Modelling Geomorphological Systems
Edited by M. G. Anderson
© 1988 John Wiley & Sons Ltd.

ERRATUM

Please note that in Chapter 13 by Frank Ahnert, on page 376, the log in
Figure 13.1 should read as follows:

$$\log L = 0.7 \log T - 2.6$$
(estimate)

Please accept our apologies for this error.

*Modelling Geomorphological
Systems*

Modelling Geomorphological Systems

Edited by

M. G. Anderson

Department of Geography,
University of Bristol

JOHN WILEY & SONS

Chichester · New York · Brisbane · Toronto · Singapore

Library of Congress Cataloging-in-Publication Data:

Modelling geomorphological systems.

   1. Geomorphology — Mathematical models.
I. Anderson, M. G.
GB400.42.M33M64 1988   551.4′072′4   87-25421
ISBN 0 471 91800 8

British Library Cataloguing in Publication Data:

Modelling geomorphological systems.
  1. Geomorphology — Mathematical models
   I. Anderson, M. G.
   551.4′0724   GB400.42.M33
ISBN 0 471 91800 8

Phototypeset by Dobbie Typesetting Service, Plymouth, Devon
Printed and bound in Great Britain by Anchor Brendon, Tiptree, Essex

To Rebecca

List of Contributors

F. AHNERT — *Geographisches Institut der RWTH, Templer-graben 55, D–5100, Aachen, W. Germany.*

M. G. ANDERSON — *Geography Department, University of Bristol, University Road, Bristol, BS8 1SS, UK.*

P. AUGUSTINUS — *Department of Earth Sciences, University of Waikato, Private Bag, Hamilton, New Zealand.*

J. BATHURST — *Department of Civil Engineering, University of Newcastle upon Tyne, Claremont Road, Newcastle upon Tyne NE1 7RU, UK.*

I. DOUGLAS — *School of Geography, University of Manchester, Manchester, M13 9PL, UK.*

W. L. GRAF — *Geography Department, Arizona State University, Tempe, Arizona, 85287, USA.*

R. D. HEY — *School of Environmental Sciences, University of East Anglia, Norwich, NR4 7TJ, UK.*

A. D. HOWARD — *Environmental Sciences Department, University of Virginia, Charlottesville, Virginia 22903, USA.*

S. HOWES — *Logica Space and Defence Systems Limited, Cobham Park, Downside Road, Cobham, Surrey, KT11 3LX, UK.*

P. F. KRSTANOVIC — *Department of Civil Engineering, Louisiana State University, Baton Rouge, Louisiana 70803, USA.*

L. J. LANE — *Southwest Rangeland Watershed Research Center, 2000 E. Allen Road, Tucson, Arizona 85719, USA.*

D. M. MARK — *Department of Geography, State University of New York at Buffalo, Francis E. Fronczak Hall, Buffalo, New York 14260, USA.*

V. G. MOON — *Department of Earth Sciences, University of Waikato, Private Bag, Hamilton, New Zealand.*

P. NADEN — *School of Geography, The University, Leeds, LS2 9JT, UK.*

G. PICKUP — *Division of Wildlife and Rangelands Research CSIRO, Central Australian Laboratory, P.O. Box 2111, Alice Springs, W.T. 5750, Australia.*

K. M. SAMBLES   *Geography Department, University of Bristol, University Road, Bristol, BS8 1SS, UK.*

M. J. SELBY   *Department of Earth Sciences, University of Waikato, Private Bag, Hamilton, New Zealand.*

E. D. SHIRLEY   *Southwest Rangeland Watershed Research Center, 2000 E. Allen Road, Tucson, Arizona 85719, USA.*

V. P. SINGH   *Department of Civil Engineering and Louisiana Water Resources Research Institute, Louisiana State University, Baton Rouge, Louisiana 70803, USA.*

R. J. STEVENSON   *Department of Earth Sciences, University of Waikato, Private Bag, Hamilton, New Zealand.*

S. TRUDGILL   *Department of Geography, Sheffield University, Sheffield, S10 2TN, UK.*

Contents

Preface

Students of geomorphology will eagerly devote themselves to laboratory and field aspects of the subject, echoing seemingly instinctive approaches to the study of landscape processes to which decades of research papers now testify. A significantly smaller proportion of students find the major driving force for their interest to lie in mathematical modelling, which by its very nature can often propel this second group beyond the activities of the former group, albeit with potentially greater uncertainty. This is not an uncommon phenomenon in the earth sciences where many students sit rather uncomfortably with mathematical formulations, and find the suggestion that they themselves 'experiment' with such formulations distinctly painful.

This text has sought to address this situation and to selectively introduce current modelling research in hillslope and river channel processes. It is hoped that the spectrum of topics covered will provide a sufficiently coherent basis for both graduate and undergraduate students, not only to see the potential range of modelling activity within hillslope and channel process investigations, but also to appreciate the modelling challenges that still remain. Not only do such potential research areas lie within the particular fields of process study, but also, and equally importantly for the subject, they relate to three further elements, namely alternative modelling approaches within a given field, the questions of model validation and verification and the scope for making computer simulation a more professional element in geomorphological investigations. These four ingredients cannot all be simultaneously satisfied in a text of this length, but it is to be hoped that increased awareness of the substantial research questions remaining will result from the reading of the chapters contained herein.

The twenty or so graduate students I have had at Bristol have all contributed to the motivation for this text. Their enthusiasm and support has made the task of editing, as well as research supervision, all the easier. In addition, I am grateful to colleagues around the world who have offered helpful advice and been kind enough to comment on specific chapters.

Malcolm G. Anderson
Department of Geography
University of Bristol

8 July 1987

Modelling Geomorphological Systems
Edited by M. G. Anderson
©1988 John Wiley & Sons Ltd.

Chapter 1

A review of the bases of geomorphological modelling

MALCOLM G. ANDERSON and KATHERINE M. SAMBLES
Department of Geography, University of Bristol

1.1 IMPORTANCE OF MODELLING IN GEOMORPHOLOGY

There now is sufficient model diversity to develop a new research methodology for geomorphological modelling. As recently as 1985, Woldenberg reviewed Chorley's 1967 model typology for geomorphology. Essentially, this typology views abstract analogue models to analyse systems, as comprising hardware, mathematical, and experimental design models. In this typology, the mathematical models (the focus of this text) were subdivided into deterministic, probabilistic, and optimization models. An analogous classification is to be found in Haines-Young and Petch (1986). In the early phases of research, such classifications are helpful in the structuring of initial ideas and approaches. This was broadly the situation in geomorphology in the late 1960s and early 1970s. However, with research developments since that time and the increase in model complexity, such classifications perhaps neither no longer reflect current research nor, more importantly, provide a comprehensive enough basis for future developments. Models are now composite in character and therefore an alternative perspective on model development may now be more appropriate than that of a classificatory approach. A further development in this field has been the relationship between data acquisition and modelling. Whilst early models were more the direct product of field investigation, the relative sophistication of the recent models allows the development of predictive elements which are not necessarily capable of direct field validation (e.g. certain recurrence interval estimations in hydrology, long-term slope development models, etc.). There has thus been a major reversal in the relative dependency between field investigations and model development over the last twenty years or so. Currently, models may be structured on a more theoretical basis, and the type of field parameterization may in fact be specified by the model itself. These two themes, the composite nature of current models and the interactive field element in model

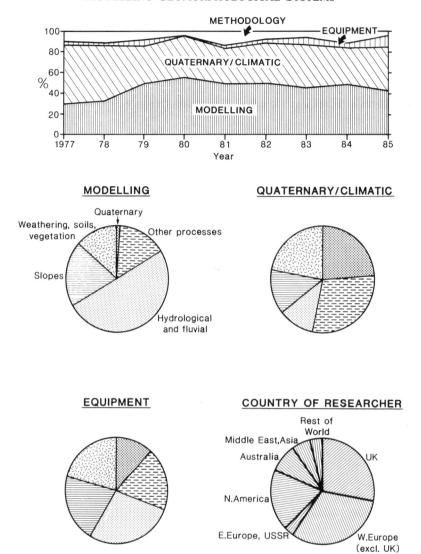

Figure 1.1. Research paper content analysis of three international geomorphology journals: *Earth Surface Processes and Landforms, Catena,* and *Zeitschrift für Geomorphologie*

parameterization and development, form the major original approach taken towards modelling both in this chapter and in the remainder of the text. These themes are augmented by evidence of recent geomorphological modelling activity, which we now briefly review.

In recent years, there has been a steady dependence upon modelling in geomorphology, as evidenced by the content of three major international geomorphological journals (see Figure 1.1(a)). Here we see that some 45 per cent of papers over the last nine years relate in some way to modelling and, for the reasons mentioned above, we have avoided any further subdivision in this figure. A breakdown of the research areas corresponding to the modelling, equipment, and other approaches (see Drake, 1977; Herwitz, 1981) together with the country of the researcher, for all papers over nine years, is given in Figure 1.1(b)), the relative dominance of hydrological and fluvial based modelling studies is clearly evident. In seeking to establish an overview, such a survey does provide an indication of modelling activity, but this could be underestimated since more subject specific journals have not been included (e.g. *Arctic and Alpine Research*; *Journal of Glaciology*). Figure 1.1(b) further illustrates the relative dominance of western European geomorphological research, as measured by these three international journals. It is thus a steady accumulation of papers relating to modelling that reinforce the need to establish a research methodology for model development that takes us beyond classification.

This chapter therefore sets out to provide a discussion on the methodology of contemporary modelling frameworks in geomorphology. From this, we generate ideas relating to model configuration and resultant uncertainty in both prediction and interpretation. The methodology of reducing modelling uncertainty is discussed and we argue that it is this area that should now be a major focus of attention in geomorphological modelling research. Finally, model transportability and future modelling needs are outlined.

1.2 BACKGROUND TO GEOMORPHOLOGICAL MODELLING

Modelling is an integral part of current geomorphological thought and methodology. Comment to this effect is found in many reviews. Few reviews, however, perhaps emphasize the perceived gulf between conceptual understanding and modelling ability as clearly as Thornes and Ferguson (1981) who reported:

'Only one British geomorphologist (M. J. Kirkby) has achieved a qualitative understanding of the relationship between form and process to match that attained by British non-geographers in closely cognate areas of earth science such as fluvial sedimentology and glacial geology.

This situation indicates a gap between our conceptual insight into landscape systems — and here our British record is good — and our ability or willingness to set up and analyse quantitative models.'

Table 1.1, in setting out some of the major themes prevalent in current geomorphology relevant to modelling research, seeks to focus attention on the more influential papers and directions as we perceive them. This is a necessary prerequisite for the establishment of a more thorough investigation of modelling research status and future directions. Table 1.1 therefore portrays current geomorphological methodology and research attitudes — it does not aim to provide a comprehensive list of *all* relevant papers. A summary of each paper is followed by a brief discussion, comprising quotes, research directives, and comments. As the table shows, methodology is concerned with all aspects of geomorphology, within which rigour in both inference (e.g. Church, 1984) and utility is receiving increased attention. However, despite this relatively sound base, few papers transpose their methodological concepts into directives for future modelling efforts. As such, the various methodologies in Table 1.1 may remain untested, unless modelling strategies are developed which facilitate articulation and, where appropriate, integration of these methodologies. The work taking us towards this goal is currently limited. Puvaneswaran and Conacher (1983) illustrate the long-term extrapolation problems associated with short-term data, and Thornes (1983) provides examples to complement his

Table 1.1. Some of the major themes prevalent in current geomorphology relevant to modelling research

Title	Summary	Applications
TIME		
LANDSCAPE SENSITIVITY AND CHANGE Author: Brunsden and Thornes (1979) Other references: Anderson and Calver (1982) Hack, (1975) Horton (1945) Schumm and Lichty (1965) Selby (1974)	Problem of extrapolation of short-term processes into longer-term (100 to 10 000 yrs) Introduces the concepts of: Constant process — characteristic form, Relaxation and characteristic form time, Pulsed and ramped inputs, Complex response of landscape, Landscape threshold.	'For any given set of environmental conditions, through the operation of a constant set of a constant set of processes, there will be a tendency over time to produce a set of characteristic landforms'. Future studies should concentrate on: Identification of process domain sequences, relaxation times and paths, progression along these paths and the sequence and magnitude of the inputs causing the changes. Theoretical discussion with no practical examples given.
APPLICABLE MODELS OF LONG-TERM LANDFORM EVOLUTION Author: Brunsden (1980) Other references: Brunsden and Kesel (1973) Davis (1899) Schumm (1978)	Two major research areas identified: (a) Boundary conditions of landform development. (b) Magnitude, frequency and sequence of environmental change and relaxation times. Amplification of Brunsden and Thornes (1979). Geomorphological time is divided into lag, relaxation and characteristic form time. Characteristics of: (a) Insensitive areas: persistent relief, convergence of form, stagnancy of development and continuous change. (b) Sensitive areas: complex landscape and episodic change.	Mainly theoretical discussion but includes some examples of relaxation times, etc. Research questions and problems envisaged stated clearly although no practical advice or examples are provided.
EXTRAPOLATION OF SHORT-TERM PROCESS	Extrapolates short-term erosion rates by overland flow into rates for or long-term	'Rarely have attempts been made to relate process data and evidence of climatic variations

DATA TO LONG-TERM LANDFORM DEVELOPMENT Author: Puvaneswaran and Conacher (1983) Other references: Conacher and Dalrymple (1978)	valley development in response to an envisaged need to relate process data to long term landform development. Problems encountered: (a) Time zero definition and identification. (b) The operation of other slope processes. (c) Temporal climatic variation causing variation in erosion intensity—could not see how to overcome this problem.	to explanations of total landform development. Processors, particularly have tended increasingly to study processes for their own sake'. Study attempts to extrapolate short-term data into long-term. Some of the resultant problems were thought to be insurmountable.
EVOLUTIONARY GEOMORPHOLOGY (1983) Author: Thornes (1983) Other references: Graf (1979) May (1974) Maynard Smith (1974) Schumm (1973) Smith and Bretherton (1972) Wise et al. (1982)	Proposes evolutionary dynamic theory especially the notions of stability, instability and bifurcation as the basis for a model of long-term geomorphological behaviour.	Need for the adoption of a dynamical approach to geomorphology and reinterpretation of processes and landscape. Examples are very specific.
'ERGODIC' REASONING IN GEOMORPHOLOGY Author: Paine (1985) Other references: Abrahams (1972) Chorley and Kennedy (1971) Craig (1982) Howard (1982) Mosley and Zimpfer (1976) Savigear (1952) Thorn (1982) Thornes (1982) Thornes and Brunsden (1977)	Review of applications and utility of ergodic reasoning. Few geomorphological studies are truly ergodic, i.e. space is rigorously substituted for time, most are location-for-time substitutions. Two categories: (a) Characteristic form models. (b) Relaxation-time models.	General review and theoretical discussion. Many examples and references. No definite conclusions reached.

continued

Table 1.1. (continued)

Title	Summary	Applications
SCALE		
ON SCALES OF INVESTIGATION IN GEOMORPHOLOGY Author: Mark (1980)	Dichotomy between 'landscape scale geomorphology' and 'process geomorphology'.	'The mainstream of "process geomorphology" has indeed learned much about the workings of various processes; however deductive statements about forms at the landscape scale have not been found, if indeed they have seriously been searched for'.
CAUSAL AND FUNCTIONAL RELATIONS IN FLUVIAL GEOMORPHOLOGY Author: Hey (1979)	Discusses causality and its implications on fluvial geomorphology, especially channel response and development. Advocates a mathematical approach using the continuum concept as this can be applied to both static and dynamic systems.	Further study should concentrate on the rates of change of variables, the governing physical properties operating at a channel section and the feedback mechanisms operating between sections. Mathematics is recommended. 'Further advances in fluvial geomorphological research are dependent on the adoption of mathematical reasoning'.
TECHNIQUES		
HARDWARE MODELS IN GEOMORPHOLOGY Author: Mosley and Zimpfer (1978)	Three types of hardware models in geomorphology: (a) Segments of unscaled reality (b) Scale models (c) Analogue models. Principles of hardware modelling. Hardware models in drainage basin morphometry, slope morphology and river morphology. Advantages and disadvantages of hardware models.	Comprehensive review.

Other references:
Foster (1975)
Kirkby (1967)
Schumm (1956)
Tanner (1960)

| RECENT TRENDS OF EXPERIMENTAL GEOMORPHOLOGY IN THE FIELD

Author:
Richter (1981)

Other references:
Anderson and Burt (1978)
Ahnert (1976)
Ahnert (1977)
Embleton et al. (1978)
Schick (1978)
Slaymaker et al. (1978)
Young (1978) | Review of quantitative research in the field in the context of process studies to:
(a) Analyse processes and determine their importance in recent relief development.
(b) Extrapolate these results to cover long time spans and larger areas.
(c) Enable process prediction, for use in applied geomorphology
Standardization required as is recognition of experiments as contributors to geomorphological theory or applied geomorphology. | A lot of the papers cited are in German and French. German terminology — 'functional geomorphology' and 'morphodynamics' — is used. |
| ON EXPERIMENTAL METHOD IN GEOMORPHOLOGY

Author:
Church (1984)

Other references:
Ahnert (1980)
Hewlett et al. (1969)
Slaymaker et al. (1980)
Southard et al. (1980)
Sugawara (1972) | Thorough discussion of experimentation in geomorphology covering its definition, conceptual basis, control and criteria, non-experimental studies, the uses of experiments and experimental catchments. Three categories of experiment, I, II and III. Only types I (direct manipulation) and III (stratified observations) qualify as genuine experiments. | Calls for a general appreciation of the difficulties of experimentation and the requirements of scientific method. |

continued

Table 1.1. (continued)

MODELS IN GEOMORPHOLOGY Author: Woldenburg (1985) Other references: Chorley (1967)	An edited collection of papers from the Binghampton Symposia series. Four groups of four papers on aspects of modelling glacial, coastal, fluvial and Martian geomorphology. Papers deal with model building and testing, computer simulation, physical analogue mathematical and verbal models. A wide range, variety and application of models and modelling.	A broad model definition: 'geomorphic models and theories are hypotheses about system form and/or process and/or behaviour'. The preface contains a model classification, which in this chapter we are suggesting could be profitably revised.
MODELLING (CH. 9 PHYSICAL GEOGRAPHY: ITS NATURE AND ITS METHODS) Author: Haines-Young and Petch (1986)	The role of models in science and in the testing of theories. Model classification on basis: 1. Is model deterministic or stochastic? 2. Is model partially or fully specified? 3. Is model a hardware or software model? Alternative to Woldenburg approach to model classification.	A strict model definition is used: 'Models are devices used to make predictions'. The conclusions that 'models are only usefully considered to be mechanisms by which predictions are made' is perhaps a little restrictive. The importance of model error in inducing theory error is recognized.

REVIEWS

WHITHER GEOMORPHOLOGY? Author: Williams (1978)	A brief summary of Australian geomorphology and its future directions. Three types of geomorphology prevail in Australia: — Tectonic evolution — Quaternary geomorphology	Geomorphology is seen as a branch of geodynamics. 'Process studies alone will never allow us to devise unequivocal models of landscape evolution'. — Process studies past and present Plate tectonics are an important influence on Australian geomorphology.

THEORIES, PARADIGMS, MAPPING AND GEOMORPHOLOGY Author: St-Onge (1981) Other references: Baker and Pyne (1978) Higgins (1975) Strahler (1952) Twidale (1977)	Identifies the lack of general theory in geomorphology today in comparison to Davis's time. The functions and attributes of a theory are discussed. The need for mapping in geomorphology and a view of landforms/processes within their regional area context is identified.	Calls for a new all-encompassing geomorphological theory (or paradigm) to halt the risk of fragmentation of geomorphology as a whole, as well as giving a sound theoretical base to the discipline.
FLUVIAL GEOMORPHOLOGY— PROCESS EXPLICIT OR IMPLICIT? Author: Gregory (1985a)	A review of a year's work in fluvial geomorphology. Divided into: (a) Documenting processes, i.e. identifying and measuring processes. (b) Amplifying processes, i.e. consolidating and building on existing research. (c) Questioning processes, i.e. questioning/examining existing accepted concepts. (d) Extending processes, e.g. identification of process thresholds. (e) Applying processes, i.e. applied geomorphology.	See the annual reviews in 'Progress in Physical Geography' for other geomorphological research areas.
THE NATURE OF PHYSICAL GEOGRAPHY Author: Gregory (1985b)	Chapters 8 and 10 are the most relevant dealing with physical geography from 1970–1980 and in the future. The current and future methodology, status and role of academic physical geography is discussed.	General, broad based discussion, covering all of physical geography from 1850–1980s.

continued

Title	Summary	Applications
TRENDS AND DIRECTIONS IN HYDROLOGY Author: Burges (1986) Other references: Moore et al. (1986) O'Loughlin (1986)	Introduces a special issue of Water Resources Research devoted to trends and directions in hydrology. The concluding remarks highlight aspects of hydrology that have been neglected, i.e. hydrologic interactions under 'normal climatic conditions, where substantial changes have been, or will be made to — vegetal cover — topography — groundwater states — soil water chemistry	Future work envisaged as resembling that of Moore et al. (1986) and O'Loughlin (1986) — at that scale and flexibility.
DILETTANTISM IN HYDROLOGY: TRANSITION OR DESTINY? Author: Klemes (1986) Other references: Box (1976) Braben (1985) Dumitrescu and Nemec (1979) Klemes (1983) Yevjevich (1968)	Calls for the creation of hydrology as a separate science in its own right. Misuse of mathematics — 'mathemistry'. Five directions for future research are identified: 1. More knowledge of climatology, meteorology, geology and ecology. 2. Inclusion of other forms of energy in hydrological models other than kinetic and potential energy of water. 3. Examination of the relevance of the river basin notion. 4. New measurement methods. 5. Study thermodynamics, geochemistry, soil physics and plant physiology.	A useful discussion which raises a lot of points as to the future of hydrology. 'For a good mathematical model it is not enough to work well. It must work for the right reasons'. 'Hydrologists are concerned not about the validity of their hydrology but their numbers'. 'The possibility that a wrong concept may well produce a reasonable number'. 'Mathematics has been used to redefine a hydrologic problem rather than solve it'.
THE HYDROLOGY OF TOMORROW Author: Kundzewicz (1986) Other references: Naef (1980) Rodriguez-Iturbe and Valdes (1979)	Considers the hydrology of the past ten years, the present and the future, with an emphasis on certain topics within hydrology. The role and status of mathematical models in hydrology is discussed.	Future development should concentrate on: (i) the recognition and description of hydrological processes. (ii) Water management (assessment of water resources and improvement in hydrological forecasting). The years 1970–1985 did not bring too many illuminating conceptions in hydrological research.

theoretical discussion on evolutionary geomorphology. Neither of these approaches, however, coincide with our perceived need to generate specific modelling strategies to validate the methodologies in Table 1.1.

There are, however, selected models that have the potential to address this important issue and are themselves illustrative of the current modelling capability in geomorphology. Table 1.2 illustrates four state-of-the-art models which have very significant attributes with regard to the specified need to establish methodological–modelling interaction. Four major features characterize these models. They are all physically based, have clearly definable assumptions, facilitate the modelling of processes that could not be undertaken in the field and, finally, allow for predictions to be made forward (Kirkby, 1985a,b) and backward (Boulton et al., 1984) in time. As such, models of this form have the potential to answer the research (methodological) questions posed in such papers as that of Brunsden (1980) (Table 1.1).

However, to undertake this satisfactorily presupposes a model structure of which one is conceptually certain. The problem of methodological acceptance is firmly rooted back in problems of model conceptual uncertainty. For example, for each of the models illustrated, there are alternatives to the submodels actually employed. Figure 1.2 illustrates this condition in which conceptual uncertainty in model formulation is illustrated in general terms. From this figure, the submodels may be replaced by a number of alternatives (structural uncertainty), and if data uncertainty is included in parameter specification too, then an almost prohibitively large number of model alternatives become available for validation (both structurally and in terms of numerical internal and external validation). Thus we now need to be able to develop modelling search paths that are parsimonious.

Certain ingredients of this complexity (Figure 1.2) have been discussed by earlier workers. Leimkuhler (p. 69, 1982), for example, made three observations regarding (then) future model development:
— Models will have to grow transparently in their structure as well as their data
— They will have to be more disaggregated and detailed to incorporate more and more aspects of the real system, and
— They will not be constructed as single-purpose models which run for some time and are thrown away afterwards, but will be applied, run and further developed for a longer time period.

In addition, he reminds us of Comte's Law: 'As the degree of complexity of a system rises (i.e. the better a model is suited to reflect reality), the range of possible generalizations decreases (i.e. there are fewer meaningful aggregates)', although as Beven (1985) has observed, this should not deter us from the problems of tackling more complex modelling of complex systems.

Physically based models in particular need to be viewed very critically in respect of both their 'explanation' and future development. The potential dangers of such model development resulting from a narrow and strict deductive concept of explanation are particularly real. As Rescher (1962) argues, such a view 'may be buttressed by fond memories of what explanation used to be like in nineteenth-century physics'. It is for this reason that we consider the conceptualization of model development should increasingly be discussed and presented in the form of Figure 1.2.

Table 1.2. Four state-of-the-art models

(a) Physically based hydrology models

Objectives:

Forecasting the effects of land use change, spatially variable inputs, and forecasting for pollutants and sediments and ungauged catchments

(Beven, 1985, p. 407)

Model structure:

Description of model in Jonch-Clausen (1979).
Physically based distributed model with spatial distribution of catchment parameters achieved in the horizontal (grid network) and vertical (column of horizontal layers in each grid square).

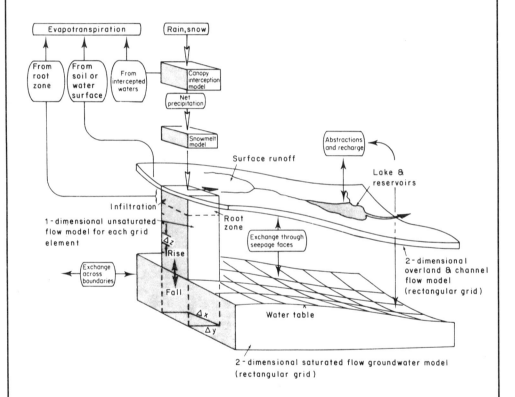

(Reproduced by permission of the Institute of Hydrology)

Potential geomorphological applications relate to sediment transport and solute modelling (see Burt, 1986).

Great expenditure on programming, computer resources data preparation and field experimentation.

(See Table 13.1 in Beven (1985) for parameter needed for 1 grid square).

Very restricted internal validation of distributed models—but then in any case there is restricted knowledge on the actual internal processes.

Table 1.2. *(continued)*

References:

Bathurst (1986a)
Bathurst (1986b)
Beven and O'Connell (1982)
Rohdenburg *et al.* (1986)

(b) **Modelling evolution of regolith-mantled slopes**

Objectives:

To explore relationships between rock type, soil and slope form (Kirkby, 1985a).

Model structure:

Interactions considered are shown below:

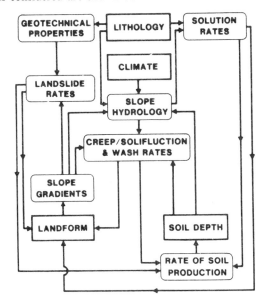

An integrated slope evolution model. Function relationships used: soil weathering (Kirkby, (1977); soil thickness and transport rates (Ahnert, 1964); models for soil creep and solifluction (Kirkby, 1984). Lithology represented as a sequence of parallel strata.

See, too, Kirkby (1985b) for a slope profile model. Profile evolution modelled in three parts: for the weathering profile, the organic profile and the inorganic profile with nutrient cycling. Here certain significant processes are neglected (e.g. organic complexing and physical clay translocation).

References:

Ahnert (1964)
Freeze (1986)
Kirkby (1977)
Kirkby (1984)
Kirkby (1985a)
Kirkby (1985b)

continued

Table 1.2. *(continued)*

(c) **Modelling Debris Flow Mobilization**

Objectives:

To determine the effects of groundwater flow on hillslope failure and liquefaction (Iverson and Major, 1986).

Model structure:

An analytical solution for the limiting stable slope angle of a semi-infinite, homogeneous, isotropic cohesionless soil mass subject to steady, uniform Darcian seepage is developed resulting in a time- and space-dependent model of downslope, landslide motion.

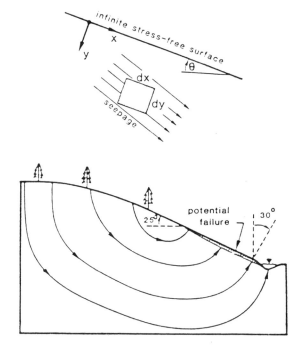

(Copyright by the American Geophysical Union.)

Seepage is confined to x–y plane. Model is valid only for small departures from steady-state landslide behaviour. A principal result from the model is that, with all other factors constant, minimum slope stability occurs when seepage is directed such that $\theta = 90° - \phi$ (see figure).

References:

Iverson (1983)
Iverson (1986a)
Iverson (1986b)
Iverson and Major (1986)

continued

Table 1.2. *(continued)*

(d) **Glaciological modelling of ice sheets**

Objectives:

Development of a model to explore the effect of net mass-balance pattern, basal boundary conditions and subglacial topography on the size and shape of ice sheets (Boulton *et al.*, 1984).

Model structure:
Steady plane isothermal flow solution to construct ice sheet profiles. Ice sheet coordinate system utilized:

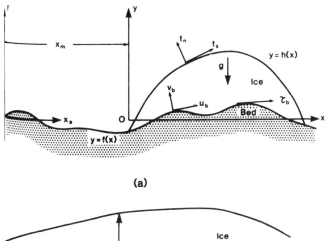

(a)

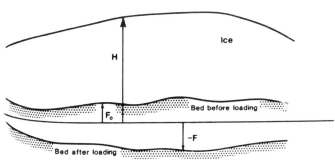

(b)

Reconstructions of Quaternary ice sheets in Northern England undertaken by Boulton *et al.* (1984). (Reproduced from Journal of Glaciology, Vol. 30, 1984, p. 140 by courtesy of the International Glaciological Society.)

Problems/restrictions/assumptions:

Finite slope topography excluded.
Basal drainage is negligible.
Temperature variation excluded.

References:

Boulton *et al.* (1984)

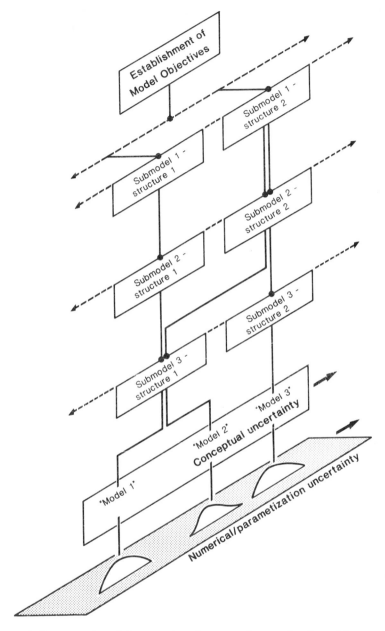

Figure 1.2. Representation of the generation of conceptual uncertainty in model generation

1.3 MODEL UNCERTAINTY

From Figure 1.2, it can be seen that a major requirement for both current and future modelling efforts must firstly be the identification of uncertainty (conceptual and data/parameter induced) and, secondly, the reduction of such uncertainty. In methodological terms, this viewpoint is really in its infancy, although there are four areas that can be identified that provide appropriate entry points.

Given the permutations of model construction (Figure 1.2) it will become increasingly important to review each stage of process modelling to identify the precise *modelling assumptions* that are made. For example, we must be able to distinguish between those cases where processes are omitted on the basis of pre-existing evidence (e.g. Armstrong, 1980, in a slope and soil simulation study) and those where processes are excluded to facilitate the development of the model (i.e. processes that are not easily quantifiable and physically or statistically manipulatable: see Hickin, 1984). However, such 'processes' as vegetation and climate are frequently employed (usually out of necessity) as indicators of past geomorphological conditions (post-diction), but are incorporated to a much lesser extent when considering future conditions (see Anderson and Burt, 1980, p. 14).

We must be able to evaluate the assumptions not only within a single model, but eventually also to be able to establish groups of methodologically similar models based upon assumption criteria. In this context, it is useful to note the progress made in catchment hydrology modelling. Howes (1985) reviewed the range of methodologies used to incorporate variability within hydrological model structures. She cited numerous models falling into five separate methodologies: lumped and semilumped (spatial variability excluded), semidistributed, geometrically distributed, and probability distributed (in which spatial variability is incorporated). Currently, such a parallel in a geomorphological context is much harder to achieve due to the relatively small number of available model structures. Here we are echoing the earlier observation of Thornes and Ferguson (1981) in section 1.2 above.

As attempts are made to include submodels of complex processes, such as vegetation, there are problems of (a) modelling the process itself, and (b) ensuring the correct representation of submodel interaction (discussed further in section 1.4). Rigorous testing becomes a more stringent requirement as submodels from other areas are drawn into geomorphological models if we are to avoid the 'possibility that a wrong concept may well produce a reasonable number' (Klemes, 1986). Kundzewicz (1986) extends this point in his review of the progress of scientific hydrology. Constraints, he argues, exist in the development of models due to the interrelationships between conceptualization and scale. He continues:

'there is a discrete spectrum of levels of scale with a set of physical laws valid at any particular scale but which may differ between levels. The laws at a higher level can be obtained from the laws at a lower level with the help of ''reduction of the category of multitude to one of singleness'' (Klemes, 1978, p. 352) i.e. through averaging or integration. Hydrology embraces a relatively small number of levels of scale whose extension, i.e. developing the interface with neighbouring subject areas (hydraulics, meteorology, climatology), could induce progress. This extension is possible in two directions, viz. upwards and downwards (Klemes, 1983). An example of upwards conceptualization is the passage from the point (hydrodynamic) scale to the river reach (hydrological) scale. Instead of considering every point of the river reach modelled via the partial differential equations of mathematical physics (hydrodynamic scale), an external approach may be used in which an ordinary differential equation for the reach is obtained by lumping the spatial representation. *The disadvantages of upwards conceptualization are the consequences of the incomplete knowledge of the phenomenon on a lower level. Such gaps are typically filled with the help of unchecked or simplified assumptions that may weaken the physical sense of the whole construction* (Klemes, 1983).' (Reproduced by permission of Blackwell Scientific Publications Limited.)

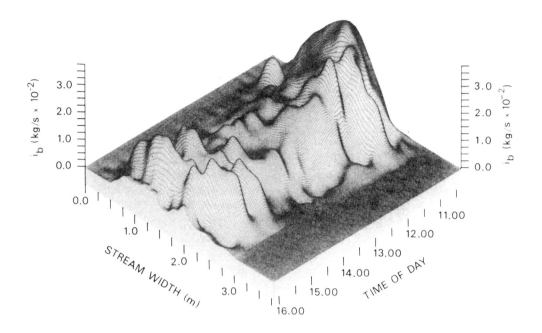

Figure 1.3. Spatial and temporal variations in the pattern of bedload transport in a study reach (after Gomez, 1983, reproduced by permission of John Wiley & Sons, Ltd)

In this context, we view the status of geomorphological modelling to be similarly placed. There are, however, clear illustrations of the importance of identifying and reexamining basic assumptions in geomorphological modelling that can currently be cited. Gomez (1983) illustrates the complex spatial and temporal variations in bedload transport (Figure 1.3). This variation in transport rates was inconsistent with assumptions of existing bedload-transport rate–stream power relationships which exclude the effect of bed armouring. Church (1981) provides a further illustration in discussing an error in Horton's slope function, in which he observes that 'it appears that the critical capacity for verification of results which lies at the heart of scientific method has not been sufficiently appreciated'.

A second area of model uncertainty relates to that of *large magnitude, low frequency events*. Most geomorphological processes are characterized by relatively low rates of activity and long elapsed times for cumulatively significant events to occur. Geomorphologists must thus somehow seek the means to determine the factors which integrate the effects of landform change (Church, 1980). Anderson and Calver (1977) have alluded to this problem in general terms:

'At any one time, landscape form depends on the overall effect of past frequency distributions of sizes of geomorphic events and on the order of the more recent events. The time elapsed since a particular event is of relevance, since, even if features do last the mean recurrence interval of their formative event, their degree of interim degradation can be of importance to landscape appearance in the short term. When the landscape is viewed in terms of the establishment of dynamic equilibria over medium-length time spans, it is the probability distribution of occurrences of events of particular magnitudes over that span that relates to the general prevailing form, while the particular sequence of events define the degree of oscillation about that form. In the very long term it appears that the sequence of geomorphic events *per se* is of comparatively little importance to landscape form: rather, the landscape progresses through a series of forms under the continued activity of events of different magnitudes, whose probabilities of occurrence may themselves be seen to vary over time owing to external causes, of which landscape degradation itself may, of course, be one'. (Reproduced by permission of The Institute of British Geographers).

This point has been further noted by Pickup and Rieger (1979), where they observe 'a channel is more likely to be a product of a whole range of flows and of the sequential nature of the flows. Furthermore, the response of a channel to discharge is not constant, but varies according to the frequency of scale of the discharge'. The low frequency, high magnitude event in channel studies having a short time base, renders both the nature and results of change relatively more obvious than similar events of other processes where the time base is significantly longer (e.g. glacial retreat). Chorley *et al.* (1984) summarize this complexity by depicting drainage basin landforms as comprising a nested hierarchy of differing sensitivity and recovery, so that at any time differently adjusted landforms (I–IV) may result from a given initial change in process (Figure 1.4). Thus alluvial channels have a quick recovery time, whilst divides have a greater permanency in the landscape. In general terms, it may never be possible to fully resolve this area of model uncertainty. In addition, in a specific context, an investigation may be operating within the relict of a high magnitude, low frequency event, which the investigator may have no means of identifying.

The above elements of uncertainty (model assumption and magnitude/frequency) relate to the *inputs* of model design and to data acquisition. Two further elements of uncertainty that can be illustrated relate to model *output*. Here there are three type situations. Firstly, where the investigator is in possession of the results against which the model is to be tested. The choice of technique by which a comparison or evaluation of this type is undertaken need not always be obvious. For example, when comparing predicted and observed hydrographs, there are several moment measures that can be employed, each one relating to a specified characteristic of the hydrograph (see Anderson and Howes, 1986, p. 181). The second situation relates to that where a short period of observations (e.g. 1 year) provides the basis of long-term (say 10-year) predictions. Konikow and Patten (1985, p. 264) outline a situation in which a 40-year calibration did not provide a reliable basis for predicting groundwater levels for a 10-year period (see, too, Person and Konikow, 1986). Finally, certain geomorphological models deal with time spans which mean that validation of any style is difficult in terms of the final model output. However, it may be that results from selected submodel elements can be falsified. Therefore, development of long-term geomorphological models is reliant upon the verification of falsification of selected submodel components, and not the end result, unless such a result is logically discountable. For example, in the long-term modelling of landforms, if hillslope/channel interaction appears 'correct' there is no basis for assessing the predicted scale of the interaction (see Armstrong, 1980).

Any geomorphological modelling investigation has to take on board the elements of modelling uncertainty we have reviewed above. We have stressed these areas in an attempt to focus modelling attention more firmly on these important issues. Composite model structures (Figure 1.2) have great potential to provide as a vehicle for reducing such

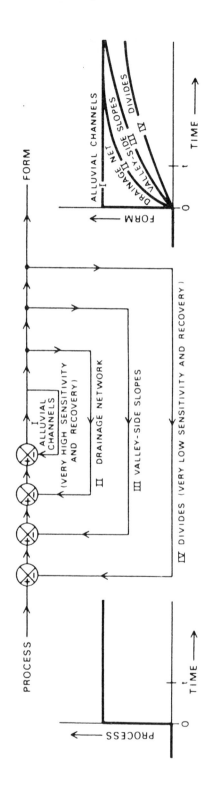

Figure 1.4. Drainage basin landforms depicted as a rested hierarchy of differing sensitivity and recovery. See text for complete explanation (after Chorley *et al.*, 1984, reproduced by permission of Methuen and Co. Ltd.)

uncertainty. This is despite the acknowledged inclusion of uncertainty that strictly accompanies such modelling schemes. Taken overall, validation of submodels (in the context of the overall model structure) allows the identification of uncertainty and simultaneously provides the means of potentially reducing uncertainty. The time base for modelling is a critical ingredient in the context of both the type of uncertainty we need to reduce, and of the associated validation approaches that may be available to the investigator.

1.4 MODEL TIME BASE

Notwithstanding the question of uncertainty raised in the previous section, a general aim of geomorphological modelling must remain as that of long-term landscape simulation. Various authors have addressed this issue; for example, Puvaneswaran and Conacher (1983) have stated that 'rarely have attempts been made to relate process data and evidence of climatic variation to explanation of total landform development'. Thus there are very few models that undertake long term landscape modelling. However, Table 1.3 illustrates four principal time bases that modelling can operate on. In addition, varying spatial scales can be examined. For example, the study by Andrews and Mahaffy (1976) is of interest since this models the varying spatial extent of an ice sheet through time (Figure 1.5). Boulton et al.'s (1984) paper extends this work in providing an estimation of climatic parameters operating at the time of maximum/minimum ice. The emphasis of contemporary geomorphological modelling, however, generally emphasizes the shorter time base. Such schemes may eventually provide the submodel components of longer term models—for example, Iverson (1986b) has developed a model for debris flow mobilization, i.e. a scheme that has effectively introduced a significant time element into studies of slope instability by the consideration of post-failure dynamics. In a parallel study, O'Loughlin (1986) has

Table 1.3. Four principal time bases for modelling

Time basis	Author	Model basis	Scale and objective
Current	Reynaud (1973)	Bed friction obeys Coulomb's Law of solid friction. Glen's non-linear creep law. Model combines both internal deformation *and* basal sliding.	Valley glacier scale
Minutes	Li *et al.* (1976)	Finite difference approach to model sediment outflow hydrographs with implications for the effect of slope shape.	Hillslope scale
10^3 years	Andrews and Mahaffy (1976)	3-D ice model based on assumption that ice sheet moves under its own weight by simple shear strain parallel to the geoid. Tests the hypothesis of instantaneous glacierization.	Continental scale
Long term	Kirkby (1971)	Continuity equation–solutions parameterized by empirical process relationships.	Hillslope scale

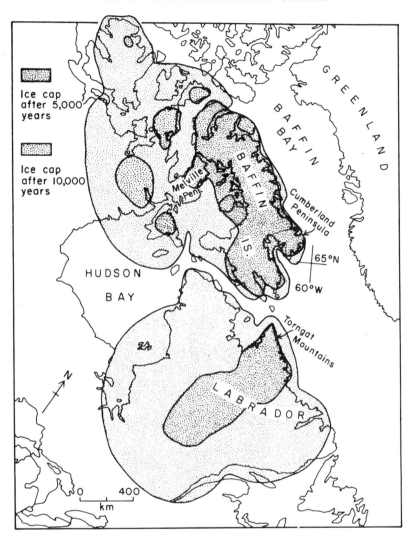

Figure 1.5. The location of the Laurentide Ice Sheet after 10 050 years of growth as predicted by a three-dimensional numerical ice flow model (after Andrews and Mahaffy, 1976, reproduced by permission of Academic Press, Inc.)

modelled the effect of variable forest clearing on catchment hydrology, thereby achieving a relatively flexible (if short-term) space/time interaction (Figure 1.6); see, too, Moore *et al.* (1986). Kirkby (1984), however, provides a good (and rare) example of longer term modelling of landform development—Figure 1.7. This study is particularly valuable because it utilizes Savigear's (1952) ergodic reasoning as a validation tool.

In the absence of validation possibilities, potential model inaccuracy is likely to increase. Armstrong (1980) has illustrated the evolution of soil depth (Figure 1.8) by considering weathering, soil creep, and fluvial equations (soil, creep rate) and the restricted number of parameters employed. The simulated results cannot readily be substantiated and illustrate potential problems with respect to the coupling of hillslope and channel submodels.

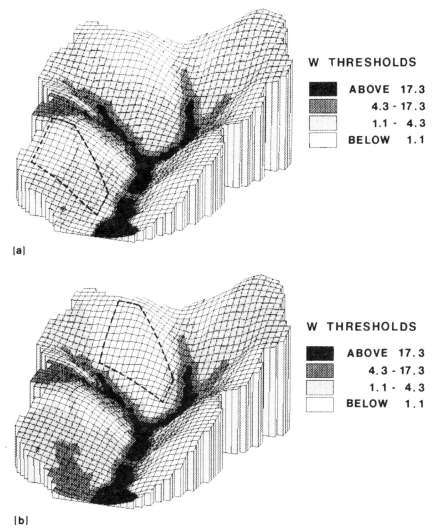

Figure 1.6. Predicted extent of saturation zones from two analyses simulating forest clearing. The polygons define different regions where the drainage flux is reduced to 25 per cent of the value over the remainder of the catchment and representing these areas where the trees were retained. (W is a wetness index.) (After O'Loughlin, 1986, copyright by the American Geophysical Union)

This point is further alluded to by Sprunt (1972) in respect of the simulated development of a third order drainage basin on an initially uniform plane (see Figure 1.9), where he remarks on an unrealistic result obtained for channel angle–basin area relationships. Kirkby (1977) has alluded more directly to this issue in the context of soil development models as components of slope models. He acknowledges 'the importance of interaction of slope profiles and processes, but could find no soil models which were compatible with the inputs and outputs of slope development models'.

These examples tend to illustrate the need to further research potential submodel interaction appropriate to the longer timescales. The difficulties we have illustrated above

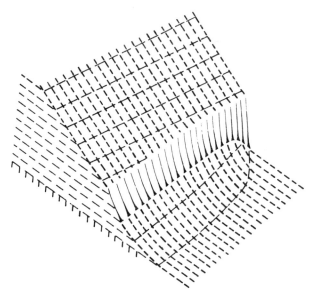

Figure 1.7. Simulation of slope development (after Kirkby, 1984, reproduced by permission of Gebrüder Bantraeger)

t = 1000

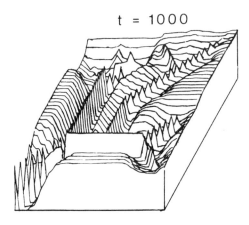

Figure 1.8. Evolution of soil depth (after Armstrong, 1980, reproduced by permission of Catena Verlag)

(which occur in long-term modelling) testify to this need, where such interaction becomes as important as the submodel processes themselves.

1.5 FUTURE MODELLING DEVELOPMENT: CONSTRAINTS AND PROSPECTS

As a theoretical basis, we may envisage geomorphological modelling to follow the style of Figure 1.2, with the associated problems of uncertainty coupled with the goal of modelling long-term development. Such modelling schemes have practical constraints of which we must be cognisant. Figure 1.10 illustrates two such limitations. Figure 1.10 shows that as

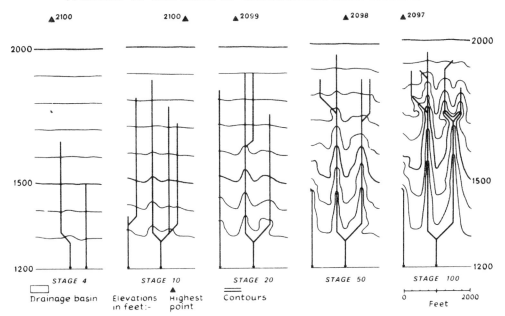

Figure 1.9. Simulation of drainage networks (after Sprunt, 1972, reproduced by permission of Methuen & Co. Ltd)

models develop in complexity, the time taken to fully understand the model increases (i.e. transportability decreases). With increased model complexity computing costs and time increase, perhaps to the point that research time availability may constrain further development. There may well be a complex relationship between academic value and the next model version (1→2 in Figure 1.10(b)) due to the external imposition of funding constraints. For example, it may well take of the order of two years to be fully familiar with certain of the state-of-the-art schemes outlined in Table 1.2. Within the usual research period of three years, the ability of the researcher to make an original contribution becomes progressively limited. This is further compounded if the researcher leaves the institution at the end of that period, since corporate memory is effectively thereby significantly reduced.

In the discussion thus far, we have sought to establish the following points:

1. The need to abandon a classificatory approach to modelling in favour of the recognition of the need to accommodate composite model structures which transcend this classification.

2. The need to further accommodate the existence of both conceptual and data induced uncertainty in modelling (Figures 1.2 and 1.3).

3. The need to recognize the different time bases in modelling and to remain cognisant of the long-term objectives of geomorphological interpretation (both post-diction and prediction).

4. Finally, the need to attempt to circumvent the potentially significant problem of time constraints on research in respect of the transportability of the state-of-the-art models (Tables 1.2 and 1.3 for example).

These general points overlay the chapters that follow in this text, from which a further set of modelling objectives can be determined.

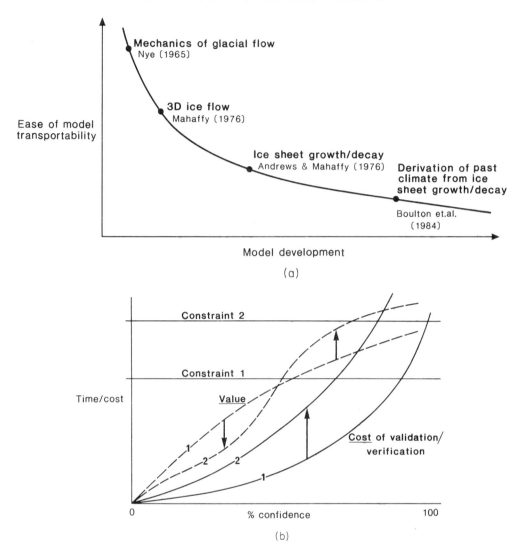

Figure 1.10. (a) Illustration of the progressive development of a model coupled to decreasing ease of transportability; (b) Model cost/value relationships changing in respect of an initial (1) and later version (2) of the model. Constraints of either time (1) or money (2) may alter the relative value achieved

There is little doubt that the most significant progress made in geomorphological theory is in the understanding of process mechanics (Kirkby, 1987). However, in the application of such theories to landscape form or other interrelated applications (e.g. land use change) our efforts are much more tentative and ill-formed, and as Kirkby observes it is here that almost everything remains to be done.

This text has sought to develop this theme in an effort to focus geomorphological inquiry into those more 'tentative' areas. In particular, it is implicitly recognized that it is now feasible to develop models (and indeed paradigms) that are not directly amenable to validation in a classical sense. We can perhaps validate elements of the process mechanics components, but we cannot validate the model. Anderson and Rogers (1987) comment on this situation in the context of hydrological distributed models:

> 'Fully distributed models have two structural levels in their design: a) the choice of processes to be included; and b) the description of the distributed processes. Thus, although there may be only three or four fully distributed models on the basis of the former structural element, there are potentially 'n' such models when considering both structural levels. In the case of these models, there are two further issues since the situation is more complex in structural terms. There are numerous other decisions such as methods of solution for the equations, determination of parameter values, specification of initial conditions and catchment sub-division that have to be made in, first, operating such models and, second, in applying them to a particular catchment.'

thereby stressing the need for a major revision in our concept of model validation.

Certain of the initial chapters in this text thus illustrate an approach towards these areas which have hitherto been only tentatively tackled, and where classical validation is the most difficult—catastrophe theory (Chapter 2) and equilibrium concepts (Chapter 3). These two chapters relate to what has been termed high level theories, being theories less related to individual data and intended to have a general and wide applicability—as Graf observes in Chapter 2, catastrophe theory can be a useful input for the construction of generalizations. Continuing this theme are two chapters (4 and 13) that attempt to develop further concepts relating to landscape form. Chapter 4 attempts to further channel network theory by a consideration of networks with forks and lakes and to argue the core role networks play in simulation of erosion and landscape evolution. Chapter 13 complements this network modelling review, by a consideration of landform change modelling. Here Ahnert argues that eksystematic (internally induced) changes may be introduced into functional-geomorphological process–response models of landform evolution.

At the process level of modelling river dynamics, considerable recent progress has been made. This is evidenced by Chapters 5, 6, and 8. Chapter 5 reviews empirical rational and dynamic models of channel morphology which currently provide the basis for explanatory modelling in which the bed topography feeds back to the hydraulic conditions via convective acceleration effects caused by shoaling/deepening, and in which the spatial fields of shear stress and bed material transport are mapped at the within-reach scale.

Channel flow models (Chapter 6) provide a further example of recent increased sophistication in modelling, highlighting new and significant data acquisition needs to both calibrate and validate—a point of even greater relevance when it comes to sediment transport modelling (Chapter 8).

Chapters 5–8 emphasize the explicit data needs relating to parameters needed to fit a model, input data needed to maintain the simulation and state variables describing selected conditions which may provide the basis for model calibration and verification. Whilst Bathurst (Chapter 6) states these separate requirements, it is clear that this division must be increasingly recognized by geomorphologists.

It is vital that the development of hillslope modelling proceeds at a rate that at selected stages is able to provide coherence for river channel form and process modelling in the context of landform modelling. Integration in the final analysis may serve as both stimulus and check on modelling approaches taken in each respective field, and highlight areas of relative omission in a way that too close a process mechanical view within any one field may not provide.

One such area of relative omission is that of rock slope processes in regard both to initial stages of weathering and subsequent controls (Chapter 12). Such areas must be seen as of potential significance as we attempt to move to a higher degree of spatial integration (see erosion cell mosaic concept by Pickup in Chapter 7). Although not explicitly covered in this text, Anderson and Richards (1987) have highlighted the need for modelling post-failure mechanisms and for the incorporation of long-term weathering effects to be included in models of long-term slope evolution, in both the spatial and temporal contexts.

Having highlighted thematic elements, a major topic of discussion throughout the text is the relative efficiency of physically based modelling. Whilst such modelling is still advancing in concept, several authors highlight the need for cautious optimism here, and some propose stochastic modelling to be given some prominence (either implicitly or explicitly—see Singh *et al.*, Chapter 9). This debate will continue given the spectrum of modelling applications, but the major feature that should be dwelt upon is the following set of proposals which may be derived from the chapters in this text, specifically the need in future geomorphological modelling research to:

1. Maintain, and develop further, higher level theories in geomorphology (see Chapters 1–4, 13 and 14).
2. Develop models in areas of relative model omission (see Chapters 11, 12, and 15).
3. Develop models which couple systems and the treatment of spatially integrative modelling (see Chapters 7, 8, and 11).
4. Continue development of process mechanics in established modelling areas and develop further appropriate data structures for validation as well as model operation (see Chapters 5–8, and 7, 9, and 10).
5. Make geomorphological models more generally available, and thus explore more professionally than hitherto, aspects of computer simulation and coding. This is seen as a major element to enhance the status of geomorphological modelling, which if it were to occur, would parallel developments in availability of models in hydrology and other earth science areas (see Chapter 15).

REFERENCES

Abrahams, A. D. (1972). Environmental constraints on the substitution of space for time in the study of natural channel networks. *Geol. Soc. Am. Bull.*, **83**, 1523–1530.

Ahnert, F. (1964). Slope retreat as a function of waste cover thickness—some preliminary computer orientated models. *Annals Assoc. of Am. Geog.*, **54**, 412.

Ahnert, F. (1976). Brief description of a comprehensive 3-dimensional process–response model of landform development. *Zeitschrift für Geomorphologie*, **25**, 29–49.

Ahnert, F. (1977). Some comments on the quantitative formulation of geomorphological processes in a theoretical model. *Earth Surface Processes*, **2**, 191–201.

Ahnert, F. (1980). A note on measurements and experiments in geomorphology. *Zeitschrift für Geomorphologie, SB.*, **35**, 1–10.

Anderson, M. G. and Burt, T. P. (1978). Towards more detailed field monitoring of variable source areas. *Water Res. Res.*, **14**, 1123–1131.

Anderson, M. G. and Burt, T. P. (1980). Methods of geomorphological investigation. In Goudie, A. (Ed.), *Techniques in Geomorphology*, George Allen and Unwin, London, 3–21.

Anderson, M. G. and Calver, A. (1977). On the persistence of landscape features formed by a large flood. *Trans. Inst. Brit. Geogrs.*, **2**, 2, 243–254.

Anderson, M. G. and Calver, A. (1982). Exmoor channel patterns in relation to the flood of 1952. *Proc. Ussher Soc.*, **5**, 362–367

Anderson, M. G. and Howes, S. (1986). Hillslope hydrology models for forecasting in ungauged watersheds. In Abrahams, A. D. (Ed.), *Hillslope Processes*, Allen and Unwin, Winchester, Mass., 161–186.

Anderson, M. G. and Richards, K. S. (1987). Modelling slope stability: the complementary nature of geotechnical and geomorphological approaches. In Anderson, M. G. and Richards, K. S. (Eds.), *Slope Stability: Geotechnical Engineering and Geomorphology*, John Wiley, Chichester, 1–9.

Anderson, M. G. and Rogers, C. C. M. (1987). Catchment scale distributed models: a discussion of future research directions. *Progress in Physical Geography*, **11**, 28–51.

Andrews, J. T. and Mahaffy, A. W. (1976). Growth rate of the Laurentide Ice Sheet and sea level lowering (with emphasis on the 115 000 BP Sea Level Low). *Quaternary Research*, **6**, 167–183.

Armstrong, A. C. (1980). Soils and slopes in a humid temperate environment: a simulation study. *Catena*, **7**, 327–338.

Baker, R. V. and Pyne, S. (1978). G. K. Gilbert and modern geomorphology. *Am. J. Sci.*, **278**, 97–123.

Bathurst, J. C. (1986a). Physically-based distributed modelling of an upland catchment using the Systeme Hydrologique Europeen. *J. Hydrol.* **87**, 79–102.

Bathurst, J. C. (1986b). Sensitivity analysis of the Systeme Hydrologique Europeen for an upland catchment. *J. Hydrol.*, **87**, 103–123.

Beven, K. (1985). Distributed models. In Anderson, M. G. and Burt, T. P. (Eds.), *Hydrological Forecasting*, Wiley, Chichester, 405–435.

Beven, K. J. and O'Connell, P. E. (1982). On the role of distributed models in hydrology. *Institute of Hydrology, Report 81*, Wallingford, England.

Boulton, G. S., Smith, G. D., and Morland, L. W. (1984). The reconstruction of former ice sheets and their mass balance characteristics using a non-linearly viscous flow model. *J. of Glaciology*, **30**, 140–152.

Box, G. E. P. (1976). Science and statistics. *J. Am. Stat. Assoc.*, **71**, 791–799.

Braben, D. W. (1985). Innovation and academic research. *Nature*, **316**, 401–402.

Brunsden, D. (1980). Applicable models of long term landform evolution. *Zeitschrift für Geomorphologie*, SB, **36**, 16–26.

Brunsden, D. and Kesel, R. H. (1973). The evolution of a Mississippi river bluff in historic time. *J. Geol.*, **81**, 576–597.

Brunsden, D. and Thornes, J. B. (1979). Landscape sensitivity and change. *T.I.B.G.*, **4**, 463–484.

Burt, T. P. (1986). Slopes and slope processes: progress report. *Progress in Physical Geography*, **10**, 547–562.

Burges, S. J (1986). Trends and directions in hydrology. *Water Resources and Research*, **22**, 15–58.

Chorley, R. J. (1967). Models in geomorphology. In Chorley, R. J. and Haggett, P. (Eds.), *Models in Geography*, Methuen, London, 59–96.

Chorley, R. J. and Kennedy, B. A. (1971). *Physical Geography: A Systems Approach*, Prentice-Hall, London.

Chorley, R. J., Schumm, S. A., and Sugden, D. E. (1984). *Geomorphology*, Methuen, London.

Church, M. (1980). Records of recent geomorphology events. In R. A. Cullingford, D. A. Davidson, and J. Lewin (Eds.), *Timescales in Geomorphology*, John Wiley, Chichester, 13–29.

Church, M. (1981). Horton's slope function. *Earth Surface Processes and Landforms*, **6**, 199–201.

Church, M. (1984). On experimental method in geomorphology. In Burt, T. P. and Walling, D. E., (Eds.), *Catchment Experiments in Fluvial Geomorphology*, Geo Books, Norwich.

Conacher, A. J. and Dalrymple, J. B. (1978). Identification, measurement and interpretation of some pedogeomorphic processes. *Zeitschrift für Geomorphologie*, SB, **29**, 1–9.

Craig, R. G. (1982). The ergodic principle in erosional models. In Thorn, C. E. (Ed.), *Space and Time in Geomorphology*, London, George Allen and Unwin.

Davis, W. M. (1899). The geographical cycle. *Geogr. J.*, **14**, 481–504.

Drake, L. D. (1977). Depositional fabrics in basal till reflect alignments during transportation. *Earth Surface Processes*, **2**, 309–317.

Dumitrescu, S. and Nemec, J. (1979). Hydrology—a look back and a look forward. In *Three Centuries of Scientific Hydrology*, Unesco–WHO–IAHS, Paris, 16–22.

Embleton, C., Brunsden, D., and Jones, D. K. C. (Eds.) (1978). *Geomorphology, Present Problems and Future Prospects*, Oxford University Press, Oxford.

Foster, J. E. (1975). Physical modelling techniques used in river models. In *Symposium on Modelling Techniques. New York Am. Soc. Civil Engs.*, 540–559.

Freeze, R. A. (1986). Modelling interrelationships between climate, hydrology and hydrogeology and the development of slopes. In Anderson, M. G. and Richards, K. S. (Eds.), *Slope Stability: Geotechnical Engineering and Geomorphology*, 381–403.

Gomez, B. (1983). Temporal variations in bedload transport rates: the effect of progressive bed armouring. *Earth Surface Processes and Landforms*, **8**, 41–54.

Graf, W. L. (1979). Catastrophe theory as a model for changes in fluvial systems. In Rhodes, D. D. and Williams, G. P. (Eds.), *Adjustments of the Fluvial System*, Dubque, Kendall Hunt, 13–32.

Gregory, K. J. (1985a). Progress report: fluvial geomorphology—process explicit and implicit? *Prog. in Phys. Geog.*, **4**, 414–424.

Gregory, K. J. (1985b). *The Nature of Physical Geography*, Edward Arnold.

Hack, J. T. (1975). Dynamic equilibrium and landscape evolution. In Melhorn, W. N. and Flemal, R. C. (Eds.), *Theories of Landform Development*, Binghampton, N.Y., 87–102.

Haines-Young, R. H. and Petch, J. R. (1986). *Physical Geography: its Nature and its Methods*, Harper and Row, London, 230 pp.

Herwitz, S. R. (1981). Landforms under a tropical wet forest cover on the Osa Peninsula, Costa Rica. *Zeitschrift für Geomorphologie*, **25**, 259–270.

Hewlett, J. D., Lull, M. W. and Reinhart, K. G. (1969). In defence of experimental watersheds. *Water Resources Research*, **5**, 306–316.

Hey, R. D. (1979). Causal and functional relations in fluvial geomorphology. *Earth Surface Processes*, **4**, 179–182.

Hickin, E. J. (1984). Vegetation and river channel dynamics. *Canadian Geographer*, **XXVIII**, 111–126.

Higgins, R. S. (1975). Theories of landscape development: perspective. In Melhorn, W. M. and Flemal, R. C. (Eds.), *Theories of Landform Development*, Binghampton State University of New York Publications in Geomorphology, 1–26.

Horton, R. E. (1945). Erosional development of streams and their drainage basins: hydrophysical approach to quantitative morphology. *Bull. Geol. Soc. Am.*, **56**, 275–370.

Howard, A. D. (1982). Equilibrium and time scales in geomorphology: application to sandbed alluvial streams. *Earth Surface Processes and Landforms*, **7**, 303–325.

Howes, S. (1985). *A mathematical hydrological model for the ungauged catchment*, Unpublished Ph.d. thesis, University of Bristol, 404 pp.

Iverson, R. M. (1983). A model for creeping flow in landslides—discussion. *Bull. Assoc. Eng. Geol.*, **20**, 455–458.

Iverson, R. M. (1986a). Unsteady nonuniform landslide motion, 1. Theoretical dynamics and the steady datum state. *J. Geol.*, **94**, 1–15.

Iverson, R. M. (1986b). Dynamics of slow landslides: a theory for time-dependent behaviour. In Abrahams, A. D. (Ed.), *Hillslope Processes*, Allen and Unwin, Winchester, Mass., 297–317.

Iverson, R. M. and Major, J. J. (1986). Groundwater seepage vectors and the potential for hillslope failure and debris flow mobilization. *Water Resources Research*, **22**, 1543–1548.

Jonch-Clausen, T. (1979). Systeme Hydrologique Europeen: a short description. *SHE Report 1, Danish Hydraulic Institute*, Horsholm, Denmark.

Kirkby, M. J. (1967). Measurement and theory of soil creep. *J. Geol.*, **75**, 359–378.

Kirkby, M. J. (1971). Hillslope process–response models based on the continuity equation. *Inst. Br. Geogr. Spec. Publ.*, **3**, 15–30.

Kirkby, M. J. (1977). Soil development models as a component of slope models. *Earth Surface Processes*, **2**, 203–230.

Kirkby, M. J. (1984). Modelling cliff development in South Wales: Savigear reviewed. *Zeitschrift für Geomorphologie, N.F.*, **28**, 4, 405–426.

Kirkby, M. J. (1985a). A model for the evolution of regolith-mantled slopes. In Woldenburg, M. J. (Ed.), *Models in Geomorphology*, Allen and Unwin, Binghampton Series.

Kirkby, M. J. (1985b). A basis for soil profile modelling in a geomorphic context. *Journal of Soil Science*, **36**, 97–121.

Kirkby, M. J. (1987). Papers on theoretical geomorphology: introduction. In Gardiner, V. (Ed.), *International Geomorphology 1986, Part II*, 1, 2, John Wiley, Chichester.

Klemes, V. (1978). Physically based stochastic hydrologic analysis. *Adv. in Hydroscience*, **11**, 285–356.

Klemes, V. (1983). Conceptualization and scale in hydrology. *J. Hydrol.* **65**, 1–23.

Klemes, V. (1986). Dilettantism in hydrology: transition or destiny? *Water Resources Research*, **22**, 9, 177S–188S.

Konikow, L. F. and Patten, E. P. Jr. (1985). Groundwater forecasting. In Anderson, M. G. and Burt, T. P. (Eds.), *Hydrological Forecasting*, Chichester, John Wiley, 221–270.

Kundzewicz, Z. W. (1986). The hydrology of tomorrow. *Hydrological Sciences Journal*, **31**, 223–235.

Leimkuhler, K. (1982). Some methodological problems in energy modelling. In Cellier, F. E. (Ed.), *Progress in Modelling and Simulation*, Academic Press, London, 61–73.

Li, R. M., Simons, D. B., and Carder, D. R. (1976). Mathematical modelling of overland flow for soil erosion. *National Soil Erosion Conf., Purdue University, Lafayette*, 25–26.

Mahaffy, M. A. W. (1976). A numerical three-dimensional ice flow model. *J. Geophys. Res.*, **81**, 1059–1066.

Mark, D. M. (1980). On scales of investigation in geomorphology. *Canadian Geographer*, **XXIV**, 1, 81–82.

May, R. M. (1974). Biological populations with non-overlapping generations. Stable points, stable cycles and chaos. *Science*, **186**, 645–647.

Maynard Smith, J. (1974). *Models in Ecology*, Cambridge University Press, Cambridge, 146 pp.

Moore, I. D., Mackay, S. M., Wallbrink, P. J., Burch, G. J., and O'Loughlin, E. M. (1986). Hydrologic characteristics and modelling of a small forested catchment in Southeastern New South Wales. Pre-logging condition. *J. of Hydrology*, **88**, 307–335.

Mosley, M. P. and Zimpfer, G. L. (1976). Explanation in geomorphology. *Zeitschrift für Geomorphologie, NF*, **20**, 381–390.

Mosley, M. P. and Zimpfer, G. L. (1978). Hardware models in geomorphology. *Prog. in Physical Geog.*, **2**, 438–461.

Naef, F. (1980). Can we model the rainfall–runoff process today? *Hydrological Sciences Bulletin*, **26**, 281–289.

Nye, J. F. (1965). The flow of a glacier in a channel of rectangular, elliptic or parabolic cross section. *J. Glaciology*, **5**, 661–690.

O'Loughlin, E. M. (1986). Prediction of surface saturation zones in natural catchments by topographic analysis. *Water Resources Research*, **22**, 5, 794–804.

Paine, A. D. M. (1985). Ergodic reasoning in geomorphology: time for a review of the term? *Prog. in Phys. Geog.*, **9**, 1–15.

Person, M. and Konikow, L. F. (1986). Recalibration and predictive reliability of a solute–transport model of an irrigated stream–aquifer system. *J. Hydrology*, **87**, 145–165.

Pickup, G. and Rieger, W. A. (1979). A conceptual model of the relationship between channel characteristics and discharge. *Earth Surface Processes*, **4**, 37–42.

Puvaneswaran, P. and Conacher, A. J. (1983). Extrapolation of short-term process data to long-term landform development: a case study from southwestern Australia. *Catena*, **10**, 321–337.

Rescher, N. (1962). The stochastic revolution and the nature of scientific explanation. *Synthese*, **14**, 200–215.

Reynaud, L. (1973). Flow of a valley glacier with a solid friction law. *J. of Glaciology*, **12**, 65, 251–258.

Richter, G. (1981). Recent trends of experimental geomorphology in the field. *Earth Surface Process and Landforms*, **6**, 215–219.

Rodriguez-Iturbe, I. and Valdes, J. B. (1979). The geomorphologic structure of hydrologic response. *Water Resources Research*, **15**, 1409–1420.

Rohdenburg, H., Diekkruger, B., and Bork, H-R. (1986). Deterministic hydrological site and catchment models for the analysis of agroecosystems. *Catena*, **13**, 119–137.

Savigear, R. A. G. (1952). Some observations on slope development in South Wales. *Trans. Inst. Brit. Geogr.*, **18**, 31–52.

Schick, A. P. (1978). Field experiments in arid fluvial environments: considerations for research design. *Zeitschift für Geomorphologie, SB*, **29**, 22-28.

Schumm, S. A. (1956). Evolution of drainage systems and slopes in badlands at Perth Amboy, New Jersey. *Geol. Soc. Am. Bull.*, **67**, 597-646.

Schumm, S. A. (1973). Geomorphic thresholds and complex response of drainage systems. In Morisawa, M. (Eds.), *Fluvial Geomorphology*, Publications in Geomorphology, State University of New York, Binghampton, 299-310.

Schumm, S. A. (1978). Geomorphic thresholds: the concept and its applications. *Trans. Inst. Brit. Geogrs.*, **4**, 485-515.

Schumm, S. A. and Lichty, R. W. (1965). Time, space and causality in geomorphology. *Am. J. Sci.*, **263**, 110-119.

Selby, M. J. (1974). Dominant geomorphic events in landform evolution. *Bull. Inst. Ass. Engng. Geol.*, **9**, 85-89.

Slaymaker, O., Dunne, T., and Rapp, A. (1980). Geomorphic experiments on hillslopes. *Zeitschrift für Geomorphologie SB*, **35**, 5-7.

Slaymaker, O., Rapp, A., and Dunne, T. (1978). Field instrumentation and geomorphological problems. *Zeitschrift für Geomorphologie, SB*, **29**.

Smith, T. R. and Bretherton, F. P. (1972). Stability and the conservation of mass in drainage basin evolution. *Water Resources Research*, **8**, 1506-1527.

Southard, J. B., Boguchwal, L. A., and Romea, R. D. (1980). Test of scale modelling of sediment transport in steady unidirectional flow. *Earth Surface Processes*, **5**, 17-23.

Sprunt, B. (1972). Digital simulation of drainage development. In Chorley, R. J. (Ed.), *Spatial Analysis in Geomorphology*, Methuen, London, 317-389.

St-Onge, D. A. (1981). Theories, paradigms, mapping and geomorphology. *Canadian Geographer*, **XXV**, 307-315.

Strahler, A. N. (1952). Dynamic basis of geomorphology. *Ball. Geol. Soc. Am.*, **63**, 923-938.

Sugawara, M. (1972). Difficult problems about small experimental basins and necessity of collecting information on large basin. In *Symposium on the results of research in representative and experimental basins, Wellington, New Zealand, 1-8 December 1970*, International Association of Scientific Hydrology Publication No. 96, 393-397.

Tanner, W. F. (1960). Helical flow, a possible cause of meandering. *J. Geophysical Res.*, **65**, 993-995.

Thorn, C. E. (1982). *Space and Time in Geomorphology*, George Allen and Unwin, London.

Thornes, J. B. (1982). Problems in the identification of stability and structure from temporal data series. In Thorn, C. E. (Ed.), *Space and Time in Geomorphology*, George Allen and Unwin, London.

Thornes, J. B. (1983). Evolutionary geomorphology. *Geography*, **68**, 225-235.

Thornes, J. B. and Brunsden, D. (1977). *Geomorphology and Time*, Methuen, London.

Thornes, J. B. and Ferguson, R. I. (1981). Geomorphology. In Wrigley, N. and Bennett, R. J. (Eds.), *Quantitative Geography*, Routledge and Kegan Paul, London, 284-293.

Twidale, C. R. (1977). Fragile foundations: some methodological problems in geomorphological research. *Revue de Geomorphologie Dynamique*, **26**, 85-95.

Williams, M. (1978). Whither Australian geomorphology? *Australian Geogr.*, **14**, 63-64.

Wise, S. M., Thornes, J. B., and Gilman, A. (1982). How old are the badlands? A case study from southeast Spain. In Bryan, R. and Yair, A. (Eds.), *Badland Geomorphology and Piping*, Geobooks, Norwich, 259-277.

Woldenburg, M. J. (1985). *Models in Geomorphology*, George Allen and Unwin, Binghampton Series.

Yevjevich, V. (1968). Misconceptions in hydrology and their consequences. *Water Resources Research*, **4**, 225-232.

Young, A. (1978). A twelve-year record of soil movement on a slope. *Zeitschrift für Geomorphologie, SB*, **29**, 104-110.

Modelling Geomorphological Systems
Edited by M. G. Anderson
©1988 John Wiley & Sons Ltd.

Chapter 2

Applications of catastrophe theory in fluvial geomorphology

WILLIAM L. GRAF
Department of Geography, Arizona State University

2.1 INTRODUCTION

Catastrophe theory is a part of mathematical topology which characterizes the spatial and temporal structures of change in general systems. Because geomorphology deals with spatial and temporal changes in the forms and processes of earth surface systems, catastrophe theory offers a potentially useful language for expressing theories that explain changes in geomorphological systems. Catastrophe theory alone is neither catastrophic nor theoretical. The label derives from translations of the original French texts in which catastrophe refers to a mathematical–topologic surface that depicts a change structure. Catastrophe theory is not a theory in the scientific sense because it does not explain processes but merely describes them. Despite the misleading label, catastrophe theory provides the geomorphologic scientist with a tool for theory construction. The purpose of this chapter is to explore the fundamental aspects of catastrophe theory and to demonstrate their applicability to theory construction in fluvial geomorphology. The following pages represent a refined version of a previous discussion of the subject (Graf, 1979b).

The nature of change in fluvial geomorphic systems commands increasing research attention. During the development of explanations in fluvial geomorphology after Horton's (1932, 1945) introduction of hydrophysical methods, researchers placed great emphasis on equilibrium conditions in fluvial systems (Leopold, Wolman, and Miller, 1964). In the engineering community, explanations depend largely on the assumption of stability and regime theory, wherein processes exhibit mutual interdependency and mutual adjustment (Shen, 1971). More recent work has emphasized the nature of change in geomorphic systems and the adjustments between states of equilibrium (Graf, 1977; Gregory, 1977). In some cases, fluvial systems in particular may never achieve the equilibrium envisioned by relatively simple hydraulic theory (Stevens *et al.*, 1975). Catastrophe theory may be useful in developing explanations that achieve a more satisfying balance between the focus on system stability and the focus on system change.

The application of catastrophe theory in geomorphologic problems depends on the use of three general concepts widely accepted in the science: general systems theory, equilibrium tendencies, and the concept of the interplay between forces and resistances. The precepts of catastrophe theory presuppose that the objects of study are related to each other in a general system framework. A general system is a collection of elements and their interrelationships. In geomorphology the majority of research in the past several decades has implicitly used the concept of a general system in the description of geomorphic forms and processes. Explicit definition of river systems as open systems by Strahler (1952, p. 676) gave rise to a series of works exploring the utility general systems theory in geomorphology (Chorley, 1962; Chorley and Kennedy, 1971; Huggett, 1980, 1985).

Catastrophe theory also assumes that the system under investigation tends toward some equilibrium condition with mutual adjustment of all the system elements toward some predictable arrangement of forms and processes (outlined by Gilbert, 1877, for example). The theory characterizes changes in systems as being changes of equilibrium states, so that in deterministic interpretations there is a perfect definition of solutions. Empirical approaches to catastrophe theory using geomorphic data inevitably generate more poorly defined solutions because all the observations used for input do not represent equilibrium states. In catastrophe theory applications, there is no 'quasi-equilibrium' as defined by Langbein and Leopold (1964) because values fix system states at various instances in time and they are either in equilibrium or disequilibrium arrangements.

Finally, application of catastrophe theory in geomorphology is most likely to be successful if it is established in a system of precisely defined forces and resistances. I have argued previously that understanding the resolution of forces and resistances is the key to unravelling fluvial geomorphic processes (Graf, 1979a), and analysis of forces and resistances lies at the heart of much recent geomorphic research into fluvial processes. An entire paradigm of palaeohydraulics, typified by the work of Church (1978), investigates spatial–temporal changes based on analyses of fluvial forces and particle resistances. Catastrophe theory lends itself especially well to the description of systems wherein the controlling factors are force and resistance and the responding factor under investigation is a system attribute.

2.2 DEVELOPMENT OF CATASTROPHE THEORY

Rene Thom of the French Institute for Advanced Scientific Studies created catastrophe theory as a mathematical topologic language to describe spatial and temporal changes in a four-dimensional universe (Thom, 1975). He postulated that change in time and space could be represented by geometric figures, called catastrophes, whose properties were unaffected by deformation. Using differential topology he derived a series of seven such figures which characterized all the possible changes in the four-dimensional universe. His proofs, which have not been successfully challenged, show that although various systems may have different quantitative dimensions, they share certain qualitative characteristics that can be described by his catastrophes. He contended that the qualitative similarities were more important than the dimensional differences. In a geomorphologic example, stream channels have a wide range of physical dimensions throughout the world, but all stream channels share some predictable behavioural characteristics. From the standpoint of the geomorphologist (and from most explanatory perspectives, Thom argued) the similarities in patterns of behaviour are more important than the differences in size or magnitude among the streams.

Catastrophe theory does not supply theoretical explanations. Most geomorphologists view a theory as a set of statements that explain relationships among variables (Harvey, 1969), and catastrophe theory offers no explanation. It merely provides a descriptive model with a specific language that shows the relationships among variables that interact with each other over time and space. It does not explain why these interactions occur, a task that is left to the researcher who applies this model to describe the data.

The importance of language in scientific research cannot be underestimated, because the language used in the theory determines what is real and what is not. If an observation cannot be expressed in available language, it cannot be included in theoretical explanations and therefore from a scientific perspective it does not exist (Boulding, 1956, p. 71). Most of modern geomorphology uses calculus-based mathematics as a language. Applications of calculus have been remarkably successful in characterizing geomorphologic systems, but only a short-sighted researcher would reject out of hand the opportunity to describe and analyse systems from alternative perspectives, especially perspectives which promise to provide insight into the nature of change as well as the nature of stability.

The emphasis on the qualitative rather than the quantitative aspects of the physical world represents a logical outcome of the application of topology where arrangement is more important than magnitude. Since many systems have qualitative attributes that are of interest to researchers, initial applications of the theory attacked problems of changes in qualitative form. Thom used his work to analyse the changes that occurred in life forms during evolution. Subsequently, Zeeman (1977) made extensive use of the theory in a variety of applications in the biological sciences including the description of cell division and change and the behaviour of the brain. Physicists described light caustics (geometric patterns of refracted radiation) as manifestations of catastrophes (Trinkhaus and Drepper, 1977), and engineers used the theory to predict structural failure in beams (Thompson and Hunt, 1973). Zeeman (1977) also suggested applications in the social and behavioural sciences. Wagstaff (1976) explored the potential utility of the theory for geography, while Wilson (1976, 1981) provided a complete analysis of its application to problems associated with urban geography.

Applications of catastrophe theory in the earth sciences have been largely of a speculative nature. Cubitt and Shaw (1976) and Henley (1976) used catastrophe theory as a substitute for other mathematical models in geology. Later investigations emphasized the implications of stable structures of change postulated in catastrophe theory (Slingerland, 1981; Scheidegger (1983). In fluvial geomorphology, Graf (1979b) outlined basic precepts of the theory and offered exploratory examples. Thornes (1980) applied the qualitative aspects of catastrophe theory to analysis of ephemeral streams in dryland Spain, and Graf (1982) used it to describe junction processes in semiarid streams in the Henry Mountains, Utah. Richards (1982, p. 215) used a catastrophe theory representation for channel patterns.

Scientific researchers have not reached total agreement on the utility of catastrophe theory. Although no one has successfully challenged the mathematical–topologic proofs of the theory by its originator (Thom, 1975), applications of the theory have stirred considerable debate. Wagstaff (1976) questioned whether or not the theory could increase understanding of spatial processes. Applications in social and behavioural sciences seem to have created more problems than they have solved. Many critics view catastrophe theory as a sort of intellectual toy that, once used briefly for amusement but without appreciable accomplishment, will soon be cast aside (Croll, 1976; Zahler and Sussman, 1977). Viewing the fanfare with which the theory appeared in scientific and general literature (e.g. Zeeman, 1977), Kolata (1977) characterized it as an 'emperor with no clothes.' Gregory (1985, p. 181), although intrigued by the possibilities of the theory, wondered whether or not it would be one more example of researchers' infatuation with the method rather than the issues at hand.

Since its general introduction in the mid-1970s, catastrophe theory has seen few demanding tests, but a number of workers have used it in fluvial geomorphologic research. Whether or not it will ultimately become an established component of geomorphologic theory building is not yet clear, but catastrophe theory offers new perspectives worthy of close examination. The development and implementation of any intellectual tool requires considerable time as illustrated by the long incubation periods required for the adoption by geomorphology of such fundamental principles as evolution, equilibrium, and statistical analysis. If it is ultimately accepted by the geomorphological community, catastrophe theory will also undergo a long apprenticeship.

2.3 FUNDAMENTAL PRINCIPLES OF CATASTROPHE THEORY

An algebraic interpretation of the underpinnings of catastrophe theory begins with an analysis of energy functions, potentials, and the role of equilibrium (Woodcock and Poston, 1974; Stewart, 1975; Woodcock and Davis, 1978). If a simple physical system is defined by two control (or independent) variables a and b and one response (or dependent) variable x, the three are related to each other by an energy function which defines their interactions: $E(a,b,x)$. For every ordered pair (a,b) there is a value of x that minimizes the energy function E. When E is at the minimum possible, the system is said to be at equilibrium, and algebraically the first derivative of the energy function is zero:

$$dE/dx = 0 \qquad (1)$$

A map of the values (a,b,x) which satisfy equation 1 defines a three-dimensional surface which Thom (1975) labelled a catastrophe (Figure 2.1). The surface is not perfectly smooth, but has a wrinkle or fold in the range of (a,b) values where x may adopt more than one value and still satisfy equation 1. A physical interpretation is that in the wrinkle or fold area, the system may adopt one of two equilibrium states. A third unstable equilibrium state also exists in this range of (a,b) values.

The energy function is of vital importance because it defines the nature of the catastrophe surface. For catastrophe theory applications, energy functions must be Morse functions, meaning they have local minimums. These local minimums, where the first derivative equals zero, are referred to as isolated, nondegenerate, or Morse critical points. Degenerate or non-Morse points are not stable (Gilmore, 1981, provides a more complete mathematical explanation).

At any one time, the condition of a system such as the one outlined above can be graphically plotted as a point in Figure 2.1. If the system is in equilibrium, that is with its energy function at a minimum and with the first derivative equal to zero, the point plots on the catastrophe surface. If the system is not in equilibrium and the derivative of

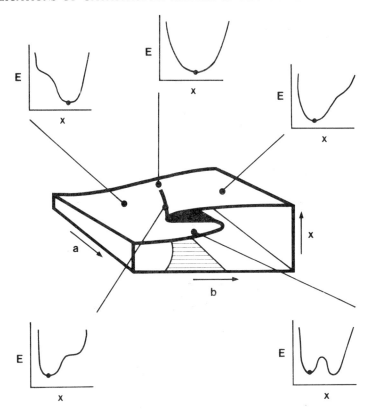

Figure 2.1. Relationships between energy function (E) and responding variable (x) shown in the inset diagrams for a variety of points on a cusp catastrophe. The point in each diagram represents the system conditions at a single time

the energy function does not equal zero, the point plots above or below the surface. Because the catastrophe surface represents a particular solution to the first derivative of the energy function, it is a mathematical singularity and is the only possible solution.

If a system is in a condition that is not an equilibrium state, the energy function operates to move the system toward the equilibrium position. This tendency represents the system potential in a physical sense. Thom's catastrophe functions included catastrophe 'germs' or energy functions which mathematically defined these potentials for systems with up to five control factors (Table 2.1). The simplest catastrophe is a fold, a system with only one control variable and one response variable. In fluvial geomorphology most systems of interest cannot be defined with only one control variable, so the fold has no direct utility for the field. Because the science is concerned with the interactions between force and resistance, however, the cusp catastrophe with its two control factors and one response variable offers some promise (Figure 2.1).

Except for the first two, the catastrophe functions outlined in Table 2.1 produce catastrophes or singularities that are too complex for representation in three-dimensional diagrams. If one or more of the dimensions of the complex catastrophes is held constant, they may be diagrammed (Figure 2.2), but if the fluvial systems considered by the

Table 2.1. Definitions of the seven basic catastrophes

Singularity	Control factors	Behaviour factors	E, Energy function	Derivative: when equal to zero defines singularity
Fold	1	1	$\frac{1}{3}x^3 - ax$	$x^2 - a$
Cusp	2	1	$\frac{1}{4}x^4 - ax - \frac{1}{2}bx^2$	$x^3 - a - bx$
Swallowtail	3	1	$\frac{1}{5}x^5 - ax - \frac{1}{2}bx^2 - \frac{1}{3}cx^3$	$x^4 - a - bx - cx^2$
Butterfly	4	1	$\frac{1}{6}x^6 - ax - \frac{1}{2}bx^2 - \frac{1}{3}cx^3 - \frac{1}{4}dx^4$	$x^5 - a - bx - cx^2 - dx^3$
Hyperbolic	3	2	$x^3 + y^3 + ax + by + cxy$	$3x^2 + a + cy$ $3y^2 + b + cx$
Elliptic	3	2	$x^3 - xy^2 + ax + by + cx^2 + cy^2$	$3x^2 - y^2 + a + 2cx$ $-2xy + b + 2cy$
Parabolic	4	2	$x^2y + y^4 + ax + by + cx^2 + dy^2$	$2xy + a + 2cx$ $x^2 + 4y^3 + b + 2dy$

Figure 2.2. An example of a high-order catastrophe surface, the hyperbolic umbilic catastrophe, shown with all but three variables held constant to facilitate illustration (Woodcock and Davis, 1978, p. 83)

geomorphologist are subject to no more than the general controls of force and resistance, the high-order catastrophes are not needed.

2.4 THE CUSP CATASTROPHE

If the system under analysis consists of two control factors and one response variable, the resulting structure of change is always a cusp catastrophe. Because geomorphologic features are the products of the controls imposed by force and resistance, the cusp is the catastrophe most likely to be useful to geomorphologic researchers. In the following discussion of the properties of the cusp catastrophe, the example of changing channel patterns illustrates the fundamental aspects of the cusp. Channel patterns are generally of three types: straight, meandering, and braided (Leopold and Wolman, 1957), although transitional forms also exist. A simple measure of channel pattern as a responding variable in a geomorphic system is sinuosity, the ratio of along-channel length to the straight-line distance between beginning and ending points of the study reach.

The controlling factors of force and resistance also have simple definitions. Force and power are not the same physical quantities, but stream power is a widely accepted measure of the ability of a stream to perform work and is therefore useful in the present context. Stream power is

$$\Omega = \gamma QS \qquad (2)$$

where Ω = total stream power (N s^{-1}), γ = unit weight of water (9,807 n m^{-3}), Q = discharge (m^3 s^{-1}), and S = energy slope (often represented by channel gradient). Resistance can also be expressed in similar units based on empirical data, but it is more commonly stated in the form of the silt/sand ratio (M) of the bed and bank materials (reviewed by Schumm, 1977, p. 115–117). Resistance may also be related to bank vegetation (Smith, 1976). When the controls of stream power and bank resistance interact, a cusp catastrophe represents the behaviour of the responding variable of channel sinuosity (Figure 2.3).

A point on the channel pattern catastrophe represents power, resistance, and sinuosity values at a given time. If the system is in equilibrium, the point lies on the surface of the figure. As values of power or resistance change, the point moves about on the surface, assuming that the system operation maintains equilibrium. Movement of the point represents change through time, so that the path of point movement dictates the nature of change in the responding variable of sinuosity. Smooth and abrupt changes are possible in the equilibrium system because changes in the control variables may force the system point to move through a smooth transition zone on the figure or across the fold, resulting in abrupt change.

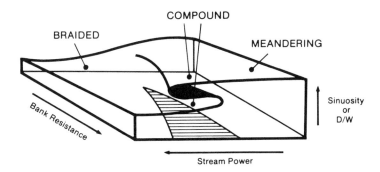

Figure 2.3. The cusp catastrophe relating the control factors of stream power and resistance with the responding variable of sinuosity to produce various channel patterns

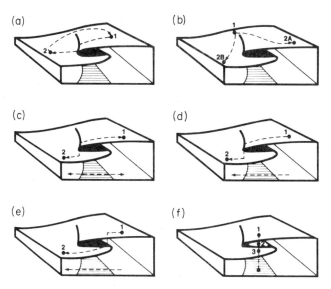

Figure 2.4. Various properties of the cusp catastrophe: (a) smooth and abrupt change; (b) divergence and convergence; (c) hysteresis; (d) Maxwell jump convention; (e) perfect delay jump convention; (f) bimodal equilibrium and the inaccessible or unstable state

As an example, suppose that at time 1 a channel is in a meandering configuration, but that at time 2 it is braided. The transition could have taken place by an increase in stream power which would have moved the point representing the system to the left on the cusp and over the fold (see Figure 2.4(a)), resulting in an abrupt decrease in sinuosity and establishment of a braided channel. Such changes occur on dryland rivers such as the Cimarron (Schumm and Lichty, 1963), the Gila (Burkham, 1972), and the Fremont (Graf, 1983). The same change, or its reverse, could take a different path, controlled by simultaneous adjustments in power and resistance in such a way as to move the point representing the system through the smooth portion of the catastrophe surface (Figure 2.4(a)). When dryland streams make the transition from braided to meandering, it is frequently by this latter route as observed on the Cimarron, Gila, and Fremont Rivers. The power adjustment derives from declining discharges, while the resistance changes result from changing vegetation communities.

Because the cusp catastrophe includes a fold, divergent and convergent behaviour are possible from the same starting or ending conditions. Adjustments in system controls may move a point representing system conditions from the apex of the fold onto either the upper or lower sheet from the same starting point. Conversely, two different systems represented on a cusp catastrophe might change in such a manner as to converge on a common condition. In the example of the channel pattern catastrophe, a channel might be in an intermediate state between meandering and braided (point 1, Figure 2.4(b)), but if the system experiences a significant increase in resistance, the point might be shifted forward on the catastrophe surface. Depending on subtle changes in stream power, the point could move to the upper sheet of Figure 2.4(b), or it could move to the lower sheet representing braided conditions (point 2B, Figure 2.4(b)). In the western United States, adjustments of this type in the twentieth century on the Gila River in Arizona (Graf, 1981) and on the Platte River in

Nebraska (Eschner *et al.*, 1983) have moved the points representing channel pattern to the upper sheet of Figure 2.4(b) and the streams have taken on meandering characteristics. The major adjustments in control factors were vegetation growth that increased resistance (Smith, 1981) and decreased discharges that reduced stream power (Kirchner and Karlinger, 1983).

The operation of many geomorphic systems includes hysteresis, an episodic reversal of trends of change with given states occurring repetitively. The cusp catastrophe represents hysteresis by the movement of the point representing system conditions back and forth across the catastrophe surface (Figure 2.4(c)). In the channel pattern example, if stream power varies over time, the system may respond by alternating between meandering and braided conditions as represented by movement of the point across the fold and then back again in response to change in control variable on the axis across Figure 2.4(c). Alternating meandering and braided channel deposits are common components of the fluvial depositional record (for example, Karl, 1976).

The abrupt change in a system represented by the movement of the point from one sheet to another in the fold area (or bifurcation set) is a jump. Three jump conventions are Maxwell, perfect delay, and random. In the Maxwell convention, the point representing system conditions moves from one surface to the other as soon as it is possible to do so; that is, immediately upon entry into the bifurcation set (Figure 2.4(e)). In the perfect delay convention, the point does not move from one surface to the other until it is about to leave the bifurcation set; that is, when there is no other option (Figure 2.4(d)). In the random convention, once the system enters the fold area the jump occurs with equal probability anywhere within the fold or bifurcation set. In a well-known hydraulic example, apparently the transition from one type of flow to another occurs by a random jump convention. In a certain bifurcation set of paired values of Reynold's numbers and friction coefficients, either tranquil or turbulent flow may occur, representing two possible equilibrium states given one set of control factors (e.g. Chow, 1959, p. 10). The jump from one set of flow conditions to the other is a random event in the bifurcation set.

The shape of the cusp catastrophe in the fold area suggests that in a certain range of the values of the control variables (the bifurcation set) three equilibrium states exist for the response variable. Two of these equilibrium states are stable, and are represented by points 1 and 3 on the upper and lower sheets as shown in Figure 2.4(f). Point 2 in Figure 2.4(f) is on an intermediate sheet and is unstable. Most research efforts disregard this 'inaccessible' region or state. The area in the bifurcation set therefore is a zone of bimodal equilibrium. In the channel pattern application, catastrophe theory implies that at certain values of stream power and resistance, channel sinuosity has one of two possible equilibrium values, and that the channel may be either meandering or braided. Compound channels which exhibit this dual behaviour occur in some streams that are returning to a meandering configuration after a large flood creates braided conditions (Graf, 1987). Compound channels may also be intentionally designed to create stream courses that are more stable than straight conduits (Keller, 1975).

2.5 CATASTROPHE DEFINITION

The major contribution of catastrophe theory is as a qualitative statement about the nature of system change. Quantitative specification of catastrophes is of less importance, but given the present emphasis on quantitative analysis in science generally and in geomorphology particularly, brief mention of quantification of catastrophes is warranted. The two possible routes to quantitative specification are deterministic (deductive) and statistical (inductive).

In the deterministic approach it is necessary to specify the energy function that drives the system variables toward equilibrium conditions, which in turn define the catastrophe surface. In geomorphology the nature of the energy functions involving force and resistance are unknown, and are not likely to be revealed in the near future. A deterministic approach to catastrophe specification is therefore not likely.

An empirical alternative is to observe numerous real world cases of ordered pairs of values for the two control factors of force and resistance and to observe the value of the response variable for each ordered pair. If catastrophe theory is a valid universal interpretation of space–time changes, and if the empirical data derive from a suitably wide range, the data themselves will reveal the outline of the catastrophe. Simple three-dimensional plots of the data reveal the form of the catastrophe surface. Because in the geomorphological world not every assemblage of values for control and response factors (a,b,x) represents a perfect equilibrium case, the resulting surface will not consist of smooth sheets but rather undulating surfaces that have bulges or sags to account for the non-equilibrium outliers.

The cusp catastrophe may be simplified for purposes of empirical definition into two sheets, partially joined but with most of each sheet separate. Trend surface analyses can define the nature of each sheet in quantitative terms and can produce a smoothed version of the data that presumably closely approximates the true equilibrium conditions of the catastrophe surface. Estimation of the two trend surfaces must be accompanied by additional analyses of data in the bifurcation set, because in that range of ordered pairs of values for the control factors (a,b) resulting values of the response variable might fall on either surface. Discriminant analysis can be used to determine the most efficient separation of the points into two exclusive groups. The points separated from each other in the discriminant analysis define the edges of the fold.

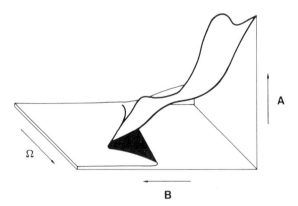

Figure 2.5. An empirical cusp catastrophe representing amount of gully or arroyo cutting in the Front Range of Colorado. Surface irregularities result from the inclusion of non-equilibrium observation points. The control factors are the stream power (Ω) of the 10-year flood and the resistance on the valley floor measured by biomass (B). The response variable is the cross-sectional area (A) of channel erosion. Data from Graf, 1979b

The concept of threshold is directly related to concepts of catastrophe theory because in both cases abrupt changes occur across a mathematically defined boundary. A major difference is in the case of thresholds as used for specific fluvial systems by Brice (1966), Patton and Schumm (1975), and Graf (1979a). The authors characterized the response of the channels under investigation as either entrenched or not entrenched. The cusp catastrophe includes the additional complexity of a response variable with ratio data, such as a measure of the depth of entrenchment or volume of material eroded from the cross-section. The inclusion of ratio data in the analysis converts the simple line threshold defined in two-dimensional space into a more complex three-dimensional fold. It may be that the specific representations by Brice, Patton, Schumm, and Graf are partial views of more complex cusp catastrophes that would be outlined with more complete data as shown in Figure 2.5.

2.6 EXAMPLES OF APPLICATION

The application of catastrophe theory to fluvial geomorphic problems includes issues related to channel patterns, arroyo excavation, sedimentation at channel junctions, ephemeral stream processes, and sediment dynamics at the basin scale. The channel pattern example outlined in the previous section of this chapter is a useful conceptual mechanism that simplifies a complex interplay of variables, but the cusp catastrophe has not been quantified. Sufficient data have been published for this relatively simple task at least at the laboratory level. Field-level research must address resistance offered by vegetation as well as particle size and cohesion.

Graf (1979b) used unit stream power and resistance in similar units as control factors with depth of channel erosion as the response variable in the construction of a quantified cusp catastrophe representing gully and arroyo development. With a graphically defined cusp, the upper sheet had undulations while the lower sheet was smooth (Figure 2.5). Smoothed surfaces resulted from a subsequent application of catastrophe theory where the control factors were the unit stream powers of two channels, a trunk stream and a tributary, while the response variable was the amount of recent sedimentation in the junction (Figure 2.6). If the trunk stream was much more powerful than the tributary, little sediment accumulated. If the tributary was more powerful (usually because of its steep gradient)

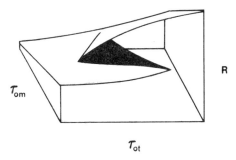

τ_{om}

τ_{ot}

R

Figure 2.6. An empirical cusp catastrophe representing the depth of sedimentation (R) at junctions in a gully or arroyo system in the Henry Mountains region, Utah. Control factors are shear stress in the main (τ_{om}) and tributary (τ_{oT}) channels during the 10-year flood. Surfaces are smooth trend surfaces. Data from Graf, 1982

much sediment accumulated in the junction. At intermediate values, mixed amounts of sedimentation resulted in the bifurcation set.

In his analysis of ephemeral streams in Spain, Thornes (1980) emphasized the stability of the change structure in catastrophe theory. He pointed out that adjustments in the water and transport functions of the ephemeral streams were susceptible to description using catastrophe theory.

J. Phillips (personal communication, 1986) has applied the precepts of catastrophe theory in the analysis of basin-wide sediment budgets for streams in the southeastern United States. Catastrophe theory may be useful in describing the sometimes gradual, sometimes abrupt adjustments in the throughput of sediment in these piedmont streams which control the influx of sediment to estuaries. Phillips found that contributions of catastrophe theory appear to be limited to mere description, adding little in the way of informative insight.

2.7 CONCLUSIONS

Many of the general characteristics of catastrophe theory provide advantages as well as disadvantages in fluvial geomorphic applications. For example, catastrophe theory provides a new perspective on the operations of systems and stimulates the researcher to search for explanations different from those suggested by previous approaches. The newness of the theory is also a powerful disadvantage, however. Recently proposed, incompletely explored, and subject of much debate, catastrophe theory has become available to geomorphology at a time when the science is at the end of three decades of expansion. A review of geomorphic literature suggests that the science is expending research capital in consolidation efforts more than in opening new theoretical territory.

An advantage of catastrophe theory is that it requires the reduction of research problems to the essence of an analysis of forces, resistances, and responses. As a result, the solutions to many different problems are more likely to merge into generalizations and to form cohesive theories. On the other hand, the reduction of geomorphic controls to force and resistance components is not necessarily common or easily accomplished. In many cases, the connection between fundamental physical controls and complex systems observed in the real world has yet to be established.

The operation of control factors that impart to a system a potential to drive the system toward some equilibrium is an inherently pleasing concept for geomorphic applications, but the precise manner of operation of such energy functions remains beyond the theoretical reach of the science. No one has yet written comprehensive and specific energy functions for complex geomorphic systems, and the science still must deal with its mechanics at a restricted scale of analysis.

Catastrophe theory deals with systems that tend in their operation toward an equilibrium whereby there is a balance between force and resistance. The concept of equilibrium pervades geomorphic research and makes it somewhat compatible with catastrophe theory. Some geomorphic systems rarely if ever achieve equilibrium, however, because of their complexity or because the forces which drive them are sporadic or unstable. Some dryland rivers, for example, appear not to tend toward a particular equilibrium condition because they do not experience sustained flows that possess enough power to alter their forms (Stevens *et al.*, 1975; Graf, 1981). The mutual adjustment that occurs in humid-region streams is absent, so that it would not be possible to empirically define the catastrophe surface.

The emphasis on the qualitative in catastrophe theory is also both a benefit and a bane. The qualitative aspects make communication of the theory and its applications easy, and in a field science like geomorphology, field impressions are easily translated into catastrophe theory. But geomorphology has recently undergone a quantitative revolution (Salisbury, 1971; Gregory, 1985) and at present emphasizes quantitative analysis. Catastrophe theory is not likely to contribute to this aspect of the science, but rather will provide more support for the theory-building aspects.

Catastrophe theory suggests that change in all systems with two control factors and one response variable behave in the same way, represented by a cusp catastrophe. Therefore all geomorphic systems controlled by force and resistance will also behave in the same way. This stability of change structure is one of the most valuable contributions of catastrophe theory, because it implies that the fold in the cusp exists, even if a limited data set does not reveal it. Beyond this concept the generality of the theory may mean that it can contribute little else.

The scientific objective of generalizations has the practical promise of predictive capability. Catastrophe theory can provide useful input for the construction of generalizations, and it can provide a sophisticated language for theory building. Quantitative specification of the location of the fold may be useful in predicting the likelihood of impending rapid changes.

Catastrophe theory offers the geomorphologist an additional language for the intellectual exercise of research. Whether or not it ultimately comes to common use in the science depends on its success in solving the problems important to geomorphology. By the mid-1980s there have been few attempts to use catastrophe theory, a circumstance likely to continue as long as other methods solve the problems of the present paradigms. When the unsolved puzzles mount to a level unacceptable to the science, catastrophe theory warrants a complete test before researchers discard it as the 'emperor with no clothes.'

REFERENCES

Boulding, K. E. (1956). *The Image: Knowledge in Life and Society*, University of Michigan Press.

Brice, J. C. (1966). Erosion and deposition in the loess-mantled Great Plains, Medicine Creek drainage basin, Nebraska. *US Geol. Surv. Prof. Pap.*, **352-H**.

Burkham, D. E. (1972). Channel changes of the Gila River in Safford Valley, Arizona, 1846–1970. *US Geol. Surv. Prof. Pap.*, **655–G**.

Chorley, R. J. (1962). Geomorphology and general systems theory. *US Geol. Surv. Prof. Pap.*, **500–B**.

Chorley, R. J. and Kennedy, B. A. (1971). *Physical Geography: A Systems Approach*, Prentice-Hall.

Chow, V. T. (1959). *Open Channel Hydraulics*, McGraw-Hill.

Church, M. (1978). Paleohydrologic reconstructions from a Holocene valley fill. In Miall, A. D. (Ed.), *Fluvial Sedimentology*, Canadian Society of Petroleum Geologists, 743–772.

Croll, J. (1976). Is catastrophe theory dangerous? *New Scientist*, **70**, 630–632.

Cubitt, J. M. and Shaw, B. (1976). The geological implications of steady-state mechanisms in catastrophe theory. *Mathematical Geol.*, **8**, 657–662.

Eschner, T. R., Hadley, R. F., and Crowley, K. D. (1983). Hydrologic and morphologic changes in channels of the Platte River Basin in Colorado, Wyoming, and Nebraska: a historical perspective. *US Geol. Surv. Prof. Pap.*, **1277–A**.

Gilbert, G. K. (1877). Report on the geology of the Henry Mountains. *US Geogr. and Geol. Surv. of the Rocky Mountain Region*.

Gilmore, R. (1981). *Catastrophe Theory for Scientists and Engineers*, Wiley.

Graf, W. L. (1977). The rate law in fluvial geomorphology. *Amer. J. Sci.*, **277**, 178–191.

Graf, W. L. (1979a). Development of montane arroyos and gullies. *Earth Surf. Proc.*, **4**, 1–14.

Graf, W. L. (1979b). Catastrophe theory as a model for change in fluvial systems. In Rhodes, D. D. and Williams, G. P. (Eds.), *Adjustments in the Fluvial System*, Kendall-Hunt, 13–32.

Graf, W. L. (1981). Channel instability in a braided sand-bed river. *Water Resources Res.*, **17**, 1087–1094.

Graf, W. L. (1982). Spatial variation of fluvial processes in semi-arid lands. In Thorn, C. E. (Ed.), *Space and Time in Geomorphology*, George Allen and Unwin.

Graf, W. L. (1983). Downstream changes in stream power in the Henry Mountains, Utah. *Ann. Assoc. Amer. Geogr.*, **73**, 373–387.

Graf, W. L. (1987). Geomorphic evidence of floods along arid-region rivers. In Baker, V. R., Kochel, R. C., and Patton, P. C. (Eds.), *Flood Geomorphology*, Wiley, in press.

Gregory, K. J. (Ed.) (1977). *River Channel Change*, Wiley.

Gregory, K. J. (1985). *The Nature of Physical Geography*, Edward Arnold.

Harvey, D. (1969). *Explanation in Geography*, St. Martin's.

Henley, S. (1976). Catastrophe theory models in geology. *Mathematical Geol.*, **8**, 649–655.

Horton, R. E. (1932). Drainage basin characteristics. *Amer. Geophys. Un. Trans.*, **13**, 649–655.

Horton, R. E. (1945). Erosional development of streams and their drainage basins: hydrophysical approach to quantitative morphology. *Geol. Soc. Amer. Bull.*, **56**, 275–370.

Huggett, R. (1980). *Systems Analysis in Geography*, Clarendon Press.

Huggett, R. (1985). *Earth Surface Systems*, Springer-Verlag.

Karl, H. H. (1976). Depositional history of Dakota Formation (Cretaceous) sandstone, southwestern Nebraska. *J. Sedimentary Petrol.*, **46**, 124–131.

Keller, E. A. (1975). Channelization: environmental, geomorphic, and engineering aspects. In Coates, D. R. (Ed.), *Geomorphology and Engineering*, Dowden, Hutchinson, and Ross, 115–140.

Kirchner, J. E. and Karlinger, M. R. (1983). Effects of water development on surface-water hydrology, Platte River Basin in Colorado, Wyoming, and Nebraska upstream from Duncan, Nebraska. *US Geol. Surv. Prof. Pap.*, **1277–B**.

Kolata, G. B. (1977). Catastrophe theory: the emperor has no clothes. *Sci.*, **287**, 350–351.

Langbein, W. B. and Leopold, L. B. (1964). Quasi-equilibrium states in channel morphology. *Amer. J. Sci.*, **262**, 782–794.

Leopold, L. B. and Wolman, M. G. (1957). River channel patterns — braided, meandering and straight. *US Geol. Surv. Prof. Pap.*, **282–B**.

Leopold, L. B., Wolman, M. G., and Miller, J. P. (1964). *Fluvial Processes in Geomorphology*, W. H. Freeman.

Patton, P. C. and Schumm, S. A. (1975). Gully erosion, northern Colorado: a threshold phenomenon. *Geol.*, **3**, 88–90.

Richards, K. (1982). *Rivers: Form and Process in Alluvial Channels*, London, Methuen.

Salisbury, N. E. (1971). Threads of inquiry in quantitative geomorphology. In Morisawa, M. (Ed.), *Quantitative Geomorphology: Some Aspects and Applications*, State University of New York at Binghamton, 9–60.

Scheidegger, A. E. (1983). Instability principle in geomorphic equilibrium. *Zeit. für Geomorph.*, **27**, 1–19.

Schumm, S. A. (1977). *The Fluvial System*, New York, John Wiley and Sons.

Schumm, S. A. and Lichty, R. W. (1963). Channel widening and flood plain construction along Cimarron River in south-western Kansas. *US Geol. Surv. Prof. Pap.*, **352–D**.

Shen, H. W. (Ed.) (1971). *River Mechanics*, Water Resources Publications.

Slingerland, R. (1981). Qualitative stability analysis of geologic systems, with an example from river hydraulic geometry. *Geol.*, **99**, 491–493.

Smith, D. E. (1981). *Riparian Vegetation and Sedimentation in a Braided River*, MA Thesis, Department of Geography, Arizona State University.

Smith, D. G. (1976). Effect of vegetation on lateral migration of anastomosed channels of a glacier meltwater river. *Geol. Soc. Amer. Bull.*, **87**, 857–860.

Stevens, M. A., Simons, D. B., and Richardson, E. V. (1975). Non-equilibrium river form. *Amer. Soc. Civil Eng., J. Hydr. Div.*, **101**, 557–566.

Stewart, I. (1975). The seven elementary catastrophes. *New Scientist*, **68**, 447–454.

Strahler, A. N. (1952). Dynamic basis of geomorphology. *Geol. Soc. Amer. Bull.*, **63**, 923–938.

Thom, R. (1975). *Structural Stability and Morphogenesis: An Outline of a General Theory of Models*, W. A. Benjamin.

Thompson, J. M. T. and Hunt, G. W. (1973). *A General Theory of Elastic Stability*, Wiley.

Thornes, J. B. (1980). Structural instability and ephemeral channel behavior. *Zeit. für Geomorph. Suppl.*, **36**, 233–244.

Trinkhaus, H. and Drepper, F. (1977). On the analysis of diffraction catastrophes. *J. Phys.*, **A10**, 1–16.

Wagstaff, J. M. (1976). Some thoughts about geography and catastrophe theory. *Area*, **8**, 319–320.

Wilson, A. G. (1976). Catastrophe theory and urban modeling: an application to modal choice. *Env. and Plan.*, **8**, 351–356.

Wilson, A. G. (1981). *Catastrophe Theory and Bifurcation: Applications in Urban and Regional Geography*, University of California Press.

Woodcock, A. E. R. and Davis, M. (1978). *Catastrophe Theory*, E. P. Dutton.

Woodcock, A. E. R. and Poston, T. (1974). *A Geometrical Study of the Elementary Catastrophes*, Springer-Verlag.

Zahler, R. S. and Sussmann, H. J. (1977). Claims and accomplishments of applied catastrophe theory. *Nature*, **269**, 759–763.

Zeeman, E. C. (1977). *Catastrophe Theory: Selected Papers, 1972–1977*, Addison-Wesley.

Modelling Geomorphological Systems
Edited by M. G. Anderson
©1988 John Wiley & Sons Ltd.

Chapter 3

Equilibrium models in geomorphology

ALAN D. HOWARD

Department of Environmental Sciences, University of Virginia,
Charlottesville, VA 22903, USA

3.1 INTRODUCTION

The terms *equilibrium* and *steady state* imply a particular type of relationship between observable quantities. In this chapter this relationship will be defined and illustrated with several examples of geomorphic models that assume equilibrium. Since most readers already have an intuitive grasp of equilibrium, they may find it surprising that the topic is controversial and requires careful discussion both of the operational definition of equilibrium and of criteria for evaluation of equilibrium or disequilibrium in natural systems. Examples are offered below from simple natural and mathematical systems, and the pitfalls or limitations in the use of equilibrium models are discussed.

3.1.1 Equilibrium Defined

The first hurdle is one of definition. The terms equilibrium and steady state have been given a variety of imprecise, overlapping, and sometimes contradictory definitions (see, for example, the discussions by Chorley and Kennedy, 1971, Ch. 6, and Allen, 1974). Howard (1982) offered a definition that conforms to the spirit of prior usage while offering operational rules capable of quantitative testing. This definition is paraphrased and somewhat amended below as a series of propositions followed by interpretive corollaries and clarifications.

Equilibrium refers to a type of temporal relationship between one or more external variables, or inputs, and a single internal variable, or output, that has the following characteristics:

1. Changes in the inputs must cause measurable changes in the output either immediately or after a finite time. This eliminates the trivial case of inputs that have no effect upon the output.

2. The value of the output at a given time is related by a single-valued, temporally invariant functional relationship to the value(s) of the input(s) at the same time, within a consensual degree of accuracy.
3. The functional relationship should be capable of repeatable testing, either experimentally or observationally.
4. An equilibrium relationship may be limited to certain ranges of the input values and/or to certain rates of change of the input values.

The heart of the definition is the single-valued functional relationship, which incorporates both the notion of time independence in the relationship between inputs and outputs and the possibility that equilibrium can be maintained as both inputs and the outputs change through time (this is analogous to the concept of reversible reactions in chemical equilibria). The concept of an allowable margin of error in the functional relationship moves equilibrium from an unattainable ideal to a useful and quantitatively testable tool.

Although the above definition is fairly self-contained, a number of implications and clarifications need to be explored.

Equilibrium makes no necessary statement as to the chain of cause and effect. The terms input and output suggest a causal dependence, but this is not intended. Inputs and outputs are defined merely as quantities that can be measured more-or-less simultaneously and repeatedly. Similarly, the terms external and internal variables are not meant to imply physical separation or one-way fluxes of mass or energy. In other words, equilibrium as proposed here implies a 'black-box' approach to system identification. In general, an investigator is concerned with the question of cause and effect, and tries to make the external, or input variables the causes and the internal, or output variables, the effect. However, feedback relationships often make such distinctions difficult or impossible. Furthermore, identification of cause and effect generally requires information and hypotheses supplemental to the determination of equilibrium.

Equilibrium is not a property of the system as a whole. No natural system can ever exhibit single-valued functional relationships between all measurable properties at all physical and temporal scales. For example, the velocity of water in a canal may be essentially uniform and steady when measured by a traditional current meter, and related in a predictable manner to the flow depth and the cross-sectional characteristics and gradient of the canal. However, when the flow is measured by hot-wire anemometry, the small-scale rapid turbulent fluctuations lose a single-valued relationship to depth and canal characteristics (although a statistical relationship still applies).

Much of the confusion in the discussion of equilibrium in geomorphic systems arises from the attempt to identify physical systems as being in equilibrium as opposed to exhibiting disequilibrium or possibly 'inherent instability'. In a broad sense, this issue involves our propensity to confuse models and measurements with reality. For example, we generally feel that the 'velocity' of water in a channel flow is a well-defined property of the system. However, from the preceding paragraph, it is apparent that velocity depends not only upon the system but also upon how it is measured, and in particular, upon the characteristic physical and temporal scales of the measuring procedure. Because of these considerations, the definition states that equilibrium criteria must be applied separately to each internal, or output, variable, whereas several external variables may be involved for each internal variable. In general, all temporally varying external variables that may appreciably affect the output variable should be included if a functional relationship is to be identified.

Although equilibrium is manifested by a temporally invariant functional relationship, temporal change in internal and external variables is necessary to identify the relationship and test equilibrium. Thus equilibrium does not imply either unchanging inputs or output. In fact, a completely static relationship is indecisive about the sensitivity of the output to changes in the input. As will become clearer below, the occurrence or lack of an equilibrium relationship depends upon the rate at which the output variable can adjust, or respond, to changes in the input variable.

The selection of input and output variables and their methods of measurement are often determined by a desire to find an equilibrium relationship. All measurements involve temporal and areal/volumetric averaging. The selection of an appropriate scale of averaging is generally influenced by the characteristic response time of the output variables to changes in input as well as other pragmatic or theoretical concerns. Mismatching of spatiotemporal scales for measurement of input and output parameters, or measurements that are made at timescales much longer or much shorter than the characteristic response time will generally complicate or obliterate any possible equilibrium relationships. Returning to the previous example, if velocity is measured by 'instantaneous' hot-wire anemometer readings, no simple hydraulic relationships are likely to be discovered relating the velocity to channel form variables and depth measurements made at larger physical–temporal scales. The importance of timescales to equilibrium concepts is further explored in the following examples, and has been discussed by several authors, including Thornes and Brunsden, 1977; Brunsden, 1980; Cullingford *et al.*, 1980; and Howard, 1982.

3.2   EQUILIBRIUM CONCEPTS AND EXAMPLES

In the following sections the above definition will be applied to various types of system models. The first type of system will be a simple deterministic linear system. The equilibrium concept will then be extended to more complicated systems that involve non-linear relationships, spatial extension, and/or stochastic inputs.

3.2.1   Equilibrium in Deterministic Linear Systems

Howard (1982) illustrated the definition of equilibrium by considering a simple linear system in which an ouput, or response, $y(t)$, is a weighted average of past values of a single input, $x(t)$. The weighting function is assumed to be a negative exponential, so that:

$$y(t) = \lambda \int_0^\infty x(t - \tau) \, e^{-\lambda \tau} \, d\tau \tag{1}$$

where $T = 1/\lambda$ is the characteristic relaxation time of the system (the relaxation time is also sometimes defined as the time, T', such that $e^{-\lambda T'} = 0.5$ (Waide and Webster, 1976)). The conditions for equilibrium and the dynamics of such a system can be illustrated by considering the output response to simple inputs such as step changes, sinusoids, and impulses.

The output response to a step change in the input at time t_0 from a value of C_1 to C_2 is, for $t > t_0$:

$$y(t) = C_2 + (C_1 - C_2) \, e^{-\lambda(t - t_0)}, \tag{2}$$

assuming $x(t) = C_1$ for all $t < t_0$.

The inputs to many natural systems are quasiperiodic (with, for example, a yearly cycle). Such inputs can be represented conceptually as a sum of sinusoidal inputs with different frequencies superimposed upon an average value. For a single frequency component with an average value:

$$x(t) = \alpha \sin\{\omega t\} + C, \text{ so that} \tag{3}$$

$$y(t) = A \sin\{\omega t - \theta\} + C, \text{ where} \tag{4}$$

$$A = \alpha\beta/(\beta^2 + 1)^{1/2} \text{ and } \theta = \arctan 1/\beta.$$

Also, $\beta = \lambda/\omega$ and it is assumed that the input $x(t)$ has been maintained for all past time.

For this type of input the system response is a constant value with a superimposed sinusoidal component that is a delayed and damped replica of the input. The ratio A/α is conventionally termed the magnitude ratio, and θ is the phase shift (see Pickup and Rieger, 1979). These, in turn, are functions of the parameter β (termed here the *response ratio*) that relates the input frequency to the system relaxation time. The magnitude ratio and phase shift are plotted as a function of the response ratio in Figure 3.1.

Another type of simple input is a linear trend:

$$x(t) = C + \delta t, \tag{5}$$

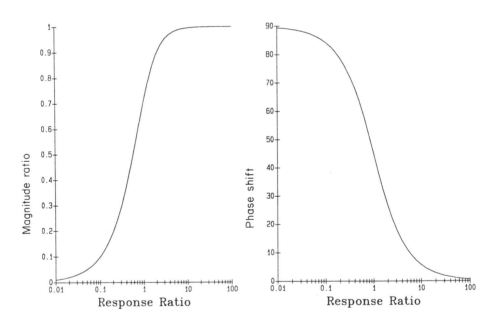

Figure 3.1. The magnitude ratio (A/α) and phase shift (θ, in degrees) plotted as a function of the response ratio (β) for a linear system with finite memory

which has the response:

$$y(t) = C + \delta t - \delta/\lambda. \tag{6}$$

A final simple input is an impulse of magnitude M occurring at time $t = t_0$:

$$x(t) = C + M\Delta(t - t_0), \tag{7}$$

where $\Delta(t - t_0)$ is a unit impulse. The system response for $t > t_0$ is:

$$y(t) = C + M\lambda \, e^{-\lambda(t - t_0)}. \tag{8}$$

The response of our example linear system to an arbitrary input can generally be modelled as a linear combination of such simple inputs.

3.2.2 Criteria for Equilibrium

Equilibrium occurs when the value of $y(t)$ is sufficiently close to the value that it would have if the input variable(s) were constant over an infinite time ($x(t) = C$). This *ultimate* equilibrium value, $Y(t)$, for this system is:

$$Y(t) = \lambda \int_0^\infty C \, e^{-\lambda\tau} \, d\tau = C \tag{9}$$

However, as is evident from this equation, an infinite time is required to reach the ultimate equilibrium, but a *consensual* equilibrium can be defined that permits a specified departure from ultimate equilibrium. In this paper the unmodified term equilibrium refers to such a consensual equilibrium.

Two possible definitions of consensual equilibrium for the example system are discussed here. The first definition requires that the response be within a given fraction of its ultimate value, $Y(t)$, such that:

$$|Y(t) - y(t)| \leqslant \epsilon |Y(t)|, \tag{10}$$

where ϵ is the consensual magnitude of acceptable deviation.

For a step change in input (Equation 2) equilibrium will occur for $t \geqslant t_e$, where:

$$t_e = t_0 - \ln\{(\epsilon |C_2|)/|C_1 - C_2|\}. \tag{11}$$

However, if $t_e \leqslant t_0$ (that is, if $\epsilon |C_2| \geqslant |C_1 - C_2|$), then equilibrium is considered not to be disturbed by the step change. For a sinusoidal input the deviation of the input and output signals also varies sinusoidally, so that a modification of equation 10 is proposed that uses the maximum deviation during an input cycle as a criterion:

$$\mathscr{M}|y(t) - Y(t)| \leqslant \epsilon \mathscr{A}|Y(t)|, \tag{12}$$

where $\mathcal{M}$ is the maximum, and $\mathcal{A}$ the average value of their arguments over one input cycle. For the input of equation 4 this criterion yields equilibrium for:

$$|\alpha \sin\{\theta\}| \leqslant \epsilon |C|. \tag{13}$$

For small values of the response ratio $\sin\{\theta\} \approx 1$, so that equilibrium will occur only for small amplitude sinusoidal components. For very small response ratios the output responds only to the average value of the input, that is, the high frequency components are filtered out. In such cases it may be warranted to choose a different input variable which is a temporally averaged value of the original variable, since equilibrium might then occur. This prefiltering should have a characteristic response ratio of the same order of magnitude as the system response ratio. Howard (1982) suggested that cyclical components that elicit little system response due to their high frequency can be ignored with regard to the occurrence of equilibrium; the approach recommended here of redefining the input variable by filtering the original variable is conceptually simpler and more consistent.

For some purposes it may be expedient to define equilibrium differently than suggested in equation 10 or equation 12. For example, for systems whose mean value is near zero, an absolute deviation may be appropriate:

$$|Y(t) - y(t)| \leqslant \epsilon, \tag{14}$$

whereas for step changes a proportional response may be informative

$$|Y_2 - y(t)| \leqslant \epsilon |Y_2 - Y_1|, \tag{15}$$

where Y_1 and Y_2 are the initial and final equilibrium response values, respectively. If $y(t)$ and $Y(t)$ are finite and range through several orders of magnitude, then the logarithms of the inputs and output may be used in testing for equilibrium.

The selection of an equilibrium criterion and the limit of consensual error will depend upon the purpose of the study and the nature of the system being examined.

3.2.3   Systems with Stochastic Components

The description of many systems must incorporate random or stochastic components. The implication of randomness to equilibrium concepts depends upon the nature of the randomness and the manner in which it enters into the relationship between inputs and the output.

For some systems the stochastic component can be considered to be part of the input signal, or as an additional, perhaps uncontrolled input signal. For example, random rainfall is an input to synthetic streamflow synthesis.

The status of random inputs relative to equilibrium depends upon the purpose and design of the model. In some cases the random component can be treated as an additional input, or as a component of an input with additional deterministic aspects. Treatment of such random components is conceptually no different than the simple input signals discussed above, in that disturbance from equilibrium depends upon the magnitude of the random component, and the temporal characteristics of the randomness. In particular, the likelihood of disturbance from equilibrium depends upon the strength of the random signal in various frequency ranges. Stochastic components that change slowly compared to the system response time are unlikely to cause departure from equilibrium. Components that change over timescales similar to the response time may cause disequilibrium if they are strong, and very high frequency stochastic inputs will affect the output through an averaged response. Mathematical modelling of system response to stochastic inputs is fairly tractable in the case of linear systems, and is discussed in several texts on stochastic processes, such as Papoulis (1984).

However, in cases where prediction is a goal of the model (rather than, say, identification of the response characteristics of the system), any random input component of large enough magnitude to affect the relationship between the deterministic inputs and the output would be considered to cause disequilibrium.

In some cases the stochastic component may be modelled as being part of the internal functioning of the system, that is, the response of the system to a well-defined input involves some level of unpredictability in the output. One source of such unpredictability might be uncertain knowledge of the initial state of the system, or lack of information on system inputs prior to experimentation or observation. This type of unpredictability, if sufficiently large, does indicate lack of equilibrium, since the criterion is a single-valued relationship between inputs and the output. Alternatively, the system may be very complicated and our model an inadequate description of system response. Complicated responses to simple inputs, for example, thresholds or oscillatory behaviour, generally imply that equilibrium concepts are inapplicable. However, the same physical system when viewed from another perspective, using different input and/or output variables, might exhibit equilibrium. An example from models of meandering streams is discussed below.

Errors in measurement of inputs and/or the output are another source of randomness. In some cases the measurements may appreciably disturb the functioning of the system. Because these are superimposed upon the system being modelled, the scientist would generally prefer to consider such random elements to be irrelevant to the question of equilibrium in the target system. However, and particularly in non-experimental situations, it may be difficult to refine the accuracy and precision of measurement. In fact, it may be impossible to distinguish measurement errors from random components of system inputs or system behaviour, with a resulting uncertainty about presence or absence of equilibrium.

Random components that are due to measurement errors in inputs or outputs have an important distinction from random inputs that affect the output via the system response. The system response transforms the random inputs in the same manner as deterministic inputs. Systems with 'memory', such as the averaging response of equation 1, cause correlation between successive output values, even if the inputs are uncorrelated. Kirkby (1987) shows that this introduces a type of Hurst effect in system response.

Systems that are areally very complex, and are therefore impossible to measure adequately, may be modelled as having spatial randomness. For example, permeability in an aquifer is often modelled as having random components at a spectrum of spatial scales from individual grains to aquifer-wide. However, the permeability can usually be considered to be temporally constant, so that it is possible to model steady-state response to constant recharge. Other geomorphic systems are not spatially fixed (such as meandering streams), so that areal randomness (e.g. of bank resistance) causes temporal randomness.

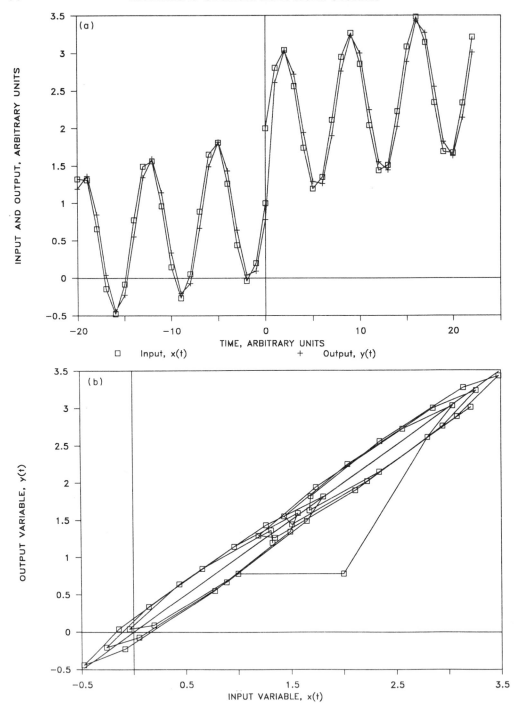

Figure 3.2. Output response for a linear system with finite memory and short relaxation time to an input with a combination of a trend, cyclical fluctuation, and step change: (a) Input and output as a function of time; (b) Output plotted as function of input, showing ultimate equilibrium response (straight line)

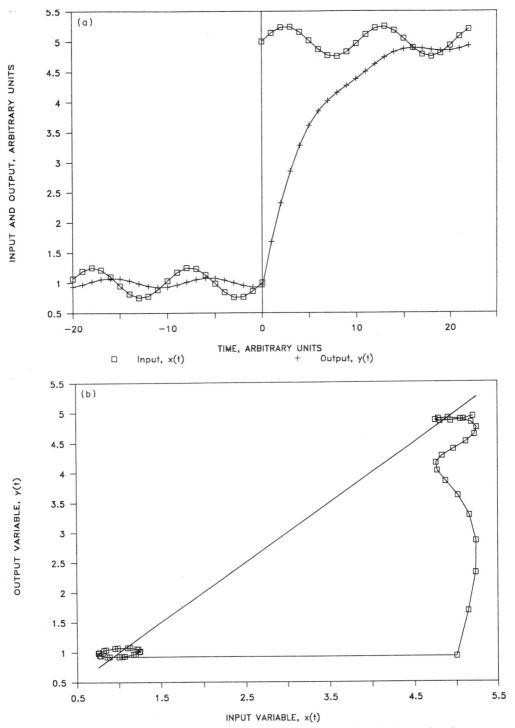

Figure 3.3. Output response to a large step input. See Figure 3.2 for further explanation

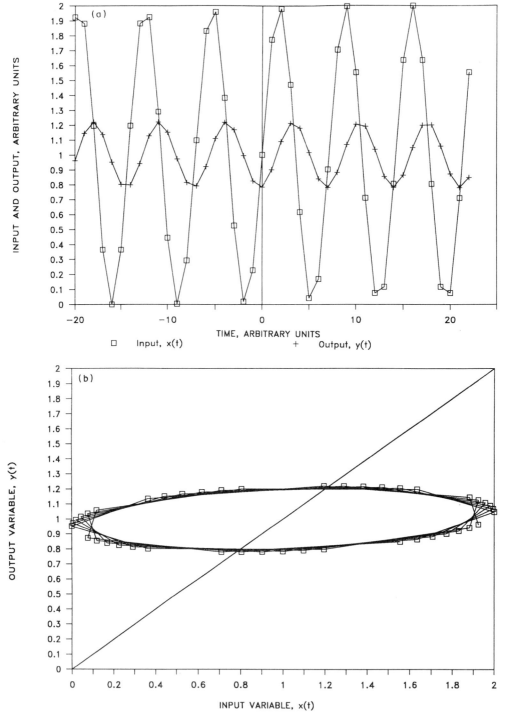

Figure 3.4. Output response to high-frequency cyclical input. See Figure 3.2 for further explanation

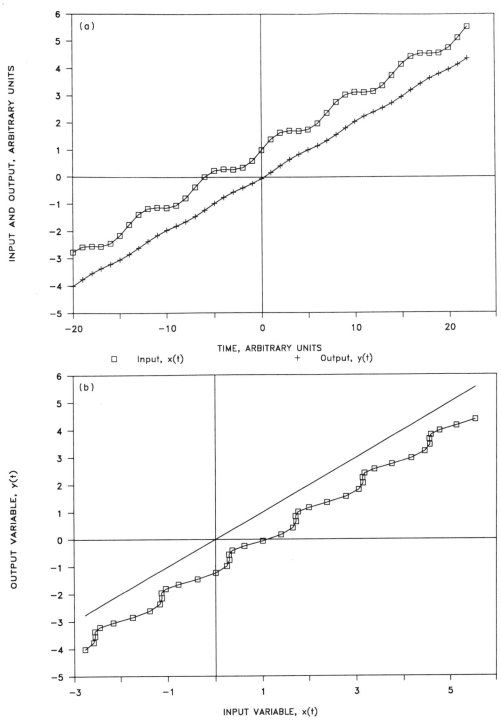

Figure 3.5. Output response to a rapid trend in input. See Figure 3.2 for further explanation

3.2.4   Identification of System Response and Equilibrium

Equilibrium and non-equilibrium behaviour of the simple linear system presented above is illustrated in Figures 3.2–3.5. Figure 3.2(a) shows a system input consisting of a sinusoidal component with a superimposed trend and a step change at time $t = 0$. A short relaxation time for the system has been chosen, so that the output closely follows the input and a plot of the input versus the output (Figure 3.2(b)) shows a nearly linear relationship with the only major deviation occurring immediately after the step change (in this and Figures 3.3b, 3.4b, and 3.5b the ultimate equilibrium value is the straight line of equality between input and output). In Figure 3.3 a large step change is coupled with a slow relaxation time, so that the plot relating input and output is complicated and multivalued. Figure 3.4 illustrates the amplitude damping and phase lag that occur when the system relaxation time is slow compared to the frequency of the input. The input is not strongly related to the input and strong hysteresis is shown in the input–output plot (Figure 3.4(b)). An input characterized by a rapid trend (Figure 3.5(a)) creates a bias in the relationship between the input and output (Figure 3.5(b)).

The simplest scenario for examining equilibrium behaviour occurs when observations of inputs and output are paired but are not sequentially linked. This arises commonly in experimental situations where the inputs are controlled and fixed at constant values and the output readings are taken when all transients have damped out. Each input and output pair result from a separate experimental observation. Many hydraulic and sediment transport relationships have been developed from such experimentation. The equilibrium behaviour of the system can therefore be determined if the experiment is well-defined (lack of uncontrolled inputs, suitable range of input variation, measurements with acceptable errors, sufficient length of system operation under constant input to assure equilibrium, etc.) and if the system is well-behaved (lack of oscillatory or threshold response within the range of input variation, lack of hysteresis, etc.). The main limitation of such experimentation is lack of characterization of transient system response.

Many non-experimental field data in geomorpholgy are analysed in a similar manner. Most hydraulic geometry relationships and morphometric studies fall in this category. Most, however, do not have the observational controls that permit the assertion of controlled, constant inputs and lack of transient responses. As a result, most such relationships exhibit large degrees of scatter. The difficulties of morphometric studies are well known, but a brief review is important to examine the implications for determination of equilibrium.

The causes of scatter are difficult to determine in morphometric and similar studies. One source may be transient response to changing inputs, such as is shown in Figures 3.3–3.5. Uncontrolled inputs are another source, as are measurement errors. If a random input or measurement error component were superimposed upon the responses of Figures 3.3–3.5 and the temporal relationships between observations were obscured, it would be difficult to distinguish between the effects of random components and the transient response to deterministic inputs. Thresholds and oscillatory response are also possibilities. Knowledge of the temporal evolution of a system in many cases permits more definitive modelling of sources of scatter, as will be discussed below.

Two assumptions are commonly made in morphometric studies. One is that underlying trends in the relationship of inputs and the output define the equilibrium response. The trends are commonly identified by statistical regression techniques. The second assumption is that equilibrium behaviour for temporal changes in an input variable can be defined by investigation of spatially separated, but similar systems that exhibit different values of the input variable; in short, this is the technique of the 'substituting of space for time'. These assumptions can be misleading if there exists covariant response of an uncontrolled variable to variations of the input variable. Similarly, the similar spatial systems investigated to determine equilibrium may be conjointly varying in response to a systematic change in the input variables that occurred prior to the sampling. For example, trends in inputs or recent step changes can cause a systematic bias in the response (Figures 3.2 and 3.5). An example of such historical influences is the large increases in channel width that occur in some semiarid streams following major floods (Schumm and Lichty, 1963). Hydraulic geometry studies in such basins will give different relationships if done before versus immediately after such a flood event. Good experimental design can minimize covariant uncontrolled inputs, and historical information about process variations from observations or the sedimentary record can help eliminate or control for large biases.

Considerably greater possibilities for system identification and prediction are possible if the temporal response of the output to changes in the input(s) can be observed. A wide range of time series analytical tools may be applied to determine the relationships between inputs and outputs, including analysis for trends, cross-spectral analysis, and ARIMA and transfer function models discussed by Box and Jenkins (1976). For example, the simple linear system of equation 1 can be represented by the differential equation:

$$y(t) + \frac{1}{\lambda} \frac{dy}{dt} = x(t), \tag{16}$$

which in turn can be expressed as a transfer model difference equation:

$$y_t(1 + \frac{1}{\lambda \, \Delta t}) - \frac{1}{\lambda \, \Delta t} \, y_{t-1} = x_t, \tag{17}$$

where Δt is the time increment, and the subscripts t and $t-1$ refer to the present and past observation times. Box and Jenkins (1976) detail methods for estimating the coefficients of such difference equations based upon time series observations of x and y, or more generally of the inputs and output. Such models will also estimate a stochastic component that is the unfit residual to the difference model, giving a measure of the degree of fit of the model to the observed process. These time series approaches can distinguish between scatter in the relationship between inputs and outputs that is due to deterministic departures from equilibrium (e.g. Figures 3.3–3.5) from those due to random components. The time series estimation procedures can be used to test the applicability of theoretical models to observed system behaviour or in a more general exploratory mode to identify the most parsimonious mathematical model that adequately describes the system behaviour. The time series approach is, in fact, more general than strictly equilibrium models since transient response also can be estimated.

3.2.5 Examples of Geomorphic Equilibrium Models

Several classes of models that employ equilibrium assumptions or illustrate requirements for equilibrium will be discussed below. Most of these are non-linear models that pertain to spatially extended systems. Some models involve stochastic components. Despite the more complex nature of these models when compared to the previous deterministic linear system, equilibrium behaviour is still a useful concept.

Equilibrium and Grade in Alluvial Streams

Howard (1982) provided an extended discussion of the conditions under which alluvial streams can approach an equilibrium between channel gradient and the hydraulic regime (the water discharge and the size and amount of sediment supplied from upstream). Howard simulated the response of alluvial stream profiles to various temporal histories of variation in the hydraulic regime, such as step changes in discharge or sediment load, pulse inputs of sediment, and sinusoidally varying discharges or sediment loads. He showed that alluvial channels will tend to return to equilibrium following a disturbance such as a pulse input or step change, and provided estimates of how the timescale for reestablishment of equilibrium depend upon the hydraulic regime and the length of the stream or stream network. In a manner similar to the linear system discussed above, cyclical regime changes having a long period compared to the response time of the stream do not cause disequilibrium. Also similarly, stream gradients do not appreciably respond to very high frequency regime changes. Howard also investigated the response of a stream to changes in base level and the timescales of readjustment.

The gradients of dentritic stream systems respond fairly uniformly throughout the network to changes in regime or base level, whereas in long, unbranched streams the gradient response to hydraulic regime tends to propagate downstream, and base level changes propagate upstream. Howard also found that Mackin's (1948, p. 471) classic definition of a graded stream as 'one in which, over a period of years, slope [gradient] is delicately adjusted to provide, with available discharge and with prevailing channel characteristics, just the velocity required for the transport of the load supplied from the drainage basin. The graded stream is a system in equilibrium . . .' is applicable to alluvial stream channels if 'a number of years' is replaced by 'a period of time commensurate with the relaxation time of the gradient'. That is, the timescale of measurement or averaging of input variables should be commensurate with the response time of the system if equilibrium relationships are to be tested or investigated.

Dynamic Equilibrium in Landform Evolution

Hack (1960, 1965, 1975) elaborated the concept of dynamic equilibrium in landform evolution. His hypotheses were never presented in axiomatic or mathematical form. However, the core of his papers is the assumption that *if* (1) a land area undergoes a constant rate of uplift, and (2) geomorphic processes (as affected by climate) remain constant, *then* the geometry of landforms attains a steady-state. Hack then further asserted that certain areas in the southeastern United States fulfilled these conditions for dynamic equilibrium. Dynamic equilibrium is not conceptually different from the general use of equilibrium in this paper, so that the term 'dynamic' is unnecessary baggage. The present discussion will not address the question of applicability of such equilibrium to any given area, but rather clarify the conditions under which equilibrium would occur and present examples of quantitative models of such equilibrium.

Some clarification of the definition given above is necessary, in that the output variable is not clear. The temporal independence of landforms can be made more precise by making the output variable the surface gradient at a particular location. Since the working definition of equilibrium restricts the application of equilibrium to a single output variable which is presumably measured at a given location, the hypothesis also involves the assertion that equilibrium is satisfied at all spatial locations simultaneously. Quantitative dynamic models of slope and channel evolution have been developed in recent years that can be utilized to investigate the geometric properties of equilibrium landforms. A recent example is a paper of F. Ahnert (1987), in which two dimensional slope profile simulation models incorporating process assumptions about mass-wasting and slope wash erosion are started from arbitrary initial conditions and iterated through time under assumptions that the base of the slope erodes at a constant rate. An example of this type of model is shown in Figure 3.6, where it is assumed that creep is the sole erosional process and that the rate of creep

Figure 3.6. Evolution of a two-dimensional slope eroded solely by creep. Initial profile is level, and lower end of slope is eroded at a constant rate. Only the slope surface is shown for each profile

is proportional to the sine of the slope angle. Similarly, Figure 3.7 shows a slope profile in which two processes act: creep and surface runoff. Runoff erosion is assumed to occur at a rate proportional to the shear stress exerted by the runoff on the slope surface. The latter profile shows the inflection point and concave lower slope that is sculpted primarily by runoff (this concave zone typically contains rills and gullies on actual slopes). Such models have limited direct application to natural slopes due to simplified process assumptions, two-dimensional profiles, and the restrictive equilibrium assumptions outlined above. However, they serve several important purposes. Firstly, they demonstrate that fairly realistic assumptions about slope processes are compatible with the development of equilibrium slopes. The temporal and spatial feedbacks tend to create smooth slopes and gradual approaches to equilibrium that lack 'pathological' behaviour such as thresholds, oscillation, metastable states or unbounded change. Secondly, such models give an estimate of the length of time

Figure 3.7. Evolution of a two-dimensional slope eroded partially by creep and partially by slopewash. Creep is the dominant process near the divide and slopewash predominates on the lower slope, with the changeover of relative importance occurring near the slope inflection

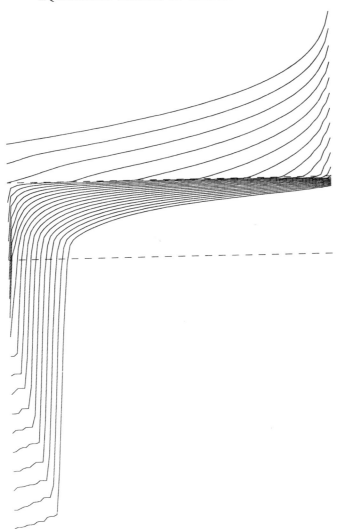

Figure 3.8. Evolution of a bedrock stream channel whose lower end downcuts at a constant rate. The discharge is assumed to increase downstream, so that the initial profile is concave. Bedrock resistance is assumed to be uniform except for the layer between the dashed lines, whose resistance is four times greater. Note that the variable resistance makes the spatial–temporal pattern of gradients and erosion rates non-uniform. The stream profiles exhibit slight wiggling or oscillation at small spatial scales that are an artifact of the simulation which would be removed by shorter temporal increments

(relaxation time) or total amount of erosion that is necessary to approach to a desired approximation of ultimate equilibrium (Ahnert, 1987). The spatial and temporal variation in erosion rates are a good index of the degree of approach to equilibrium. Finally, the equilibrium models help us to understand how the balance of processes controls the geometry of landforms. For example, in the simulations involving both creep and surface wash erosion, the drainage density (or its reciprocal, the constant of channel maintenance) functionally depends on location of the inflection point on the slope. Therefore such slope models can elucidate the process and material parameters that control spatial and temporal variations in drainage density.

The geologic materials being eroded at any given location must remain constant in erosional susceptibility in order for equilibrium to occur. When erosion progresses through layered rock, erosion rates and surface gradients vary through time. An example is shown in Figure 3.8, which is a simulation of channel erosion in which it is assumed that a bedrock-floored channel with discharge increasing downstream is being eroded at a rate proportional to the average boundary shear stress and inversely proportional to the rock resistance (Howard and Kerby, 1983). In this simulation a resistant rock layer inhibits erosion rates (and produces low gradients) above the layer, and produces very high erosion rates and steep gradients below and in the lower portion of the bed. Because the locus of exposure of the resistant bed migrates upstream, channel gradients are clearly not temporally constant.

The extension of such models from illustrating general concepts of slope erosion to application to specific natural slopes is problematic, as is suggested by the wealth of discussion that the Hack papers engendered. The first problem is the applicability of the process assumptions to the natural situation; this is a thorny issue beyond the scope of this paper. A very serious problem is the obvious lack of observation of temporal landform evolution that would be a direct test of equilibrium. Such direct testing could only be undertaken in very rapidly evolving landscapes, such as some badlands. Most of Hack's arguments in the southeastern United States were based upon substitution of space for time; the difficulties of this approach have been mentioned above.

Aside from temporal changes caused by the exposure of differing rock types, changes in climate are most likely to cause disequilibrium. Short-term climatic changes, and even individual storms, modify slopes and channels locally. This suggests that the definition of slope and channel gradients that would be tested for long-term equilibrium should involve spatial and temporal averaging at a scale equivalent to the natural relaxation timescale. In fact, some landforms exhibit threshold behaviour such as landslides at short timescales. Informative examples occur on the steep slopes of basaltic rocks on the Hawaiian Islands. Many slopes on the tradewind sides of the islands are very steep, with gradients averaging more than 60°. The abundant rain and easy weatherability of the basalt encourage rapid plant growth and soil development. But whenever an appreciable soil thickness develops, soil avalanches occur, triggered by the frequent intense rains, and the process begins anew. Therefore, over short timescales the slope behaviour is clearly episodic, but long-term average erosion rates and overall slope profiles may possibly maintain equilibrium.

The assumption of constant rate of tectonic uplift is probably untestable in most areas. However, a more tractable assumption may be substituted: the hypothesis that the regional rate of base level lowering has remained constant through time. In areas spatially and vertically remote from the oceanic base level, erosion rates and the overall relief are essentially unaffected by relative land–sea changes over timescales less than several million years. In such remote areas the timescales for readjustment of slope and headwater channel form are shorter than the timescales affecting overall relief. This should be true in areas where the local relief is much smaller than the elevation above base level. Thus individual basins should be undergoing erosion rates that are areally and temporally fairly uniform. But these erosion rates may differ significantly in contiguous basins if they drain through much different paths to the ocean. An example is the considerable relief difference across the Blue Ridge Scarp in southern Virginia and North Carolina (Hack, 1973).

In conclusion, the concept of dynamic equilibrium in landforms is of problematic applicability to natural landscapes, but it is an important and informative concept in theoretical modelling.

Equilibrium Models of Aeolian Dune Form and Migration

In contrast to the dynamic equilibrium of large regions, the application of equilibrium hypotheses to the form of sand dunes is better substantiated and process modelling is more realistic. Theoretical and empirical studies have concentrated primarily upon barchan dunes (Howard *et al.*, 1978; Howard and Walmsley, 1985; Wipperman, 1986), although some efforts have been made on field measurements and sand transport modelling on longitudinal dunes (Tsoar, 1983b) and other transverse dune forms (Tsoar, 1983a; Lancaster, 1985; Livingstone, 1986).

The barchan dune is an isolated dune forming on flat desert surfaces in essentially unidirectional winds. These dunes have been observed to migrate over long distances with little change in size or shape. Therefore, to a first approximation the barchan dune is a sedimentary landform on which the spatial pattern of erosion and deposition are balanced to maintain a constant geometry (so long as the coordinate system is considered to follow the migrating dune). This maintenance of form is of considerable help in testing theoretical models of dune evolution in that form equilibrium implies a specific areal variation in relative erosion and deposition rates over the dune that can be compared with model predictions. Furthermore, the barchan dunes migrate rapidly enough in most deserts that migration rates have been related to wind climate and dune size. The equilibrium assumption for natural barchan dunes requires that temporal variations in wind speed can be accounted for by an equivalent constant wind speed and that direction fluctuations are small. Because different wind speeds will, in general, be associated with somewhat different dune geometries, this may result in somewhat biased predictions. Some natural dunes will obviously satisfy these requirements more closely than others.

Howard *et al.* (1978) measured the areal variation in wind speed and wind direction over a natural and a model barchan and used sediment transport relationships corrected for slope effects to predict areal variations in sediment erosion and deposition rates. The predicted rates agreed in broad pattern with the rates predicted by the assumption of equilibrium. The limiting factor in this modelling approach was that the predictions were limited to the particular dune geometry for which wind measurements were taken, so that it is difficult to assess the reasons for the residuals in the prediction of erosion/deposition rates. Such residuals could have arisen from inappropriate model assumption about sediment transport mechanics, from errors in measurement of wind speed and dune geometry, and from the effects upon dune form of the naturally varying wind speeds and directions (a unidirectional oncoming wind of a fixed velocity was assumed in the modelling).

A more comprehensive modelling effort requires prediction of near-surface wind speeds and directions for a dune form of arbitrary geometry. This permits a more general simulation model in which the wind field is calculated followed by estimation of erosion and deposition are according to the sediment transport assumptions (thereby modifying the dune form), and the iterations of wind calculation and erosion/deposition are continued until an equilibrium form is obtained. Such an approach has recently been used by Howard and Walmsley (1985) and Wipperman (1986), with encouraging results. Such models require efficient algorithms for calculating the wind field as well as a powerful computer. The Howard and Walmsley simulations were troubled by numerical instability but nonetheless showed that a rounded pile of sand subject to a unidirectional wind began to take a barchanoid form. The Wipperman simulations were more successful, progressing to a realistic barchan form, but they employed a somewhat arbitrary scheme of sediment redistribution. When these models are perfected, it will be possible to test the implication of model assumptions and parameter variations on dune form, which should give insight into the factors controlling the size and shape of natural dunes, including the effects of variation of wind speed and direction.

In the dune simulation models, the occurrence of equilibrium offers a powerful test of model assumptions. Once the flow and sedimentary portions of the model are validated, the model can be utilized in more general, dynamic simulations.

Equilibrium Model of Saltation

Aeolian processes offer another example of equilibrium modelling at a shorter temporal and spatial scale that illustrate the coupling of deterministic and stochastic elements. Over the past 15 years several models of the process of sediment transport by saltation have been developed. A recent model by Ungar and Haff (1987) illustrates the necessity for assumption of one or more types of equilibrium in order to make theoretical modelling tractable. They make an elegant statement of their rationale for simplified steady-state modelling.

'The full theory of saltation is too complicated to have yet yielded to analysis, but from a simplified theory like that presented here it can clearly be seen how different parts of the saltation mechanism fit together. In particular one is interested in divining in a consistent way the self-limiting nature of the process. Such a schematic model, however, should not be judged too critically on its ability to make accurate numerical predictions, since they inevitably depend upon finer details which can be present only in a full theory. In as much as such a theory is lacking at the present time, the only way to achieve accuracy in prediction is through the use of semi-empirical formulae. The wholly empirical method has its place in practical applications, but it obtains its power at the expense of understanding. Therefore it is desirable to eschew that approach in favor of the more informative but less accurate model described in detail below . . .' (Ungar and Haff, 1987, p. 290).

One difficulty in developing a general analytical model of saltation arises from the intimate interaction between the near-surface wind and the saltating particles, such that the wind profile near the surface is modified from the usual logarithmic profile by the saltation process (Bagnold, 1973). Even more difficult to model is the impact of saltating grains on the surface and the reinjection of grains into the air.

The primary constraint that Ungar and Haff utilize in constructing their equilibrium model is that the saltation process be self-replicating, that is, on the average, the number and trajectories of grains ejected from the surface and accelerated by a constant wind shear should be reproduced following impact with the surface. Ungar and Haff present a general mathematical formulation of this constraint, but their model utilizes a particularly simple scenario that satisfies this condition. In particular, based upon empirical observations that impacts generally produce a single high-velocity rebound into the saltation population and a number of low-velocity grains that constitute the creep load, they assume that each impact generates a single new saltating grain with a velocity determined by the impact velocity, and that the population of saltating grains are all ejected with the same velocity, with the underlying assumption that the behaviour of an average or typical grain is representative of the population. Furthermore, they assume that ejection of grains in steady-state saltation is due solely to grain impact and not affected by surface wind forces.

The model employs several additional assumptions, such as a steady wind, the applicability of a mixing-length turbulent flow model, neglect of lift and spin forces on particle trajectories, lack of collisions or other direct interactions between saltating grains, a flat and unrippled surface, *inter alia*. However, these are not critical to the discussion of equilibrium aspects of the model, so they will not be discussed further.

The wind profile and saltation load for a given wind shear, particle density, and particle size are calculated by an iterative procedure subject to empirical constraints on the average saltation height and impact threshold (cessation of motion) shear velocity. Despite the rather oversimplified assumptions, the model predicts near-surface wind velocities similar to those actually observed and a reasonable dependence of transport flux upon shear velocity.

Models of Stream Meandering

Meandering is a quasi-oscillatory response of a stream that can occur in a constant hydraulic regime (e.g. the experimental studies of Friedkin, 1945). The temporal record of channel position or curvature at a given location is clearly not an equilibrium response. Furthermore, meandering in natural streams also involves complicated spatial variations because the meanders are somewhat irregular. This irregularity can be modelled by stochastic elements. Thus it may seem surprising that meandering can be an equilibrium response given a stable hydraulic regime for a suitable choice of input and output variables.

Models that employ stochastic or random elements are largely uninformative about single instances, but they become deterministic in predicting the properties of large samples and populations. For example, the kinetic theory of gases envisions random particle collisions, but it explains the pressure, volume, and temperature law for gas behaviour. In a geomorphic context, Shreve's (1966) theory of stream networks hypothesizes that any individual stream network of a given size (number of first-order tributaries) drawn randomly is equally likely to exhibit any of the topologically distinct network configurations for that size. Nevertheless, as the size of a sampled network increases, or as the number of sampled stream networks of a given size increases, averaged properties such as bifurcation, length, and area ratios tend to approach stable (equilibrium) values (Shreve, 1966, 1969).

A similar situation occurs for the quasi-oscillatory behaviour of meandering. The size and shape of individual meander loops is subject to wide variance, but size and shape statistics collected over large samples of loops approach stable population distributions if the environmental controls over meandering (the hydraulic regime and valley characteristics) remain constant along the sampled stream so that the statistics are stationary (Howard and Hemberger, in preparation; O'Neill, 1987). The change in spatial scale from output variables describing individual loops to variables measuring averages over many loops is an example of selecting the appropriate scale for testing for equilibrium.

Some progress has also been made in modelling of meandering in streams. Ferguson (1976) proposed a stochastic differential equation model for generating meandering stream patterns:

$$\theta + \frac{2h}{k}\frac{d\theta}{ds} + \frac{1}{k^2}\frac{d^2\theta}{ds^2} = \epsilon, \tag{18}$$

where θ is the local direction angle of the stream path, s is the location along that path, k is a wavenumber parameter, h is a shape, or damping parameter, and ϵ is a random forcing function, assumed to be normal Gaussian noise (Ferguson, 1976). Thus the important elements of the model are an oscillatory component, a 'memory' of past θ values which decays proportionally to h, and a random forcing component. The model can also be expressed as a second-order autoregressive model, and time series techniques can be used to estimate, or fit, the model parameters to natural streams as well as to test for spatial non-stationarity in meander characteristics. Howard and Hemberger (in preparation) and O'Neill (1987) show that the Ferguson model generates statistical meander properties similar to those of natural streams. However, the Ferguson model is not a kinematic model, since the stream pattern is generated sequentially downstream and migration behaviour is not simulated. Thus it is a model which is applicable only in 'explanation' of population meander characteristics.

More recently, kinematic models of meander movement have been proposed (Parker, 1984; Howard, 1984; Howard and Knutson, 1984; Ferguson, 1984). These models are based upon models of flow in curved channels and assume that bank erosion is proportional to bank shear stress. Such models can be used in a predictive sense for a given channel segment. They also can be used to predict population values of meander statistics (Howard and Hemberger, in preparation). Interestingly, present versions of these kinematic models are less successful than the disturbed periodic model in predicting population meander statistics (Howard and Hemberger, in preparation). The models show that the sporadic occurrence of cutoffs is as, or more, important than variations in bank resistance in creating irregularity in meander loops.

3.3 DISCUSSION

The definition of equilibrium given above suggests that the primary indication of equilibrium behaviour is a temporally invariant, single-valued relationship between the output and one or more input variables, with the provision that the relationship can manifest a consensual degree of variation. There are several ways in which the presence or absence of equilibrium may be of interest to the scientist. If the behaviour of the system is not well known, disequilibrium behaviour helps to reveal the characteristics of the system (that is, to illuminate the black box). If the goal is prediction, lack of equilibrium limits forecasting skills. Sometimes the system is well enough understood that it could be theoretically modelled, but lack of appropriate field data (or high cost of obtaining the data or running the model) prohibits application of the general models. In this case, the theoretical model can be examined to determine under what conditions an assumption of equilibrium is appropriate. In other cases, such as the saltation model, theoretical models may only be tractable if some type of equilibrium is assumed. In such models the relevance or accuracy of the equilibrium assumption needs to be determined. Testing of systems (experimentally or observationally) under controlled conditions can reveal the equilibrium behaviour of the system even if dynamic modelling is not possible.

Choice of variables characterizing inputs and outputs for natural systems depends upon the timescales of change and relaxation times of external parameters and system responses. Inputs or system behaviour that fluctuates rapidly relative to the timescale of interest are commonly assumed to be representable by an average, and possibly equilibrium value. For example, hydrogeochemical models of watersheds commonly assume chemical equilibrium. Similarly, long-term models of climatic change assume short-term thermal equilibrium at the Earth's surface. Models of channel scour and stream profile evolution in alluvial streams assume steady-state sediment transport and a dominant discharge that is a weighted average of natural discharges (Howard, 1982). Similarly, a model of valley development by groundwater sapping assumes effective groundwater flow can be represented by steady-state flow (Howard, in press). Such assumptions can, of course, be misleading or biased, and require justification.

Systems responding to slowly-changing inputs can often be assumed to remain in equilibrium. The most critical timescales of change of inputs are those commensurate with the response time of the output variable. Large trends or cyclical inputs occurring over this intermediate timescale cause disequilibrium, as do large step changes or impulses in inputs that have occurred recently as compared to the output response time.

A physical system can be examined at a variety of timescales depending upon the problem at hand. The selection of input and output variables varies with the timescale of interest, as does the likelihood of equilibrium behaviour. As pointed out by Schumm and Lichty (1965) and Schumm (1985), the role of certain components of the physical system change from independent variables (inputs) to dependent variables (outputs), or *vice versa*, as the time scale of interest changes.

Equilibrium concepts are generally not applicable to physical systems that exhibit oscillatory, threshold, hysteretic, multivalued, or strongly unpredictable responses to constant or slowly-changing inputs. Schumm (1973) suggests the term *complex response* to cover most such cases. Discussion of such system behaviour is beyond the scope of this paper, but it has been treated by many authors, including Schumm (1973), Graf (Chapter 2, this volume), Howard (1980), Thornes (1980, 1983), and Huggett (1985). However, as was discussed in the example of river meandering, a change in time or spatial scales and selection of appropriate variables may make a system showing complex response over a given timescale be predictable or in equilibrium for longer or shorter time-scales.

REFERENCES

Ahnert, F. (1987). Approaches to dynamic equilibrium in theoretical simulations of slope development. *Earth Surface Processes and Landforms*, **12**, 3–15.

Allen, J. R. L. (1974). Reaction, relaxation and lag in natural sedimentary systems: general principles, examples, and lessons. *Earth Science Reviews*, **10**, 263–342.

Bagnold, R. A. (1973). *The Physics of Blown Sand and Desert Dunes*, London, Chapman and Hall.

Box, G. P. and Jenkins, G. M. (1976). *Time Series Analysis: Forecasting and Control*, Oakland, Holden-Day, 575 pp.

Brunsden, D. (1980). Applicable models of long term landform evolution, *Annals of Geomorphology, Supplementband*, **36**, 16–25.

Chorley, R. J. and Kennedy, B. A. (1971), *Physical Geography, a Systems Approach*, London, Prentice-Hall, 370 pp.

Cullingford, R. A., Davidson, D. A., and Lewin, J. (Eds.) (1980). *Timescales in Geomorphology*, New York, Wiley, 372 pp.

Ferguson, R. I. (1976). Disturbed periodic model for river meanders. *Earth Surface Processes*, **1**, 1079–1086.

Ferguson, R. I. (1984). Kinematic model of meander migration. In *River Meandering*, New York, American Society of Civil Engineers, 952–963.

Friedkin, J. F. (1945). *A Laboratory Study of the Meandering of Alluvial Rivers*, US Army Waterways Experiment Station, Vicksburg, Mississippi.

Graf, W. L., this volume.

Hack, J. T. (1960). Interpretation of erosional topography in humid temperate regions. *American Journal of Science*, **258A**, 80–97.

Hack, J. T. (1965). Geomorphology of the Shenandoah Valley, Virginia and West Virginia, and origin of the residual ore deposits. *US Geological Survey Professional Paper*, **484**, 84 pp.

Hack, J. T. (1973). Drainage adjustment in the Appalachians. In Morisawa, M. (Ed.), *Fluvial Geomorphology*, Binghamton, New York, Publications in Geomorphology, State University of New York, 51–69.

Hack, J. T. (1975). Dynamic equilibrium and landscape evolution. In Melhorn, W. N. and Flemal, R. C. (Eds.), *Theories of Landform Development*, New York, Allen & Unwin, 87–102.

Howard, A. D. (1980). Thresholds in river regimes. In Coates, D. R. and Vitek, J. D. (Eds.), *Thresholds in Geomorphology*, London, Allen & Unwin, 227–258.

Howard, A. D. (1982). Equilibrium and time scales in geomorphology: application to sand-bed alluvial streams. *Earth Surface Processes and Landforms*, 7, 303–325.

Howard, A. D. (1984). Simulation model of meandering. In *River Meandering*, New York, American Society of Civil Engineers, 952–963.

Howard, A. D. (in press). Case study: model studies of groundwater sapping. In Higgins, C. G. (Ed.), *Groundwater Geomorphology*, Geological Society of America DNAG volume.

Howard, A. D. and Hemberger, A. T. (in preparation). Multivariate characterization of meandering.

Howard, A. D. and Knutson, T. R. (1984). Sufficient conditions for river meandering: a simulation approach. *Water Resources Research*, **20**, 1659–1667.

Howard, A. D. and Kerby, G. (1983). Channel changes in badlands. *Bulletin of the Geological Society of America*, **94**, 739–752.

Howard, A. D., Morton, J. B., Gad-el-Hak, M., and Pierce, D. (1978). Sand transport model of barchan dune evolution. *Sedimentology*, **24**, 307–338.

Howard, A. D. and Walmsley, J. L. (1985). Simulation model of isolated dune sculpture by wind. *Proceedings of the International Workshop on Physics of Blown Sand*, Department of Theoretical Statistics, University of Aarhus, Denmark. *Memoirs No. 8*, **2**, 377–391.

Huggett, R. J. (1985). *Earth Surface Systems*, New York, Springer-Verlag, 270 pp.

Kirkby, M. J. (1987). The Hurst effect and its implications for extrapolating process rates. *Earth Surface Processes and Landforms*, **12**, 57–67.

Lancaster, N. (1985). Variations in wind velocity and sediment transport on the windward flanks of desert sand dunes. *Sedimentology*, **32**, 581–595.

Livingstone, I. (1986). Geomorphological significance of wind flow patterns over a Namib linear dune. In Nickling, W. G. (Ed.), *Aeolian Geomorphology*, New York, Allen & Unwin, 97–112.

Mackin, J. H. (1948). Concept of the graded river. *Bulletin of the Geological Society of America*, **59**, 463–512.

O'Neill, M. (1987). *Meandering Channel Patterns—Analysis and Interpretation*, unpublished PhD Dissertation, State University of New York, Buffalo.

Papoulis, A. (1984). *Probability, Random Variables, and Stochastic Processes*, New York, McGraw-Hill, 576 pp.

Parker, G. (1984). Theory of meander bend deformation. In *River Meandering*, New York, American Society of Civil Engineers, 722–732.

Pickup, G. and Rieger, W. A. (1979). A conceptual model of the relationship between channel characteristics and discharge. *Earth Surface Processes*, **4**, 37–42.

Schumm, S. A. (1973). Geomorphic thresholds and complex response of drainage systems. In Morisawa, M. (Ed.), *Fluvial Geomorphology*, Publications in Geomorphology, State University of New York, Binghamton, 299–310.

Schumm, S. A. (1985). Explanation and extrapolation in geomorphology: seven reasons for geologic uncertainty. *Transactions, Japanese Geomorphology Union*, **6**, 1–18.

Schumm, S. A. and Lichty, R. W. (1963). Channel widening and flood-plain construction along Cimarron River in southwestern Kansas. *US Geological Survey Professional Paper*, **352-D**, 71–88.

Schumm, S. A. and Lichty, R. W. (1965). Time, space and causality in geomorphology. *American Journal of Science*, **263**, 110–119.

Shreve, R. L. (1966). Statistical law of stream numbers. *Journal of Geology*, **74**, 17–37.

Shreve, R. L. (1967). Infinite topologically random channel networks. *Journal of Geology*, **75**, 178–186.

Shreve, R. L. (1969). Stream lengths and basin areas in topologically random channel networks. *Journal of Geology*, **77**, 397–414.

Thornes, J. B. (1980). Structural instability and ephemeral channel behavior. *Annals of Geomorphology, Supplementband*, **36**, 233–244.

Thornes, J. B. (1983). Evolutionary geomorphology. *Geography*, **68**, 225–235.

Thornes, J. B. and Brunsden, D. (1977). *Geomorphology and Time*, London, Methuen, 208 pp.

Tsoar, H. (1983a). Wind tunnel modelling of echo and climbing dunes. In Brookfield, M. E. and Ahlbrandt, T. S. (Eds.), *Eolian Sediments and Processes*, Development in Sedimentology, New York, Elsevier, 247–259.

Tsoar, H. (1983b). Dynamic processes acting on a longitudinal (seif) dune. *Journal of Sedimentary Petrology*, **52**, 823–831.

Ungar, J. E. and Haff, P. K. (1987). Steady state saltation in air. *Sedimentology*, **34**, 289–299.

Waide, J. B. and Webster, J. R. (1976). Engineering systems analysis: applicability to ecosystems. In Patten, B. C. (Ed.), *Systems Analysis in Ecology*, Volume IV, New York, Academic Press, 329–371.

Wipperman, F. K. (1986). The wind-induced shaping and migration of an isolated dune—a numerical experiment. *Boundary-Layer Meteorology*, **36**, 319–334.

Modelling Geomorphological Systems
Edited by M. G. Anderson
©1988 John Wiley & Sons Ltd.

Chapter 4

Network models in geomorphology

DAVID M. MARK

Department of Geography, State University of New York at Buffalo

4.1 INTRODUCTION

Networks indicate the connections between and among objects. As such, networks are an essential component of any systems analysis, since if objects are related at all, there is an implicit network reflecting those relationships. In this paper, however, attention will be restricted to more concrete objects: networks which have a visible physical expression in the landscape, or which indicate the direct patterns of flow of water and/or debris across parts of the Earth's surface.

Stream channel networks fall into both of these categories. Thus, it is not surprising that the analysis of stream channels has been the principal area of network analysis in geomorphology. Recently, Abrahams (1984a) has provided an extensive and detailed review of the most common type of channel network: trivalent trees containing no lakes or forks (these terms will be explained below). Earlier reviews of channel networks by Smart (1972, 1978) and by Jarvis (1977) had a similar emphasis. This paper will attempt to complement those reviews by focussing attention on more abstract properties of channel networks, by including all types of channel networks, including those with forks (such as braided channels or deltas) as well as those with lakes, and by reviewing ridge networks and 'landscape networks' based on digital elevation models. Topics treated in depth by the earlier reviews cited above will be summarized here only briefly.

After reviewing graph-theoretic terminology and research on channel networks, this paper will turn to the other major network type with a physical expression in the landscape: ridges and divides. Such networks are prominent features of landscapes, but their patterns have received little attention from geomorphologists, perhaps because they are the places in the landscape where the *least* geomorphological activity occurs. Finally, this review will examine network models of processes over entire landscapes. If the land surface is divided into a large number of small areas (finite elements), then these can be connected together to form a whole-landscape network. Then, this 'drainage direction' network can be used to drive models of runoff, erosion, and landscape evolution. Algorithms for handling these networks will be presented. Completing the circle, the review ends by showing how these whole-landscape models can be used to generate channel networks directly from digital elevation models, and how this can be used to reexamine hypotheses about channel network composition and evolution.

4.1.1 Some Fundamentals of Graph Theory

Before beginning a review of network models in geomorphology, it is appropriate to provide a brief overview of fundamental terms and concepts from graph theory; some are illustrated in Figure 4.1. This review will in general follow the definitions and terms used by Harary (1969). First, a *graph* is defined as a set V of p *points* (often termed *vertices*), together with a set E of q *edges* (or *links*) connecting distinct pairs of points in V. A graph must have at least one vertex ($p > 0$); a graph may have no edges, in which case it is a *null* graph. Customarily, a graph is presented as a diagram; the graph itself is, however, the mathematical relation underlying the diagram. If all edges are *ordered* pairs, the graph may be termed a *directed* graph or simply a *diagraph*. If two vertices share an edge, they are said to be *adjacent*. The number of edges incident on (connected with) a vertex is the *degree* of that vertex; in a directed graph, each vertex has an *in-degree* and an *out-degree*.

Two graphs are said to be isomorphic if there exists a one-to-one correspondence of their vertices which preserves adjacencies. If a graph is embedded in (drawn on) a plane such that no edges intersect, then it is a *plane* graph. A graph which is isomorphic to any plane graph is said to be *planar*, even if it is drawn with crossing edges. Two graphs which are isomorphic, and whose points can be made to correspond through a continuous deformation in the plane are said to be topologically identical.

A *walk* is an alternating series of adjacent edges and vertices; a *path* is a walk in which all vertices, except possibly the first and last, are distinct. A *cycle* is a path whose first

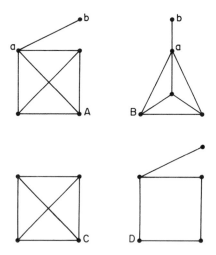

Figure 4.1. Some graph-theoretic terms and concepts: points a and b are adjacent; A and B are isomorphic; C and D are subgraphs of A (and of B); D is a spanning subtree of A (and of B); B and D are plane graphs; all four graphs are planar

and last vertices are identical. If there exists at least one path between every pair of vertices, the graph is *connected*. Then, a *tree* is a connected graph having no cycles; note that a tree with *p* vertices must have *p*-1 edges. A *rooted* graph or network just is one in which a particular vertex is singled out as the *root*. A tree rooted at a vertex of degree zero is termed a *planted* tree. Any rooted tree can be converted to a directed graph simply by directing each link toward the root.

Finally, in graph theory, the term *network* has a specific meaning: it is a graph for which there exists a function which assigns a non-negative weight (or length) to every edge. Two networks are said to be isomorphic if their underlying graphs are isomorphic; in other words, the property of isomorphism does not consider edge weights.

4.1.2 Abrahams' Review

As noted above, Athol D. Abrahams recently has provided a comprehensive and detailed review of the topic under the title 'Channel Networks: A Geomorphological Perspective' (Abrahams, 1984a). Abrahams explicitly excluded a number of topics related to channel networks from his review, and those topics will form the focus of the present review. However, it is appropriate to present an overview of Abrahams' review before providing a review of the topics he eschewed.

Abrahams' (1984a) review began by focussing on topological properties of networks without forks, lakes, or islands. Early work by Horton and other workers was integrated into the *random topology model*, a simple yet powerful model of network topology proposed by Shreve (1966) and discussed in detail below. Abrahams reviewed approaches which have been used to test the random topology model, and its implications for the arrangement of tributaries along main streams (see section 4.2.3, below).

The second major section of Abrahams' (1984a) review dealt with *length* properties of networks. Link length distributions are fundamental, since when combined with topological models they can be used to predict length properties for entire networks. The probability distributions of links lengths have been shown to differ with topologic position within the network. After a very brief review of area properties of networks, Abrahams reviewed *density* properties, computed by dividing length properties by area properties. Then, after reviewing angular properties of networks, and relating them to other network factors, Abrahams reviewed simulation models and their implications for the evolution of fluvial landscapes. Many of the themes and topics reviewed by Abrahams (1984a) are also treated in the current review; it is hoped that the perspective taken in the remainder of this review will differ from that used by Abrahams; the reader is, however, referred to Abrahams' review for detailed treatment of topics not treated in detail here.

4.2 CHANNEL NETWORKS

It is the nature of the interaction between topographic surfaces and water flow that distinct channelized flow is a common occurrence (Smith and Bretherton, 1972; Kirkby, 1980). Even in arid areas where the flow is intermittent, it is normal for a relatively permanent set of channels to form. In most such networks, lakes, islands, and braided channels are uncommon. Because of this, most studies of channel networks have explicitly or implicitly excluded such features. However, many fluvial networks include features of these types,

Table 4.1. Classification of point nodes in networks
without lakes

Inputs:	0	1	2
Outputs:			
0	—	sink (K)	sink (K)
1	source (S)	—	junction (J)
2	*	fork (F)	*

*while possible, such points are *extremely* rare, and will be
assumed not to exist.

and thus a general model of network topology must address these features. It is appropriate first to summarize a previous review of simple fluvial networks, and to review their composition; then models which can explicitly accommodate lakes will be introduced.

4.2.1   Topology of Networks Without Lakes

Basic Classification Vertices and Links

We will assume a directed network model of the channel system. Thus, the channels within a drainage basin can be represented as a set of vertices, and a set of ordered pairs of vertices (relations or connections between vertices) termed directed edges or *links*. Links will be directed according to flow direction, or net flow in the case of reversing channels in deltas and other tidal networks. Following Shreve (1966), a *channel link* is defined as an

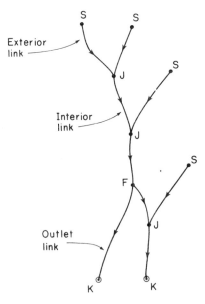

Figure 4.2. Classification of point nodes and link terminology in channel networks without lakes. Arrows indicate the flow directions. Points include sources (S), junctions (J), forks (F) and sinks (K)

Table 4.2. Classification of links in networks without lakes

Vertex type upstream:		S	J	F
Vertex type downstream:	J	exterior	interior	FJ
	F	exterior	JF	FF
	K	exterior	outlet	outlet

unbranched portion of stream channel. The vertices in the network are characterized by their in-degrees and out-degrees. This classification illustrated in Figure 4.2, and summarized in Table 4.1. In networks without lakes, the vertex at the upstream end will normally be either a *source* (point with no inputs), a *junction* (point with exactly two inputs), or a *fork* (point with one input and two outputs). The downstream end of a link can be a junction, a fork, or a *sink* (point with no outputs).

A fundamental classification of links characterizes each link by the *types* of points (vertices) at its upstream and downstream ends; secondary information may be provided by the types and other characteristics of the adjacent links. Links beginning at a source are termed *exterior* links (Shreve, 1966), and are exactly equivalent to 'first-order streams' in Strahler's (1952) modification of Horton's (1945) stream classification system. Channel links with junctions at both ends are called *interior* links (Shreve, 1966), whereas links ending in sinks may conveniently be termed *outlet* links. Other types of links do not have established names; they will be referred to by letter combinations, as shown in Figure 4.2 and Table 4.2.

Trivalent Trees

Most fluvial networks have few if any forks. In the absence of forks, channel networks are trees, rooted at their one sink or outlet. If all junctions join only two links (a reasonable assumption), then, if there are N sources, there will be $N-1$ junctions, N exterior links, and $N-1$ interior links. A directed tree rooted at a vertex of degree one, and with all other vertices of degree either 1 (sources) or 3 (junctions) is termed a *trivalent planted tree*. In a directed tree, it is possible to attach a *magnitude* (N) to each link; the magnitude of a link is simply the total number of sources upstream (Shreve, 1966). In a trivalent tree, the total number of links upstream is, of course, $2N-1$. Exterior links are, by definition, of magnitude 1.

Shreve (1966) noted that the topology of a trivalent planted tree can be uniquely defined by a string of $2N-1$ binary digits (see Figure 4.3). The string can be constructed by the following algorithm. First, begin at the basin outlet and travel upstream, turning left at every junction and reversing at every source. Then, whenever a channel link is traversed for the first time (upstream direction), write a '0' if it is an interior link, or a '1' if it is an exterior link; do nothing during downstream traversal, and turn left at the downstream end. For a complete trivalent tree, the number of 1's in the string will be one more than the number of 0's. Thus, a sequence of N 1's and $N-1$ 0's can represent a stream network if and only if the number of 1's up to some element k is greater than the number of 0's up to that point plus one, for all $k(2N-2)$. In fact, the topology of *any* planted tree containing exactly L links can be uniquely coded by a binary string of length $2L$ (de Bruijn and Morselt, 1967); the method proceeds as described above, except that a 0 is coded when any link is

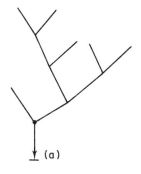

(b) 01000111011

(c) 001000010110110010 1111

Figure 4.3. (a) A magnitude-6 channel network; (b) The channel network binary string representation of (a) (after Shreve, 1966); (c) The general binary string representation of the same tree (after Bruijn and Morselt, 1967)

traversed in the 'upstream' (i.e. away from the root) direction, and a 1 is coded during the downstream traversal. Trivalent planted trees can be coded in half as many bits because their topology is so restricted.

Because the topological variety of trivalent trees is so restricted, researchers have developed link classification schemes which consider factors other than vertex type. James and Krumbein (1969) introduced a new classification for interior links of high magnitude. If the two links which merge at the upstream end of the current link are of very different magnitudes, then the low-magnitude input can be referred to as a *tributary*, with the

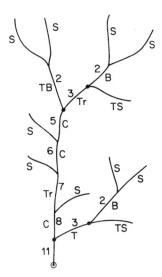

Figure 4.4. Mock's classification of link types in a magnitude-11 channel network. All source (S) and tributary-source (TB) links are of magnitude 1; magnitudes of other links are given. Interior links include bifurcating (B), tributary-bifurcating (TB), tributary (T), cis (C), and trans (Tr) links

high-magnitude input considered to be the upstream extension of the mainstream. James and Krumbein (1969) classified mainstream links as either *cis* or *trans*: in cis links, the bounding tributaries enter the main stream from the same side, whereas in trans links the tributaries join on opposite sides (see Figure 4.4).

Mock (1971) introduced a more detailed classification of links. Both interior and exterior links are classified in part according to the magnitude of link they merge with at their downstream ends; interior links also are classified according to whether or not the links merging at their upstream ends are of equal magnitude. In Mock's classification, a link is termed a *source* link if it has a source at its upstream end; a *bifurcating* link if it has two equal-magnitude inputs at its upstream end; and *tributary* if it merges at its downstream end with a link of higher magnitude. These terms are combined to form link type names; Mock referred to links which have none of these properties as *cis–trans* links. Mock's link classification scheme is illustrated in Figure 4.4 and Table 4.3.

The Random Model of Channel Network Topology

Horton (1945) presented a set of 'laws' of drainage composition; the logarithms of stream numbers, lengths, and slopes of stream length plotted as straight lines against stream order. For about two decades, these 'laws' were treated with great excitement and a certain element of awe; they appeared to show a remarkable regularity of nature. However, almost two decades later, Leopold and Langbein (1962) showed that networks produced by two different random simulation models also obeyed Horton's laws. To Leopold and Langbein, this suggests that real stream networks also have essentially random topology. In 1966, Milton stated that Leopold and Langbein's result shows that Horton's laws are irrelevant to geomorphology. In that same year, Shreve (1966) presented a model of stream topology which revolutionized the field.

Shreve (1966) introduced the concept of stream *magnitude*, the number of stream sources in a basin; this concept was discussed above. Then, the essence of Shreve's model of channel network topology is very simple: *all topologically-distinct networks of a given magnitude are equally likely to occur*. From this assumption, Shreve was able to show that networks 'obeying' Horton's law of stream numbers are extremely likely to occur (Shreve, 1967). Later Shreve (1969) added simple postulates concerning link lengths and link drainage areas. From these, a large number of channel network and drainage basin properties can be predicted (Shreve, 1969, 1975; Smart, 1972, 1978).

Shreve (1967) also noted that the probabilistic–topologic model of stream patterns can be restated in terms of the binary string representation of networks, introduced above. In the binary string for an infinite topologically-random network, the digits 0 and 1 have

Table 4.3. Mock's classification of links in trivalent networks

Upstream end: Downstream end:	source	equal N	unequal N
Joins higher N	tributary source (*TS*)	tributary bifurcating (*TB*)	tributary (*T*)
Joins equal or lower N	source (*S*)	bifurcating (*B*)	cis–trans (*CT*)

an equal probability of appearing at any position in the string. In fact, autocorrelation analysis of binary strings is an alternative way to test the random topology model, since if significant autocorrelation exists at any lag, the string (and hence the network) is non-random (Mar, 1982). However, attempts to interpret autocorrelation patterns detected in the binary strings were largely unsuccessful (D. M. Mark, unpublished research, 1982).

Testing the Random Topology Model for Trivalent Trees

Shreve (1966) presented Alfred Cayley's formula, developed in the context of chemical isomers, for determining the number of topologically-distinct trivalent planted plane trees with a given number of sources. For magnitudes 1 through 6, the numbers of distinct trees are: 1, 1, 2, 5, 14, and 42. One way to test the random topology model would be to count the numbers of networks of some fixed magnitude which fall into each distinct class, and test the hypothesis that all classes occur with equal frequency. However, the standard method for testing such a hypothesis, the Chi-squared test, requires an average expected frequency of five per class. Even for magnitude 5, this means a sample of 70 basins; for large N, this approach is impractical.

As a solution to this problem, Smart (1969) proposed the concept of an *ambilateral class* of streams. The topologically-distinct members of an ambilateral class can be obtained from each other by interchanging the subnetworks above any junction (Figure 4.5). The 14 magnitude-5 networks fall into three ambilateral classes, whereas the 42 magnitude-6

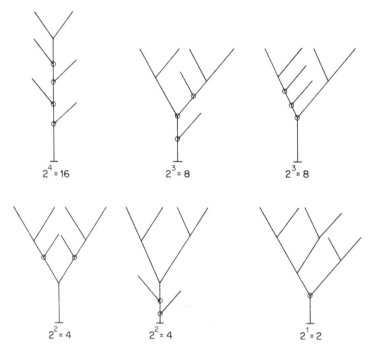

Figure 4.5. The six *ambilateral classes* for channel networks with six sources. Asymmetric junctions are circled; the number of topologically-distinct members of each ambilateral class is equal to two raised to the power of the number of such junctions

networks form six such classes. Under the random topology model, the frequency of an ambilateral class should be proportional to its number of topologically-distinct members; this in turn is 2^j, where j is the number of asymmetric junctions, places at the two merging subnetworks are not members of the same ambilateral class.

The random topology model has been tested many times in the literature, and seldom has been rejected. However, most of these tests have treated the random topology model as a *null hypothesis*. As noted by Abrahams and Mark (1986), most authors have written about Shreve's model as if it were a *general theory* from which hypotheses concerning network topology can be derived; as such, statistical tests have been biased in favour of the researchers' aims (assuming that they wished to verify the random topology model and then to use it to make further predictions). Thus, while that random topology model has been very successful in predicting network properties, much of the success may be due to the law of large numbers and to the insensitivity of average measures to minor variations (see discussion by Church and Mark, 1980). As will be discussed below, geometric considerations result in consistent and significant departures from the random topology model.

4.2.2 Geometry of Channel Networks

Link Lengths and Areas

Link lengths and areas are the most fundamental geometric properties, since if their distributions are known, they can be combined with network topology in order to predict geometric properties of larger basins. Shreve (1969) found that the assumption of a very simple link length distribution led to good predictions of many aggregate properties of drainage basins: Shreve (1969) assumed that all interior links had lengths drawn from a single population, and that exterior links have lengths drawn from another population. However, subsequent research has clearly demonstrated that different link types have different link length distributions. In fact, James and Krumbein's (1969) discovery that interior cis and trans links have different length distributions was published in the same year as Shreve's length paper. Later, Abrahams (1980) showed that T and TB interior links (see *Angular properties* below, and Table 4.3) tend to be significantly longer than CT and B links. Similarly, categories of topologically distinct exterior links have been found to have significantly different distributions. As an example, Abrahams and Campbell (1976) showed that source and tributary–source links have different length distributions. Abrahams (1984a, pp. 168–172) discussed such research in detail. While there has been much less work on link drainage areas than on link lengths (*cf.* Abrahams, 1984a, p. 172), similar tendencies may be expected for such measures.

Angular Properties

As noted by Abrahams (1984a, pp. 175–179), angular properties of networks have been the subject of considerable research. Junction angles may be of use in distinguishing drainage patterns, and also may put important constraints on other aspects of network geometry (Abrahams, 1980). Howard (1971) discussed the concept of 'optimality' in junction angles. More recently, work by Roy and Woldenberg (summarized by Roy, 1985) has derived explicit predictions of junction angles under various principles of optimality.

4.2.3   Geometric Control on Network Topology

Whereas many topological properties of natural stream networks are similar to the properties of topologically-random networks, more detailed work has revealed subtle yet consistent departures from randomness. Most of these departures can be attributed to geometric control on topology.

James and Krumbein (1969) presented one of the first empirical studies to reveal a significant departure from the random topology models in a natural network. Although the focus of their work was link length distributions, they introduced Shreve's terms *cis* and *trans* 'for same-side and opposite-side links, respectively' (James and Krumbein, 1969, p. 547). Under the random topology model, these two types of main-stream links would be expected to occur with equal frequency; this would also be predicted if tributaries enter randomly and independently on either side of the main channel. However, since a finite amount of space is needed for a tributary basin to develop, we might expect the competition for space to lead to a more-regular-than-random pattern of tributaries along each side. In the extreme, if intertributary spacing is constant and equal on both sides, then unless

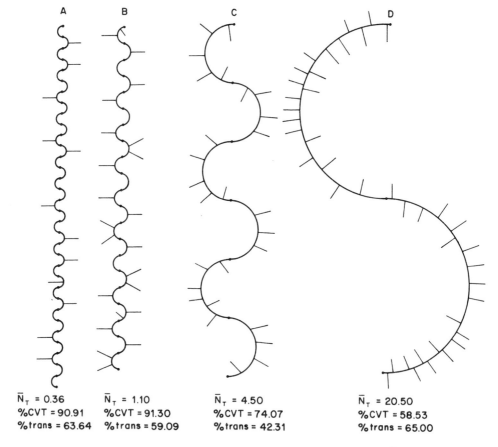

A	B	C	D
$\bar{N}_T = 0.36$	$\bar{N}_T = 1.10$	$\bar{N}_T = 4.50$	$\bar{N}_T = 20.50$
%CVT = 90.91	%CVT = 91.30	%CVT = 74.07	%CVT = 58.53
%trans = 63.64	%trans = 59.09	%trans = 42.31	%trans = 65.00

Figure 4.6. Schematic diagrams of four equally sinuous winding streams with typical $\bar{N}_T$, %CVT and %trans values (From Abrahams, 1984c)

the tributaries happen to line up exactly, *all* main-stream links would be trans links. In a study of Middle Fork, Rockcastle Creek, Kentucky, James and Krumbein (1969, p. 550) reported that 60.4 per cent of the main stream links (293 of 485) were trans links. This result is consistent with the general concept that 'space-filling' constraints lead to more-regular-than-random tributary spacing and in turn increases the proportion of trans links.

Abrahams (1984b, 1984c) has greatly elaborated the theme of geometric and hydrologic influences of network topology and geometry. In a winding valley, small tributaries tend to develop preferentially on the concave-downslope surfaces which develop on the outsides of bends; this is simply because a concave-downslope surface concentrates runoff, whereas runoff diverges on a convex slope. Intuitively, this effect should increase with the degree of valley-wall curvature. Next, the mean spacing of small tributaries should be a function of regional drainage density. Combining these two effects, Abrahams (1984b) was able to predict that there should be a non-linear relation between percentage of trans links and the mean number of tributaries per bend (see Figure 4.6). When the number of tributaries per bend is much less than one, it will be difficult to predict which bends will contain tributaries; hence, even if these all occur on the outsides of bends, we might expect an equal number of cis and trans links, or perhaps an excess of trans links due to 'space-filling' effects. Next, if there is about one tributary per bend, we would expect most main-stream links to be trans. Then, if there are several tributaries per bend, with most on the outside, there would be an excess of cis links. Finally, if the number of tributaries per bend becomes very large, the curvature might be relatively unimportant, and there would be a slight excess of trans links because of the effects identified by James and Krumbein (1969).

Abrahams (1984b) then went on to establish predictive relationships between these and other variables. For a sample of 40 winding valleys, Abrahams found that 81 per cent of the variation in the percentage of tributaries entering on the concave sides of bends could be predicted from mean number of tributaries per bend $(-)$, valley sinuosity $(+)$, and rate of bend migration $(-)$, where the signs $(+/-)$ in parentheses indicate whether the partial correlation was positive or negative. Further, Abrahams found that 71 per cent of the variation in per cent trans links could be predicted from per cent concave tributaries $(-)$, mean number of tributaries per bend $(+)$, and valley sinuosity $(-)$. Abrahams (1984c) extended this line of work by examining the influence of tributary size on tributary arrangement. He found similar effects, but larger tributaries were very often found on concave banks in part because, when they are similar in discharge to the main stream, junction-angle controls tend to reinforce this effect.

Abrahams' work on winding valleys represents a significant advance over previous channel network research, because it develops a model to predict an aspect of network topology and geometry from principles of the processes which shape the landscape. Then, the model is confirmed by careful empirical testing and statistical model-building. Informal conversations between the writer and several prominent fluvial geomorphologists indicates that Abrahams' work on winding valleys has done more than any other work to 'legitimize' network-topology research in the minds of more traditional geomorphologists. This work clearly provides a model for future work in network topology; such work is to be encouraged.

4.2.4   Networks With Forks

As noted above, a *fork* can be defined as a channel point with one input and two outputs (see *Some fundamentals of graph theory* and Table 4.1 above). Whereas these are rather rare in most channel networks, they are common in deltas, and are a major feature of braided channel systems. The topology of such networks has been examined in a small number of studies. The topology and geometry of braided channels has been studied by Howard, Keetch, and Vincent (1970) and by Krumbein and Orme (1972); Smart and Moruzzi (1972) and more recently Morisawa (1985) have examined the topology of deltas.

Smart and Moruzzi (1972) presented a model for delta channel topology. First, they showed that there is a relationship between the numbers of forks (F), junctions (J), and outlets (sinks: K). Then, there are six types of links, denoted by the node types at the upstream and downstream ends: *FF*, *FJ*, *JF*, *JJ*, *JK*, and *FK*. A random topology model assumes that the node types at the ends of a link are independent. If so, the proportion of links of a given class should equal the product of the probabilities (frequencies) of the appropriate node types. A parameter *alpha*, the ratio of J to F, was proposed as a useful measure of delta network topology. Morisawa (1985) determined the actual frequencies of link types in 20 river deltas. Of these networks, the random association of node types at link-ends could be rejected for only five (at $p = 0.05$). In these five, there was an excess of *JK* links and a corresponding deficiency of *JJ* links. Next, Morisawa (1985) used a simulation technique to produce 13 delta channel models; these networks were analysed in the same way as were the natural networks. For four of the 13 networks, the random link model could again be rejected; departures from the model also were the same as for the natural networks. Evidently, Morisawa's (1985) random growth simulation model predicts the fundamental pattern of constraints on delta channel network topology.

4.2.5   Networks With Lakes

Despite the fact that channel networks with lakes are common in glaciated terrain, and that glaciated terrain covers some thirty per cent of the Earth's land area, the topology and geometry of networks with lakes has received little attention from geomorphologists. The point terminology used above can readily be extended to include lakes. Lakes initially are classified in the same way as point vertices, designating them as *SL* (source lakes) if they have no inputs, *JL* (junction lakes) if they have two or more inputs, and KL (sink lakes) if they have no outlets. Lakes with exactly one inlet and one outlet act simply as 'connectors', and may be abbreviated as *CL*.

Lake Network Topology and Topological Randomness

In 1980, the writer drove westward across the Canadian Shield in northwestern Ontario. The disorder of the glacially 'deranged' drainage pattern there is in striking contrast to the 'orderliness' of the fluvially-eroded sandstone landscape of eastern Kentucky. And yet, as discussed above, many aspects of the drainage networks in eastern Kentucky are not significantly different from the predictions of the random-topology model. How can a drainage network be far more disorderly than a random one?

In an attempt to address this question, Mark and Goodchild (1982) developed a topologic model for drainage networks with lakes. Lakes (L) with more than one outlet are very rare, and were excluded from the model. Thus lakes are classified by whether they have 0 or 1 outlet, and by their number of inputs. Mark and Goodchild (1982) divided network vertices into two classes: 'normal points' and 'lakes'. They excluded from their model lakes with no outlet and lakes with more than one; they also excluded point forks. Thus, as with Shreve's model, the network can be represented by a planted plane tree.

Next, Mark and Goodchild (1982) represented the network topology by an *integer string* which is a generalization of Shreve's (1967) binary string (discussed in the section *Trivalent trees*, above). For consistency with that work, the symbols '0' and '1' are retained for point junctions and point sources, respectively. Then, lakes are represented by integers equal to their *number of inputs plus 2*. Thus a '2' would represent a source lake with no inlets, a '5' a lake with three inlets, etc. Mark and Goodchild (1982, p. 276) proved a necessary and sufficient condition for an integer string to represent a valid complete channel network. A network and its integer string are illustrated in Figure 4.7.

Two ways of classifying subnetworks with lakes were presented by Mark and Goodchild (1982, p. 276–7). One property is the *lake-degree set* (LDS): for a network containing exactly *L* lakes, this is simply a list of the in-degrees of those lakes. Secondly, they defined *lake-path identity sets* (LPIS), which are defined by subgraphs of the networks containing only paths draining from lakes. As an example, with two lakes and one normal source, there are exactly 15 topologically-distinct networks, with 2 LPIS and 4 LDS (see Figure 4.8).

Mark and Goodchild (1982) generalized the probabilistic–topologic model and proposed that all topologically-distinct networks with a specified set of link-type frequencies should be equally likely; in terms of the integer string, 'all substrings consisting of the same set of elements are expected to occur with equal frequency' (Mark and Goodchild, 1982, p. 279). They used observed relative frequencies of link types in a 596-link channel network from northwestern Ontario to predict frequencies of 12 subnetwork categories; frequencies were also predicted using random permutations of the original 596-element integer string (subject to validity constraints). The two sets of predicted frequencies agree closely; empirically-observed frequencies were similar. However, network types involving a mixture of lakes and 'normal' points occurred less often than predicted, whereas 'all lake' and 'no lake' subclasses were underpredicted by the random-topology model. This tendency was confirmed statistically, and appears to be due to the spatial unevenness of the lake distribution, presumably controlled by surficial geology.

Mark and Goodchild's approach assumed that the numbers of lakes of various in-degrees were given, or determined empirically. Mark (1983a) later investigated the probability distribution of lake in-degrees. Mark (1983a) suggested a model in which channel networks

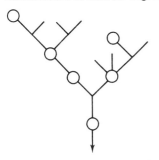

Figure 4.7. Schematic representation of a drainage network including lakes (circles) and *S*, the integer string representation of the network (From Mark and Goodchild, 1982)

$S = \{3034021011511021\}$

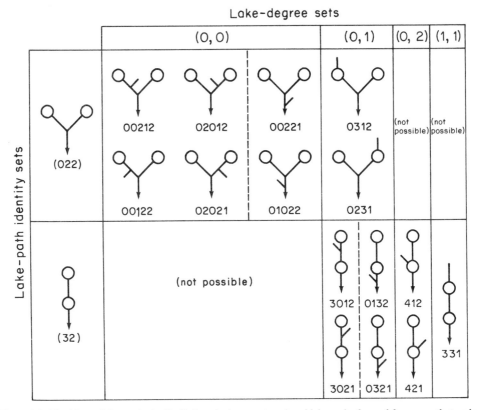

Figure 4.8. The 15 possible topologically distinct drainage networks which can be formed from exactly two lakes and one normal source (From Mark and Goodchild, 1982)

with lakes consist of a mixture of two lake populations. 'Consequent lakes' lying in glacially-scoured basins have a negative binomial distribution of in-degrees. A second population, termed 'subsequent lakes' are simply wide points on rivers, and are assumed to have exactly one inlet each. This model has three parameters, the two parameters of the negative binomial, and the mixing ratio (p). The model was tested on 15 samples from 11 glaciated regions: six river basins, one island, and eight map sheets. In all but one sample, a mixing ratio p could be found which produce a Chi-square sufficiently small that the model could not be rejected. Estimated p-values ranged from 0.042 (in a mountainous area) to 0.376 in a basin on the Canadian Shield. Mark (1983a) then measured a battery of eight geomorphometric measures for each study area, and correlated these with the lake distribution parameters. Several significant correlations were revealed, suggesting that the model parameters have some geomorphic meaning.

Link Lengths in Networks With Lakes

Mark and Averack (1984) examined the probability distribution of link lengths in networks with lakes. They divided links into eight types: exterior (E) or interior (I) links could have either lakes (L) or normal points (P) at either their upstream or downstream ends.

Three-letter codes were used; for example *ELP* denotes a link flowing from a lake with no inlets to a point junction. An asterisk (*) is used as a 'wild-card' character to indicate the class of networks having any valid symbol at each such position. Empirical data were obtained by digitizing 983 stream links, 597 from the Marchington River basin in Ontario, and 386 from Random Island, Newfoundland. Although the ocean might be considered to be topologically equivalent to a lake, links on Random Island which flowed into the ocean were given distinct codes (**O*).

First, Mark and Averack (1984) counted link frequencies in contingency tables tabulating upstream link type against downstream link type. For each sample, the hypothesis of independence could be rejected. In every case, links joining like node types (**LL*, **PP*) occurred more often that predicted, while links with mixed types were less frequent (confirming in more detail the observation made by Mark and Goodchild, 1982, for Marchington River). Next, Mark and Averack (1984) examined mean link lengths, for eight classes in Marchington River and 12 classes from Random Island. In each case, the hypothesis that all link lengths are drawn from a single population was rejected.

The mixed-gamma distribution (Abrahams and Miller, 1982) was fitted separately to each link-length class. For Marchington River, the model could not be rejected for any link class. Whereas the mixing parameters showed no particular pattern, the shape parameters of the component distributions were more interesting. *EP** links had shape parameters similar to those observed in lake-free networks; however, other links (*EL** and *I***) had shape parameters below 2; distributions with shape factors that low have been found only for interior links, and are rare (Abrahams and Miller, 1982). For Random Island, link types flowing into the ocean are interesting; EPO links had a relatively high individual shape parameter, but all other had shape parameters less than 2.8. The value of 1.26 for IPO links is the lowest reported in the literature (*cf.* Abrahams and Miller, 1982), and approaches the 1.0 which would indicate simple exponential component distributions.

4.3 RIDGE AND DIVIDE NETWORKS

Ridges and divides have received little attention from geomorphologists. Nevertheless, they form networks which often have a total length similar to that of the channels. As pointed out above, ridges represent the parts of the landscape at which there is the *least* geomorphological activity. Also, ridges often correspond with drainage divides, forming the boundaries of drainage basins. This section will begin by reviewing a topological model of divides and drainage lines introduced more than a century ago. Then, geometric models of ridge lines will be discussed.

4.3.1 Surface Networks

In his paper *On Contour Lines and Slope Lines*, English mathematician Alfred Cayley (1859) introduced a model of the topology of continuous, smooth surfaces. *Slope lines* were defined as lines of steepest descent, running perpendicular to the contours. If traced up- or down-hill, such slope lines will end at *critical points*, points at which the first derivative is zero; most points on a surface have exactly one slope line passing through them. On a continuous, single-valued function of two variables, there are three types of critical points: *peaks* (local maxima), *pits* (local minima), and *saddle points* or passes. Peaks and pits form the end-points of any number of slope lines; saddle-points are cut by two orthogonal slope-lines. One of the slope lines through a saddle point can be traced up hill to peaks at both ends; this forms what Cayley termed a *ridge line*, and is exactly equivalent to a drainage divide or watershed. The other slope line leads from one pit through the saddle to either another pit or perhaps the same one; this slope line is termed a *course line*. Course lines often, but not always, correspond with stream channels. Cayley's topologic model was later elaborated by physicist James Clerk Maxwell (1870), who provided a set of terms for the regions bounded by the lines defined above (see Figure 4.9). For regions without pits, the ridge-lines form a tree, and all course-lines lead off the study area, perhaps to the ocean.

Warntz (1966) rediscovered this work, and applied the model to the analysis of statistical surfaces representing the variation of socioeconomic data over geographic space. He later re-applied the model to topography (Warntz, 1975). Pfaltz (1976) discussed the topology of these networks, focussing on the graph which implicitly underlies them. Neither of the authors introduced terms for these features, but Mark (1979) called the network of ridge-lines and course-lines the 'Warntz network' of a surface, and the underlying graph the 'Pfaltz graph'.

Mark (1979) studied the topology and geometry of these surface networks. He found that the frequencies of different topological arrangements of ridge-lines could be predicted very well by the frequencies of minimum spanning trees for sets of randomly-distributed points representing peaks. The minimum spanning tree (MST) of a set of points is just the tree which connects the points with minimum total edge length. Peaks were distributed at random over an ellipse; by varying the elongation of the ellipse, a very good simulation of ridge-topology frequencies for different types of terrain could be obtained. Later, Mark (1981) found that, for hypothetical surface, networks which minimized total ridge length also minimized the ratio of surface to surface volume. This suggested a possible physical interpretation of the MST hypothesis. However, the reason that the MST model fits so well now appears to be much simpler, and of little geomorphic significance. Consider any

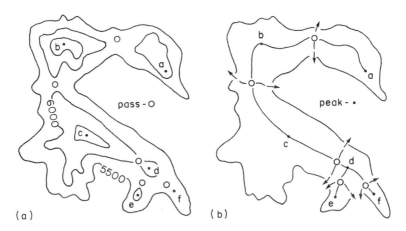

Figure 4.9. (a) Simplified contour map of a magnitude-6 hilltop, with critical points shown; (b) Critical lines forming the Warntz network of the surface (ridge lines are solid; course lines are dashed, and all lead to some pit or pits outside the bounding contour)

tree embedded in a plane; delete the links and form the minimum spanning tree of the vertices. If all of the links are about the same length, and if none of the junctions have angles less than 60°, then it is not at all surprising that the MST links simply replace the original links in almost all cases.

4.3.2   Geometric Ridges

In his famous *Report on the Geology of the Henry Mountains*, Grove Karl Gilbert (1877) provided an early quantitative model of the geometry of ridge networks in badland topography. This model seems much more promising than the one proposed by Mark (1979), but has not been studied since its introduction. Gilbert (1877, pp. 120–123) examined the pattern of secondary ridges and drainage lines along a major badland divide. He observed that where secondary ridges join the main one, the main ridge is deflected toward them, forming a zig-zag; also, at such ridge junctions, there is often a peak or local maximum, with saddles in the intervals (see Figure 4.10). Gilbert then proposed a clear and quantitative explanation for these phenomena. First, he postulated that the 'heads of the secondary drainage lines [on either side of the main ridge] are in nature tolerably defined points' (Gilbert, 1877, p. 121). Above this point, slope processes tend 'to produce an equal declivity in all directions upward from the point of convergence' (p. 121). In effect, Gilbert proposed that contours form concentric circles centred at the channel heads, and that these funnel-shaped forms are lowered and intersect with similar funnel-shaped valley-heads on the other side of the ridge. The ridge is then 'at right angles to a line connecting the two points' (Gilbert, 1877, p. 121). In fact, if all channel-heads were at the same elevation, the ridges would be exactly the boundaries of the proximal (Thiessen) polygons based on those points.

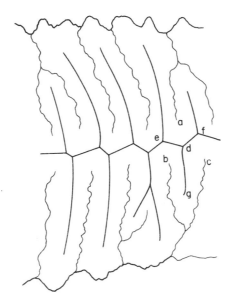

Figure 4.10. Ground-plan of a badland ridge, showing its relation to waterways. The smooth lines represent divides
(From Gilbert, 1877, Figure 58, p. 122)

This model accounts for both the zig-zag pattern of main divides and the arrangement of peaks and saddles along them, and certainly is worthy of further investigation through computer simulation.

Almost a century later, Goudie (1969) published a study of the topology of dune ridges. Goudie picked an end-node as a 'root' equivalent to a channel network outlet, and then ordered the streams using the Horton–Strahler system (Strahler, 1952). He found that these ridge networks followed 'Horton's Laws'. However, methods for defining ridges and for choosing a root node were not discussed in the paper.

Christian Werner (1972a, 1972b, 1973, 1982) also examined the topology of geometrically-defined ridges. Ridges were not defined in these papers, but were identified using a technique directly analogous to the contour crenulation method for defining channels (Werner, personal communication, 1976). In one of these papers, Werner (1972b) cut the ridge network at all passes, and analysed the topology of the small subnetworks dominated by single peaks; these are exactly the networks that would be individual channel networks and drainage basins if all heights were multiplied by -1. Werner could not reject the hypothesis that all topologically-distinct ridge trees occurred with equal frequency in the hills of Eastern Kentucky.

Werner (1972a) also showed that geometric ridge networks and channel networks are not independent. He found that pairs of adjacent first-order streams were separated by a single first-order geometric ridge in 88 of 119 cases, twice as many as a random (Poisson) model would predict. The relation has a certain stochastic component, and only approximates a geometric duality. Later, however, Werner (1973) assumed the duality to be exact, and went on to test further hypotheses regarding the relations between ridge and channel topologies. This relation was also used to deduce a relation between basin perimeter and channel factors such as length and magnitude (Werner, 1982).

4.4 LANDSCAPE FLOW NETWORKS

It is possible to consider the entire landscape to be a network. Each point receives inputs from the atmosphere and (except for peaks) from points up-slope, and then may output water and/or rock debris to other points down-slope or down-channel. In order to apply this model, it is convenient to divide the surface into a number of discrete cells or finite elements. In this section, it is assumed that these cells form a square grid in planform; triangles or other finite elements could also be used.

O'Callaghan and Mark (1984) defined the *drainage accumulation function* (D) as 'an operator which, given the drainage direction matrix [here called A] and a weight matrix W, determines a resulting matrix $[R]$ such that each element in $[R]$ represents the sum of the weights of all elements in the matrix which drain to that element' (O'Callaghan and Mark, 1984, p. 326). As an example, if each element in W is set to the unit cell area, then R will contain the contributing drainage area of every cell. Similarly, if all cells on a particular geologic formation are set to the unit area, with all others zero, then R will show the area of that rock type which drains to the individual cells; if just one element in W is set to 1, with all the rest 0, the 1's in R will indicate the path from that cell to its outlet, either a sink or the edge of the study area.

With an appropriate definition of the weight matrix W, the drainage accumulation function can be used to solve many drainage basin problems which involve a 'downstream' orientation. If all cells in W are initialized to the product of cell area and precipitation input (and this latter term could vary spatially), R will contain the water discharge passing through the cell, assuming complete runoff. Modifications of this, perhaps with a time component, could be of use in hydrograph simulation and flood prediction. If cell discharge is combined with surface gradient (derived from an elevation matrix Z), erosion for each cell can be calculated, and the matrix Z updated accordingly (Sprunt, 1972; Hugus and Mark, 1984, 1985). Thus long term erosion and landscape development could be simulated.

This surface drainage model allows the calculation to be separated into distinct modules or phases:

1. Calculation of the drainage direction matrix A, given the elevation matrix Z, and perhaps other information;
2. Algorithms for the drainage accumulation function $D(A, W)$, to yield matrix R;
3. Definition of appropriate weight matrices W, and interpretation of the resulting matrices R, in order to solve particular drainage-related problems.

Improvements to programs or procedures implementing any one of these phases could be made without altering the operation of the others.

4.4.1 The Drainage Direction Matrix

Central to most algorithms for drainage studies from digital elevation models is the assignment of drainage directions to individual grid points. While drainage directions could be computed from elevation data 'on the fly' during other processing, this would usually involve redundant calculations; instead, a *drainage direction matrix*, A, can be calculated once, and then used for all subsequent analyses. Currently, the model assumes that a grid point can drain only to exactly one neighbouring grid point. It is possible to determine a surface normal vector at every point, and project this into the horizontal plane to obtain the *aspect* at that point; then, this value can be rounded to an integer representing one of the eight directions to neighbouring cells.

Hydrologically, however, it seems more appropriate to estimate drainage direction based on the concept of steepest descent. Here, one assumes that a grid point will drain toward that neighbour which has the line of steepest descent connecting it to the grid point. If each grid cell is considered to have eight neighbours, four of these are farther away by a factor of the square root of 2. Thus in the search for the lowest neighbour, the elevation differences between diagonal neighbours and the central point should be divided by 1.41 before comparison with differences to the row and column neighbours. A point with no neighbour of lower elevation is termed a *pit*, and is given a distinctive drainage direction code. A neighbour of exactly equal elevation is not considered to be an outlet.

Pit Removal

A pit was defined as a point none of whose neighbours has a lower elevation. Such points would not drain to any of their neighbours, and thus cannot be assigned a drainage direction. At horizontal scales of, say, 10 m or greater, true pits or closed depressions are rare in natural Earth topography, being almost totally restricted to a few special geomorphic environments. Pits are fairly common in glaciated terrain, occuring as kettle holes in areas of deposition of sediment around ice blocks, or as closed rock depressions in cirques or in low-relief areas influenced by continental glaciation. In a humid environment, such pits will usually be occupied by lakes. Pits also occur in karst topography developed on limestone, in thermokarst developed over permafrost, on landslide debris, or in areas of aeolian deposition (sand dune fields). Finally, in arid or semiarid areas, low areas are often broad closed depressions which may contain ephemeral lakes or salt flats. However, in arid-land uplands and in most humid areas which were not glaciated, pits are very rare.

Pits, however, will often be found in grid-based digital elevation models. Even if the elevation is perfectly accurate at every grid point, pits may still occur due to an effect similar to aliasing: one row of the grid may have a point which falls on a channel, while in the next row downstream, the channel may pass between the grid points. Also, much gridded elevation data is now being produced using automated image correlation devices. Such data may have local, uncorrelated errors which often produce closed depressions in areas of gentle slope. Pits could be resolved and removed by altering the elevation data; alternatively, changes could be made in only the drainage direction image. In either case, pit removal could involve information about actual drainage patterns (in the form of digitized stream channels), or could simply resolve pits by a local 'flooding' procedure. Either of these could be augmented through interactive user input.

4.4.2   The Drainage Accumulation Function

In this section, several algorithms for calculating the total accumulated drainage at a point are discussed; these use the drainage direction matrix and an initial weights matrix for a study area as inputs. They assume one output per cell, except for pits which have no output; if pits are not to be treated as sinks, then pits must be removed from the drainage direction matrix, with values updated appropriately. The algorithms are completely independent of the method used to construct the drainage direction matrix, and do not require access to the original elevation data. At least four distinct algorithms are available for drainage accumulation; two of these have been discussed by Mark (1984), a third was implemented by O'Callaghan and Mark (1984), and a fourth was presented by Marks et al. (1984) and Band (1986b).

The first algorithm requires random access to both the drainage direction matrix and a weights image which eventually becomes the accumulated drainage result. Each cell is visited, and the initial weight in that cell is followed downslope to the edge of the DEM or to a pit. All cells along the path have that weight added to the accumulated weight already present. Since runoff from any cell could potentially flow entirely across the data area, the drainage directions and the runoff counters for the entire model must be available at all times.

A second approach to runoff simulation is local, and each cell is visited only once; however, this algorithm requires that the points be processed in order of decreasing elevation. If cells are processed from highest to lowest, then by the time a cell is processed, all possible inputs to that cell will have been defined (since a cell cannot receive input from one which is lower). In this approach, a sequential file is first written which contains the elevation of each point, together with the cell to which it drains; this file must then be sorted in decreasing order by elevation. A runoff vector with a length equal to the number of elevation points must be initialized to the starting weights, and then the points are processed in order. The total weight present in each cell is simply added to that already present in the cell toward which it drains; the total weight need not be carried farther at each step, since all the inputs of any cell are determined before that cell is processed.

Yet another algorithm was presented by O'Callaghan and Mark (1984). First, the number of inputs received from neighbours is calculated for each cell. Then, an iterative procedure begins. On each cycle, cells which have unresolved inputs are not processed. The weight of any cell with no unresolved inputs is added to the neighbour toward which it drains, and the number of inputs of that neighbour is decreased by one. The in-degree of a cell whose contents have been exported is set to a special value in order to indicate that no further processing is necessary. The in-degree counter is used to determine when all of a cell's inputs have been added to its value; at that time, its total drainage accumulation can be exported. On the second iteration, any cell with one input which received weight from that input on the first cycle would be processed, and eventually, the weight accumulation will be passed down to the pits or to the edges of the data area. The maximum number of iterations needed depends on the longest path from any cell, and should be proportional to the linear extent of the study area, which varies with $N^{1/2}$. Thus the expected complexity of this drainage accumulation algorithm is roughly $N^{3/2}$.

Finally, an elegant and efficient algorithm for drainage accumulation can be provided through recursion (Marks et al., 1984; Band, 1986b). Recursive procedures are ones which call themselves; such procedures are allowed in most (but not all) computer languages. Marks and others (1984) published a recursive algorithm which simply marked all cells in a basin; Band (1986b) modified the algorithm to calculate drainage areas during the recursion. The development of a recursive algorithm is similar to the construction of an inductive proof in mathematics. Consider the calculation of drainage area: the drainage area of a cell is just its own area, plus the sum of the drainage areas of neighbours draining into it. Instead of the complicated algorithm descriptions presented above, one has a very compact description:

```
procedure drainage_area (current_location):
   begin
      area : = unit_cell_area;
      for each neighbour:
         if      (the neighbour drains into the current cell)
         then    area : = area + drainage_area (neighbour);
      return (area)
   end
```

This algorithm is initiated with a call using the basin outlet as the initial location. It visits each cell exactly once, and thus the running time is a linear function of the number of cells in the basin. Notice that the procedure just keeps calling itself until it reaches a cell with no drainage inputs (usually, a cell on a ridge). With minor modification, water and sediment volumes can also be returned during the recursion, and thus the algorithm can readily be used to drive models of runoff, erosion, and landscape evolution.

4.4.3   Automated Definition of Channel Networks

The methods outlined above can be combined with a threshold for channel development to produce a synthetic channel network (O'Callaghan and Mark, 1984; Band, 1986a, 1986b). A simple drainage-area threshold, in grid cells, can produce useful channel networks (O'Callaghan and Mark, 1984; see Figure 4.11). For a more physically-meaningful definition, Band (1986a) employed a critical value of a measure proposed by Bevan and Kirkby (1979) to characterize topographic influences on runoff. DEM cells with values of $\ln(a/\tan(b))$ above some threshold are considered to be channel cells, where a is the drainage area contributing to the cell (determined by one of the methods discussed above), and b is the surface slope. With a program such as these, it is possible to generate a number of channel networks, each defined according to some (arbitrary) threshold for area of $\ln(a/\tan(b))$. Then, network properties such as link frequencies, link length distributions (by link types), junction angles, etc., can be plotted against threshold, in order to obtain objective estimates of these parameters and their sensitivity to definitions of channel sources,

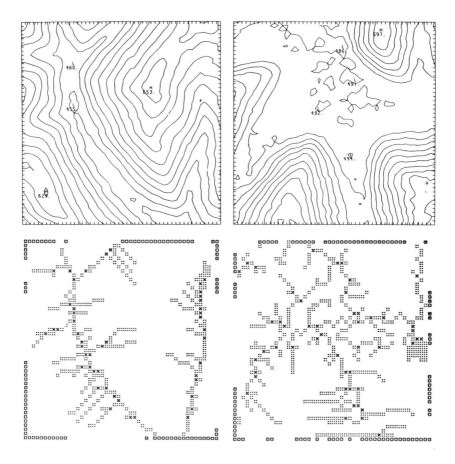

Figure 4.11. Contour maps (above) and channel networks (below) for digital elevation models from Pennsylvania (left) and Australia (right). Ticks around the margins of the contour maps indicate the grid spacing. In each case, channels are points draining 10 or more cells (after O'Callaghan and Mark, 1984, pp. 335 and 336)

a major source of operation-introduced variation in network studies. It also will be possible to obtain better information about the relationships between channel networks identified in the field and from maps (*cf*. Mark, 1983b), and to obtain large samples of network data.

4.5 SUMMARY

Networks are of paramount importance in geomorphological research. Some of the most prominent geomorphological networks are channel networks. Simple channel networks with no lakes or forks dominate in many areas, and have been studied rather intensively by geomorphologists. This review has concentrated on network types which have received less attention, including networks with forks (such as deltas or braided channels) and networks with lakes. There are other important geomorphological networks. Ridge and divide patterns have received relatively little attention, and have been examined here in more detail. Finally, flow over entire surfaces can be considered to be networks. A finite-elements approach divides land surfaces into discrete cells, and links these into a space-filling network reflecting flows of water and debris across the surface. Such networks are at the core of simulation models for runoff generation, erosion, and landscape evolution. The visible networks on the surfaces, such as the channels and the ridges, are seen to be subgraphs of the whole-surface networks. Network analysis provides a sound basis for integrated studies of landscape processes.

ACKNOWLEDGEMENTS

Portions of the work reported in this paper were developed when the writer was a Visiting Scientist with the CSIRO Division of computing research during 1983; other material included was developed under support from U.S. National Science Foundation Grant SES-8420789. The support of the both is gratefully acknowledged.

REFERENCES

Abrahams, A. D. (1980). Divide angles and their relation to interior link lengths in natural channel networks. *Geographical Analysis*, **12**, 157–171.

Abrahams, A. D. (1984a). Channel networks: a geomorphological perspective. *Water Resources Research*, **20**, 161–188.

Abrahams, A. D. (1984b). The development of tributaries of different sizes along winding streams and valleys. *Water Resources Research*, **20**, 1791–1796.

Abrahams, A. D. (1984c). Tributary development along winding streams and valleys. *American Journal of Science*, **284**, 863–892.

Abrahams, A. D. and Campbell, R. N. (1976). Source and tributary-source link lengths in natural channel networks. *Geological Society of America Bulletin*, **87**, 1016–1020.

Abrahams, A. D. and Mark, D. M. (1986). Acceptance of the random topology model of channel networks: bias in statistical tests. *Professional Geographer*, **38**, 77–81.

Abrahams, A. D. and Miller, A. J. (1982). The mixed-gamma model for channel link lengths. *Water Resources Research*, **18**, 1126–1136.

Band, L. E. (1986a). Analysis and representation of drainage basin structure with digital elevation models. *Proceedings, Second International Symposium on Spatial Data Handling*, Seattle, Washington, 5–10 July 1986, 437–450.

Band, L. E. (1986b). Topographic partition of watersheds with digital elevation models. *Water Resources Research*, **22**, 15–24.

Bevan, K. J. and Kirkby, M. J. (1979). A physically based, variable contributing area model of basin hydrology. *Hydrological Sciences Bulletin*, **24**, 43–69.

Cayley, A. (1859). On contour lines and slope lines. *Philosophical Magazine*, **18**, 264–268.

Church, M. and Mark, D. M. (1980). On size and scale in geomorphology. *Progress in Physical Geography*, **4**, 342–390.

De Bruijn, M. G. and Morselt, B. J. M. (1967). A note on plane trees. *Journal of Combinatorial Theory*, **2**, 27–34.

Gilbert, G. K. (1877). *Report on the Geology of the Henry Mountains*, U.S. Geographical and Geological Survey of the Rocky Mountain Region, Department of the Interior.

Goudie, A. (1969). Statistical laws and dune ridges in southern Africa. *Geographical Journal*, **135**, 404–406.

Harary, F. (1969). *Graph Theory*, Reading, Massachusetts, Addison Wesley.

Horton, R. E. (1945). Erosional development of streams and their drainage basins; hydrophysical approach to quantitative morphology. *Geological Society of America Bulletin*, **56**, 275–370.

Howard, A. D. (1971). Optimal angles of stream junction: geometric stability to capture, and minimum power criteria. *Water Resources Research*, **7**, 863–873.

Howard, A. D., Keetch, M. E., and Vincent, C. L. (1970). Topological and geometrical properties of braided streams. *Water Resources Research*, **6**, 1674–1688.

Hugus, M. K. and Mark, D. M. (1984). Spatial data processing for digital simulation of erosion. *Proceedings, ASP/ACSM Fall Convention*, San Antonio, Texas, September 9–14, 1984.

Hugus, M. K. and Mark, D. M. (1985). Digital simulation of erosion: a model based on geomorphic processes. *Proceedings, Sixteenth Annual Pittsburgh Conference on Modeling and Simulation*, **16**, 1, 305–309.

James, W. R. and Krumbein, W. C. (1969). Frequency distributions of stream link lengths. *Journal of Geology*, **7**, 544–565.

Jarvis, R. S. (1977). Drainage network analysis. *Progress in Physical Geography*, **1**, 271–295.

Kirkby, M. J. (1980). The stream head as a significant threshold. In Coates, D. R. and Vitek, J. D. (Eds.), *Thresholds in Geomorphology*, Binghamton Symposium, London, George Allen & Unwin, 53–73.

Krumbein, W. C. and Orme, A. R. (1972). Field mapping and computer simulation of braided-stream networks. *Geological Society of America Bulletin*, **83**, 3369–3379.

Leopold, L. B. and Langbein, W. B. (1962). The concept of entropy in landscape evolution. *US Geological Survey, Professional Paper*, **500-A**.

Mark, D. M. (1979). Topology of ridge patterns: randomness and constraints. *Geological Society of America Bulletin*, **90**, 164–172.

Mark, D. M. (1981). Topology of ridge patterns: possible physical interpretation on the 'minimum spanning tree' hypothesis. *Geology*, **9**, 370–372.

Mark, D. M. (1982). Markov dependencies in integer strings representing drainage networks: an alternative approach to testing for topological randomness (abstract only). *Abstracts, Association of American Geographers Annual Meeting*, San Antonio, Texas, April 1982, 221.

Mark, D. M. (1983a). On the composition of drainage networks containing lakes: statistical distribution of lake in-degrees. *Geographical Analysis*, **15**, 97–106.

Mark, D. M. (1983b). Relations between field surveyed channel networks and map-based geomorphometric measures in small basins near Inez, Kentucky. *Annals. Association of American Geographers*, **73**, 358–372.

Mark, D. M. (1984). Automated detection of drainage networks from digital elevation models. *Cartographica*, **21**, 168–178.

Mark, D. M. and Averack, R. (1984). Link length distributions in drainage networks with lakes. *Water Resources Research*, **20**, 457–462.

Mark, D. M. and Goodchild, M. F. (1982). Topologic model for drainage networks with lakes. *Water Resources Research*, **18**, 275–280.

Marks, D., Dozier, J., and Frew, J. (1984). Automated basin delineation from elevation data. *Geo-Processing*, **2**, 299–311.

Maxwell, J. C. (1870). On hills and dales. *Philosophical Magazine*, **40**, 421–427.

Milton, L. E. (1966). The geomorphologic irrelevance of some drainage net laws. *Australian Geographical Studies*, 4, 89–95.

Mock, S. J. (1971). A classification of channel links in stream networks. *Water Resources Research*, 7, 1558–1566.

Morisawa, M. (1985). Topological properties of delta distributary networks. In Woldenberg, M. J. (Ed.), *Models in Geomorphology*, The Binghamton Symposia in Geomorphology, International Series, No. 14, 239–268.

Moultrie, W. (1970). Systems, computer simulations, and drainage basins. *Bulletin of the Illinois Geological Society*, 12, 29–35.

O'Callaghan, J. F. and Mark, D. M. (1984). The extraction of drainage networks from digital elevation models. *Computer Vision, Graphics and Image Processing*, 28, 323–344.

Pfaltz, J. (1976). Surface networks. *Geographical Analysis*, 8, 77–93.

Roy, A. G. (1985). Optimum models of river branching angles. In Woldenberg, M. J. (Ed.), *Models in Geormophology*, The Binghamton Symposia in Geomorphology, International Series, No. 14, 269–285.

Shreve, R. L. (1966). Statistical law of stream numbers. *Journal of Geology*, 74, 17–37.

Shreve, R. L. (1967). Infinite topologically random channel networks. *Journal of Geology*, 75, 178–186.

Shreve, R. L. (1969). Stream lengths and basin areas in topologically random channel networks. *Journal of Geology*, 77, 397–414.

Shreve, R. L. (1975). The probabilistic–topologic approach to drainage-basin geomorphology. *Geology*, 3, 527–529.

Smart, J. S. (1969). Topological properties of channel networks. *Geological Society of America Bulletin*, 80, 1757–1774.

Smart, J. S. (1972). Channel networks. *Advances in Hydroscience*, 8, 305–346.

Smart, J. S. (1978). The analysis of drainage network composition. *Earth Surface Processes*, 3, 129–170.

Smart, J. S. and Moruzzi, V. L. (1972). Quantitative properties of delta channel networks. *Zeitschrift für Geomorphologie*, 16, 268–282.

Smith, T. R. and Bretherton, F. P. (1972). Stability and the conservation of mass in drainage basin elevation. *Water Resources Research*, 8, 1506–1529.

Sprunt, R. (1972). Digital simulation of drainage basin development. In Chorley, R. J. (Ed.), *Spatial Analysis in Geomorphology*, London, Methuen, 371–389.

Strahler, A. N. (1952). Hypsometric (area–altitude) analysis of erosional topography. *Geological Society of America Bulletin*, 63, 1117–1141.

Warntz, W. (1966). The topology of a socio-economic terrain and spatial flows. *Papers of the Regional Science Association*, 17, 47–61.

Warntz, W. (1975). Stream ordering and contour mapping. *Journal of Hydrology*, 25, 209–227.

Werner, C. (1972a). Channel and ridge networks in drainage basins. *Proceedings, Association of American Geographers*, 4, 109–114.

Werner, C. (1972b). Graph-theoretic analysis of ridge patterns. *International Geography 1972*, 942–945.

Werner, C. (1973). The boundaries of drainage basins for topologically-random channel networks. *Proceedings, Association of American Geographers*, 5, 287–290.

Werner, C. (1982). Analysis of length distribution of drainage basin perimeter. *Water Resources Research*, 18, 997–1005.

Modelling Geomorphological Systems
Edited by M. G. Anderson
©1988 John Wiley & Sons Ltd.

Chapter 5

Mathematical models of channel morphology

RICHARD D. HEY

School of Environmental Sciences, University of East Anglia

5.1 INTRODUCTION

In spite of their different perspectives both geomorphologists and river engineers have a common interest in predicting the shape and dimensions of alluvial channels. Traditionally geomorphologists have been concerned with the operation of river systems over millions of years; in particular erosional and depositional processes and landform evolution. Although much of their early work was descriptive, often relying on fanciful interpretations of remnant morphological features to establish a denudational chronology of a river basin or region, research over the last twenty-five years has focussed on the interaction between river mechanics and river morphology. This has produced a wealth of empirical information about the hydraulic geometry of different types of river system and a better understanding of flow and sediment transport processes in alluvial channels (Richards, 1982) which ultimately will aid the development of a physically based model to explain the morphological evolution of river systems over different space and time-scales. River engineers, in contrast, are mainly concerned with the design and implementation of flood alleviation and river improvement schemes, the training of unstable rivers and the management of river systems for water resource development purposes. Much of this work is concerned with short-term objectives, as many of these schemes only have design lives of the order of tens of years, and provide local solutions to local problems. Significantly many have caused both long-term and large-scale instability within the river system due to their effect on the sediment transport balance of the river (Raynov *et al.*, 1987). This emphasizes the need to improve stable channel design techniques and to further develop dynamic morphological models which can be used to predict the response of the river system to any proposed changes within the catchment.

This chapter reviews the different types of morphological model that are available for predicting the dimensions of stable alluvial rivers and canals and for modelling changes in channel morphology during erosion and deposition. The limitations of the various models are discussed and the research requirements to develop more deterministic modelling procedures are outlined.

5.2 STATIC MODELS OF STABLE CHANNELS

5.2.1 Variables Defining and Controlling Stable Channel Geometry

Alluvial rivers possess nine degrees of freedom since they can adjust their average bankfull width (W), depth (d), maximum depth (d_m), height (Δ) and wavelength (λ) of bedforms, slope (S), velocity (V), sinuosity (p), and meander arc length (z) through erosion and deposition. For river reaches that are in regime, implying that they do not systematically change their average shape and dimensions over a period of a few years, these can be regarded as dependent variables. In these circumstances the sediment load supplied from upstream can be transmitted without net erosion or deposition.

The principal variables controlling stable channel dimensions are the discharge (Q), bed material load (Q_s), calibre of the bed material (D), bank material, bank vegetation, and valley slope (S_v). Change in any one of these independent variables will eventually result in the development of a new regime channel geometry which is in equilibrium with the changed conditions. When stable the channel morphology will be uniquely defined by the new values of the controlling variables.

Considerable debate has centred on the designation of the dependent and independent variables over the last few years. This results from differences that occur between natural river systems and artificial channels and flumes. In the latter it is possible to prescribe some of the morphological variables, for example plan shape and width, which enables the sediment load or discharge required to maintain equilibrium to be determined (Hey, 1987). Under laboratory conditions and for artifical canals, there is, therefore, more flexibility in the choice of dependent or independent variables.

Another problem that has arisen relates to the view that channel morphology is indeterminate (Maddock, 1970). Results from flume studies are cited to show that certain combinations of designated dependent and independent variables give non reproducible results. This occurs in recirculatory flume experiments when channel slope is used as an independent variable and sediment load is considered to be a dependent variable. The maintenance of channel slope, through the adjustment of tailgate height is a discontinuous process which means that sediment load (input) is partially controlled and cannot be designated as an independent variable. This accounts for the non unique results (Hey, 1978).

For river systems there is no choice in the designation of dependent and independent variables. The variables defining the channel morphology are the response or dependent variables and the controlling or independent variables result from conditions imposed from the catchment. Channel morphology, when stable, has no control on the run off, sediment yield, valley slope, the character of the bed and bank material, or the bank vegetation.

Under regime conditions most of the controlling variables are effectively constant. The two exceptions are discharge and bed material transport as they can vary considerably through time. To overcome this difficulty Inglis (1946) suggested that a single dominant discharge could be defined which would produce the same average bankfull dimensions as a range of flows. Flume and river data indicate that this is the flow at, or about, bankfull stage. This phenomenon can be explained by considering the magnitude and frequency of sediment transport processes. Field studies indicate that the frequency of occurrence of the flow that transports most sediment in both sand and gravel-bed rivers (Andrews, 1980; Wolman and Miller, 1960) corresponds to the frequency of bankfull flow (Hey, 1975; Wolman and Leopold, 1957). With regard to bed material load, it follows that its critical value is the transport rate associated with bankfull flow (Inglis, 1949).

5.2.2 Regime Equations

'Regime theory' is the term given to empirical equations which have been developed to predict the geometry of stable alluvial channels. It is not strictly a theory.

It is not intended to review all equations that have been derived since the publication of Kennedy's classic work on canal dimensions in 1895 or Leopold and Maddock's research on rivers in 1953. Inevitably recent equations have more validity as a result of greater data availability, a better understanding of the factors controlling channel morphology and the application of rigorous statistical procedures in the analysis of the data. Most of the early equations were based on data from stable irrigation canals in India and Pakistan (Lacey, 1958; Lindley, 1919). Although many of these ignored the effect of bed and bank material type and sediment load on channel geometry, they have been successfully used on the Indian subcontinent for canal design purposes. The reason for this success, and their lack of application elsewhere, is that the bed and bank material and sediment load do not vary significantly in these canals. The constants in the regime equations will, therefore, reflect the fixed values of these unmeasured variables. When applied to channels with similar characteristics, the equations work. Application to channels with dissimilar bed and bank materials and sediment load can produce disastrous results.

Like all empirical equations care has to be exercised in their use, particularly with regard to their range of application. This should be restricted to channels which are similar to those used to derive the equations. Ideally all variables controlling channel shape and dimensions should be used in the derivation of each set of equations and, to ensure their general application, the variability within and between the variables should be maximized.

Regime equations can, for convenience, be classified on the basis of bed material size into cohesive-bed ($D < 0.0625$ mm), sand-bed ($0.0625 < D < 2$ mm) and gravel-bed equations ($D > 2$ mm). Equally they can be subdivided into mobile and quasifixed bed channels with regard to the presence or near absence of bed material transport. For this purpose silt transport rates of less than 500 ppm (parts per million by weight) are often regarded as trivial. Regime equations are listed for each type of channel which enable the width, depth and slope to be determined (metric units). Figure 5.1 illustrates the various definitions of width and depth that are used. The independent variables and their units are also listed and the range of application of each set of equations identified.

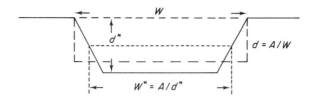

Figure 5.1. Definitions of width and depth used in regime equations
(A equals cross-sectional area)

Cohesive-bed Channels

Quasifixed Bed

Relatively little information is available on channels with cohesive beds. Simons and Albertson (1963) in their analysis of uniform water conveyance channels in alluvial material produced some tentative equations for channels with cohesive bed and banks based on data from irrigation canals in Wyoming, Colorado and Nebraska (US) and Sind (Pakistan).

Simons and Albertson's Equations

Data base and range of application:

Discharge (Q):   Bank vegetation: light–heavy
        $3.88 - 14.43\ \text{m}^3\ \text{s}^{-1}$

Sediment discharge (Q_s): $<500\ \text{ppm}$ Valley slope: 0.000063–0.000114

Median bed material size (D_{50}):   Planform: straight
        cohesive 0.029–0.36 mm

Bank material type: cohesive   Profile: uniform

Bedforms: plane   Water temperature: 21°C

Equations:

$$W^* = 3.64\ Q^{0.512} \qquad (\text{m}) \qquad\qquad (1)$$

$$d^* = 0.45\ Q^{0.361} \qquad (\text{m}) \qquad\qquad (2)$$

$$S = 0.71\ Q^{-0.343}\ (\times 10^{-6}) \qquad\qquad (3)$$

$$W = 1.11\ W^* + 1.91 \qquad (\text{m}) \qquad\qquad (4)$$

These equations have very restricted application, being based on very limited data, and should only be used for guidance. The regime slope equation is the most tentative and assumes a value of kinematic viscosity for a water temperature of 21°C. Although the original dimensionless equation, from which equation 3 was derived, indicated that slope was related to kinematic viscosity, its explicit inclusion is questionable since there was little variation in its value. In order to determine the shape of the trapezoidal section, it is necessary to calculate the top width from equation 4.

Mobile Bed

No equations are currently available.

Sand-bed Channels

Quasifixed Bed

The most comprehensive equations were derived by Blench (1969) and by Simons and Albertson (1963). Both take account of the effect of differences in the character of the bed and banks on channel morphology. Blench introduced two coefficients into his equations; a bed and a side factor to take account of these differences. Simons and Albertson, with a larger data base, including some from the US, stratified the information to produce two sets of equations, one for sand-bed and banks and one for sand-bed and cohesive banks.

Blench's Equations

Data base and range of application:
Discharge (Q): $0.03 - 2800 \, \text{m}^3 \, \text{s}^{-1}$ Bank vegetation: not specified
Sediment discharge (Q_s): $<30 \, \text{ppm}$ Valley slope: not specified
Bed material size (D): $0.1-0.6 \, \text{mm}$ Planform: straight
Bank material type: cohesive   Profile: uniform
Bed forms: ripples–dunes

Equations:

$$W^* = \frac{F_b^{0.5}}{F_s^{0.5}} Q^{0.5} \quad \text{(m)} \tag{5}$$

$$d^* = \frac{F_s^{0.33}}{F_b^{0.66}} Q^{0.33} \quad \text{(m)} \tag{6}$$

$$S = \frac{F_b^{0.833} \, F_s^{0.083} \, \nu^{0.25}}{3.63 \, g \, Q^{0.166}} \tag{7}$$

where the bed factor F_b and side factor F_s are defined by

$$F_b = \frac{V^2}{d} \quad \text{(m s}^{-2}) \tag{8}$$

$$F_s = \frac{V^3}{W} \quad \text{(m}^2 \, \text{s}^{-3}) \tag{9}$$

g being the acceleration due to gravity ($9.81 \, \text{m s}^{-2}$) and ν the kinematic viscosity ($\text{m}^2 \, \text{s}^{-1}$)

which is temperature dependent. If values of F_b and F_s cannot be obtained from measurements from similar channels in the locality, then their value can be approximated by

$$F_b = 0.58 \ D_{50}^{0.5} \text{(metric)} \qquad (10)$$

for the sand range, where D_{50} is the median particle size (mm) from sieve analysis of the bed material. The side factor is dependent on the bank material, being given by

$$\begin{aligned} F_s &= 0.009 \text{(metric) for loam of very slight cohesiveness} \\ &= 0.018 \text{(metric) for loam of medium cohesiveness} \\ &= 0.027 \text{(metric) for loam of high cohesiveness} \end{aligned}$$

Blench's method is dependent on the evaluation of bed and side factors. Equation 10 can be used to assess the former, while the latter can only be approximated. Error in F_s will cause only a small error in slope (equation 7), given its low exponent, although it will have a more significant effect on width and depth. In order to determine the cross-sectional dimensions of the channel, given values of W_* and d_*, it is necessary to make assumptions about the angle of the side slopes. Although derived for straight channels, the equations can be modified to make allowance for meandering (Blench, 1969).

Simons and Albertson's Equations

Data base and range of application:

Discharge (Q):
2.83–$11.32 \ \text{m}^3 \ \text{s}^{-1}$ (sand banks)
0.15–$2500 \ \text{m}^3 \ \text{s}^{-1}$ (cohesive banks)

Bank vegetation: light–moderate
(sand banks)
not specified
(cohesive banks)

Sediment discharge (Q_s): $< 500 \ \text{ppm}$

Valley slope: 0.000135–0.000388
(sand banks)
0.000059–0.00034
(cohesive banks)

Median bed material size (D_{50}):
0.318–$0.465 \ \text{mm}$ (sand banks)
0.06–$0.46 \ \text{mm}$ (cohesive banks)
Bank material type: cohesive/sand
Bedforms: ripples–dunes

Planform: straight

Profile: uniform

Equations:

sand banks

$$W^* = 5.72 \ Q^{0.512} \qquad \text{(m)} \qquad (11)$$

$$d^* = 0.504 \ Q^{0.361} \qquad \text{(m)} \qquad (12)$$

$$S = 0.000041 \ Q^{-0.341} \qquad (13)$$

$$W = 1.11 \ W^* + 1.91 \qquad \text{(m)} \qquad (4)$$

cohesive banks

$$W^* = 4.29 \, Q^{0.512} \qquad \text{(m)} \qquad (14)$$

$$d^* = 0.59 \, Q^{0.361} \qquad \text{(m)} \, d \le 2.1 \, \text{m} \qquad (15)$$

$$d^* = 0.44 \, Q^{0.361} + 0.674 \qquad \text{(m)} \, d > 2.1 \, \text{m} \qquad (16)$$

$$S = 0.000028 \, Q^{-0.341} \qquad (17)$$

$$W = 1.11 \, W^* + 1.91 \qquad \text{(m)} \qquad (4)$$

The equations relating to channels with sand-bed and banks are very tentative, being derived from a very small data base. In contrast considerable data were used to derive the equations for channels with cohesive banks, and these can be used with more confidence. The data points on the original graphs indicate that the width and depth relations are significantly better than the slope equation. The effect of sediment transport on channel dimensions was considered to be insignificant for transport rates < 500 ppm by weight. This contrast with Blench's assessment that values > 30 ppm had an effect on channel dimensions. Simons and Albertson also suggested that the Froude number should be less than 0.30 to maintain channel stability.

Mobile Bed Channels

Most irrigation canals in India and Pakistan operate with sediment excluders to limit transport rates in the canal system. Relatively little information is available regarding the effect of sediment load on canal morphology. On the basis of flume data, Blench's equations have been modified to take account of bed load transport.

Blench's Equations

Data base and range of application:
As for fixed-bed channels, with the exception that sediment transport rates are in range the 30–100 ppm by weight. Application in range 100–5000 ppm is more doubtful.

Equations:

$$W^* = \frac{F_{bc}^{\,0.5}}{F_s^{\,0.5}} \, Q^{0.5} \qquad \text{(m)} \qquad (5)$$

$$d^* = \frac{F_s^{\,0.33}}{F_{bc}^{\,0.66}} \, Q^{0.33} \qquad \text{(m)} \qquad (6)$$

$$S = \frac{F_{bc}^{\,0.833} \, F_s^{\,0.083} \, \nu^{0.25}}{3.63 \, g \, Q^{0.166} \left(1 + \dfrac{Q_s}{2330}\right)} \qquad (18a)$$

where Q_s is the bed load charge in ppm by weight and F_{bc} is given by

$$F_{bc} = F_b (1 + 0.012 \, Q_s) \tag{18b}$$

and F_s is defined as for fixed bed channels.

This is the only method which makes explicit allowance for the effect of sediment transport on channel dimensions. Care should be exercised in the use of this method for transport rates in excess of 100 ppm. The conversion factors used to modify the original equations are based on flume studies; they were not derived from the original field data.

Gravel-bed Channels

Considerable research has recently been carried out on the morphology of gravel-bed rivers by Kellerhals (1967), Nixon (1959), Charlton *et al.* (1978), Bray (1982), Hey (1982b), Andrews (1984), and Hey and Thorne (1986), following the pioneering work by Simons and Albertson (1963) on canals. Most ignore the effect of bed load transport on channel morphology, or assume that its value is zero or trivial. Only Kellerhals' equations actually apply for low or zero load, while Hey and Thorne were the first to explicitly include sediment load in design equations. Kellerhals' data were mainly obtained from rivers in Canada, while UK field data were investigated by Hey and Thorne.

Quasifixed bed

Kellerhals' Equations

Data base and range of application:

Dominant discharge (Q):	Bank vegetation: light
$0.03-2000 \, \text{m}^3 \, \text{s}^{-1}$	
Sediment discharge (Q_s): zero–low	Valley slope: 0.000166–0.0131
Median bed material size (D_{50}):	Planform: straight
$0.007-0.265 \, \text{m}$	
Bank material type: not specified	Profile: uniform
Bedforms: plane	

Equations:

$$W = 3.26 \, Q^{0.50} \quad (\text{m}) \tag{19}$$

$$d = 0.182 \, Q^{0.40} \, D_{90}^{-0.12} \quad (\text{m}) \tag{20}$$

$$S = 0.086 \, Q^{-0.40} \, D_{90}^{0.92} \tag{21}$$

where 90 per cent of the surface bed material is less than or equal to D_{90} as obtained by grid sampling procedures. Dominant discharge concepts were used to obtain a representative discharge value. For sites immediately downstream from lake exits an extreme flood with a high return period was used, while for sections where some bed load transport occurred discharges with a return period of between three and five years were considered representative. No information is given regarding the characteristics of the bank material, although this is likely to consist of gravel overlain by cohesive alluvium.

Mobile Bed

Hey and Thorne's Equations

Data base and range of application:

Bankfull discharge (Q):
    3.9–424 m³ s⁻¹

Bank vegetation: Type I   0% trees and shrubs
    Type II   1–5%,
    Type III   5–50%
    Type IV   >50%

Bankfull sediment discharge (as defined by Parker, Klingeman, and McClean (1982) bed load transport equation. (Q_s):
    0.001–14.14 kg s⁻¹

Median bed material size (D_{50}):
    0.014–0.176 m

Valley slope (S_v): 0.00166–0.0219

Bank material: composite, with cohesive fine sand, silt and clay overlying gravel

Planform: straight and meandering
Profile: pools and riffles

Bed forms: plane

Equations:

$$W = 3.98 \; Q^{0.52} \; Q_s^{-0.01} \quad \text{(m)} \quad \text{Vegetation type I} \quad (22a)$$

$$W = 3.08 \; Q^{0.52} \; Q_s^{-0.01} \quad \text{(m)} \quad \text{Vegetation type II} \quad (22b)$$

$$W = 2.52 \; Q^{0.52} \; Q_s^{-0.01} \quad \text{(m)} \quad \text{Vegetation type III} \quad (22c)$$

$$W = 2.17 \; Q^{0.52} \; Q_s^{-0.01} \quad \text{(m)} \quad \text{Vegetation type IV} \quad (22d)$$

$$d = 0.16 \; Q^{0.39} \; D_{50}^{-0.15} Q_s^{-0.02} \quad \text{(m)} \quad \text{Vegetation type I} \quad (23a)$$

$$d = 0.19 \; Q^{0.39} \; D_{50}^{-0.15} Q_s^{-0.02} \quad \text{(m)} \quad \text{Vegetation types II and III} \quad (23b)$$

$$d = 0.20 \; Q^{0.39} \; D_{50}^{-0.15} Q_s^{-0.02} \quad \text{(m)} \quad \text{Vegetation type IV} \quad (23c)$$

$$S = 0.087 Q^{-0.43} D_{50}^{-0.09} D_{84}^{0.84} \; Q_s^{0.10} \quad \text{Vegetation types I–IV} \quad (24)$$

Maximum depth
$$d_m = 0.20 \; Q^{0.36} \; D_{50}^{-0.56} D_{84}^{0.35} \quad \text{(m)} \quad \text{Vegetation types I–IV} \quad (25)$$

Meander arc length
$$z = 6.31 \; W \quad \text{Vegetation types I–IV} \quad (26)$$

| Sinuosity | $p = S_v/S$ | Vegetation types I–VI | (27) |

| Riffle width | $RW = 1.034\,W$ | Vegetation types I–IV | (28) |

| Riffle depth | $Rd = 0.951\,d$ | Vegetation types I–IV | (29) |

| Riffle maximum depth | $Rd_m = 0.912 d_m$ | Vegetation types I–IV | (30) |

where 50 per cent and 84 per cent of the surface bed material are less than or equal to D_{50} and D_{84} (m) respectively based on grid sampling procedures. As no information was available regarding bed load transport rates, independent estimates had to be obtained for use in the derivation of the regime equations. Before applying these equations it is important to ensure that bed load transport rates can be defined by the Parker, Klingeman, and McClean (1982) equation.

These equations can be applied to predict the plan as well as the cross sectional shape of mobile gravel-bed rivers. Allowance can also be made for the construction of pools and riffles.

Table 5.1. Degrees of freedom and governing equations* (after Hey, 1982a)

Degrees of freedom	Dependent variables	Fixed variables	Independent variables	Type of flow	Governing equations
1	V	$d, S, W,$ $d_m,$ λ, Δ, p, z	Q	Fixed bed	1 Continuity
2	V, d	$S, W, d_m,$ λ, Δ, p, z	Q, D, D_r, D_1	Fixed bed	+2 Flow resistance
3	V, d, S	$W, d_m,$ λ, Δ, p, z	$Q, Q_s, D, D_r,$ D_1	Mobile bed	Sediment +3 transport
5	$V, d, S,$ W, d_m	λ, Δ, p, z	$Q, Q_s, D, D_r,$ D_1	Mobile bed	+4 Bank erosion +5 Bar deposition
7	$V, d, S, W,$ d_m, λ, Δ	p, z	$Q, Q_s, D, D_r,$ D_1		+6 Bedforms +7 Bedforms
9	$V, d, S, W,$ $d_m, \lambda, \Delta,$ p, z	—	$Q, Q_s, D, D_r,$ D_1, S_v	Mobile bed	+8 Sinuosity +9 Riffle spacing/ meander arc length

*Average flow velocity $= V$; mean depth $= d$; channel slope $= S$; width $= W$; maximum flow depth $= d_m$; bedform wavelength $= \lambda$; bedform amplitude $= \Delta$; sinuosity $= p$; meander arc length $= z$; discharge $= Q$; sediment discharge $= Q_s$; characteristic size of bed, right and left bank sediment $= D, D_r, D_1$; valley slope $= S_v$.

5.2.3 Rational Equations

The two sets of variables defining and controlling the morphology of alluvial channels (section 5.2.1) are linked by nine physically deterministic process equations, one for each degree of freedom, which explain the adjustment mechanisms. Prescription of these equations, listed in Table 5.1, would enable the morphology of a channel to be predicted, given knowledge of the fixed and independent variables, by simultaneous solution of the appropriate process equations (Hey, 1982a). This is referred to as the rational or theoretical approach to channel design.

Three Degrees of Freedom

Solutions can currently be obtained for up to three degrees of freedom based on continuity, flow resistance and sediment transport equations. For example, it is possible to predict the slope, depth and velocity, given the cross-sectional shape of a straight, uniform and lined channel, on the basis of known values of discharge, sediment load, bed material size and bank roughness. There are many other possible combinations, particularly for canals, and it is essentially a question of deciding which three parameters need to be predicted, given the values of the remaining morphological and controlling variables (Table 5.2). The channel will be straight if the predicted or prescribed slope equals the valley slope, meandering if it is less than the valley slope and entrenched if it exceeds the valley slope. For meandering channels a different form of flow resistance equation may be required to make allowance for the effect of flow curvature on energy loss. Other possible solutions would allow combinations of discharge, sediment load or bed material size to be predicted if channel geometry is a design constraint.

Inherent in this approach is the assumption that the chosen flow resistance and sediment transport functions apply to the conditions in question and that these

Table 5.2. Possible combinations of dependent and independent variables based on continuity, flow resistance and sediment transport equations

Dependent variables	Independent variables
V, W, d	$S, Q, Q_s, D,$ stone lined non erodible banks, fixed side slope
V, S, W	$d, Q, Q_s, D,$ stone lined non erodible banks, fixed side slope
V, d, S	$W, Q, Q_s, D,$ stone lined non erodible banks, fixed side slope
W, d, S	$V, Q, Q_s, D,$ stone lined non erodible banks, fixed side slope
V, Q, W	$S, d, Q_s, D,$ stone lined non erodible banks, fixed side slope
V, Q, d	$S, W, Q_s, D,$ stone lined non erodible banks, fixed side slope
V, Q, S	$W, d, Q_s, D,$ stone lined non erodible banks, fixed side slope
Q, W, d	$S, V, Q_s, D,$ stone lined non erodible banks, fixed side slope
Q, Q_s, V	$S, W, d, D,$ stone lined non erodible banks, fixed side slope
Q, Q_s, W	$S, d, V, D,$ stone lined non erodible banks, fixed side slope
Q, Q_s, d	$S, W, V, D,$ stone lined non erodible banks, fixed side slope
Q, Q_s, S	$W, d, V, D,$ stone lined non erodible banks, fixed side slope
Q, Q_s, D	$W, d, V, S,$ stone lined non erodible banks, fixed side slope

equations are accurate. Ideally each equation used should be theoretically based and have general application. Unfortunately this is rarely the case and most have an empirical component that can restrict their range of application. Sediment transport equations can be notoriously inaccurate, particularly in gravel-bed rivers where sediment supply can limit transport rates and bed armouring occurs, which can lead to serious design errors. Before applying process based equations for design purposes, local checks should be made to verify their validity. This can be achieved either individually by checking each process equation or, collectively, by predicting the dimensions of existing stable channels in the locality.

Four Degrees of Freedom

Fixed-bed Channels

For fixed-bed channels in coarse non-cohesive material. Lane (1955) developed tractive force design concepts which enable the width, depth, slope and velocity to be determined. Essentially it is a question of maintaining the erosive forces within the channel below the resistive forces of the material on the channel perimeter. Four equations are required: (1) continuity, (2) flow resistance, (3) critical shear stress for the entrainment of the bed material, and (4) critical shear stress for the entrainment of the bank material.

Normally a Manning–Strickler equation (equation 31) is used to determine the resistance function (n) on the basis of the median bed material size (D_{50} m).

$$n = 0.047 \, D_{50}^{0.166} \tag{31}$$

The critical condition for the initiation of bed material movement can be obtained from the Shields entrainment function which takes account of both lift and drag forces. For coarse non-cohesive uniform material (D) this is given by

$$\frac{\tau_c}{g(\rho_s - \rho) \, D} = 0.056 \tag{32}$$

where τ_c is the critical shear stress ($= \gamma dS$), g the acceleration due to gravity, ρ the fluid density, ρ_s the sediment density, d the mean flow depth, and S the slope. This reduces to

$$D = 10.8 \, dS \tag{33}$$

on the assumption that $\rho = 1.0$ tonne m^{-3} and $\rho_s = 2.65$ tonne m^{-3}. However, for graded material, the critical Shields entrainment function can be as low as 0.03, based on the median grain size (D_{50}), due to differential exposure and packing of the surface material (Neill, 1968; Church, 1972). In which case

$$D_{50} = 20.2 \, dS \tag{34}$$

The critical shear stress on the bank (τ_{sc}) depends on the critical value on a plane surface (τ_c), the angle of repose (ϕ) of the material and the bank slope (θ) and is obtained by balancing forces on the grain.

$$\frac{\tau_{sc}}{\tau_c} = \cos \theta \sqrt{(1 - \frac{\tan^2 \theta}{\tan^2 \phi})} \tag{35}$$

If it is assumed that the shear stress at any point on the channel perimeter is given by γdS, which implies uniform flow and no secondary circulation to cause momentum transfer, then it follows that

$$\frac{d}{d_m} = \cos \theta \sqrt{(1 - \frac{\tan^2 \theta}{\tan^2 \phi})} \tag{36}$$

which reduces to

$$d = d_m \cos(\frac{x \tan \phi}{d_m}) \tag{37}$$

since $\tan \theta = -dd/dx$. The term inside the bracket is expressed in radians and x measures the distance from the channel centre where $d = d_m$. This enables the $d - x$ curve, defining the cross sectional shape of the channel, to be determined given the values of d_m and ϕ.

On the basis of these four governing equations, the width, depth, slope and velocity can be determined given the bed material size, which defines the angle of repose, and the discharge. If the predicted slope (S) is steeper than the local valley slope (S_v), then to transmit the prescribed discharge, it is necessary to incorporate a segment of constant depth, d_m, in the centre of the channel. This is best achieved by reversing the calculation procedure, letting $S = S_v$ and calculating the discharge capacity of the channel for comparison with the design requirement. If the former is less than the latter, then the channel needs to be widened, maintaining d_m.

Mobile-bed Channels

Attempts to develop the rational method for predicting the width of mobile-bed alluvial channels depends on the formulation of equations to define bank erosion and deposition. Laursen (1958), Henderson (1963), Ackers (1964), Chitale (1966), Charlton *et al.* (1978) and Hey (1978) all mentioned the need to develop a bank competence equation to define the erosional aspects of width adjustments. Only Hey (1978), and more recently Chang (1982), have identified the need for an equation to determine width reduction by deposition.

Although it is not possible to define equations for bank erosion and deposition for channel design purposes, it is possible to outline their mode of operation. Channel widening occurs when the shear stress at the bank exceeds the erosional resistance of the bank sediment. As a result both flow depth and bank shear stress are reduced. This process will continue until there is a balance between the shear stress and the erosional resistance of the bank.

Deposition, and a reduction in channel width, will occur when the longstream supply of sediment in the zone adjacent to the bank exceeds the transport capacity of the flow. As width/depth ratios are reduced by lateral accretion, the flow depth and bank shear stress are increased until the incoming load can be transmitted. In a meander bend both erosion and deposition can operate simultaneously at opposite sides of the channel; deposition on the point bar triggering erosion on the outer bank and *vice versa*.

The factors likely to affect the mass failure of alluvial banks include slope geometry, as affected by surface erosion and toe scour, seepage, infiltration, surcharge loading, tension cracking and vegetation. The type and scale of failure varies with soil type, in particular whether they are non-cohesive, cohesive, or composite, and considerable research has been carried out to determine factors of safety for each type of bank. These approaches have been fully reviewed by Thorne (1982) and Hey and Tovey (1987). To incorporate a bank stability equation into channel design, considerable more research is required on the geotechnical properties of the bank material in association with local flow and pore water conditions.

Bank deposition, resulting from the construction of lateral and point bars, is related to local sediment transport processes and the shear stress distributions. In meander bends, where point bar development can significantly reduce channel width, measured shear stresses indicate that at high flows maximum values occur across the toe of the point bar (Bathurst *et al.*, 1979; Bathurst 1979; Dietrich *et al.*, 1979). Although this is the zone of maximum sediment transport in the cross-section, deposition occurs due to a downstream decrease in shear stress. A number of physically based models have been developed for predicting the equilibrium flow, sediment transport and bed topography in meander bends (Engelund, 1974; Allen, 1977; Zimmerman and Kennedy, 1978). Engelund's (1974) theory is the most comprehensive and it has been shown to simulate field observations quite successfully (Bridge, 1977; Bridge and Jarvis, 1982). The model assumes that the net rate of transverse sediment transport is zero and a steady state bed condition prevails. Although point bars migrate over time due to lateral deposition, it was argued that a steady state condition is justified because lateral migration is small compared with the speed of advance of bed load particles. However, in order to develop a dynamic model of bar construction over longer time periods, it will be necessary to incorporate non steady-state bed/bar conditions.

The lack of suitable physically based equations to define width adjustments has led to the development of approaches based on energy and power principles. Effectively they rely on a variational argument in which the maximum or minimum of some parameter is sought.

Gilbert (1914) initiated this type of approach when he suggested that channels adjust either to maximize bed load transport, given the bed material size, slope and discharge, or to carry a given load on the lowest possible slope. For rivers, where sediment load and discharge are imposed, this implies that the width of the channel adjusts to minimize the slope to transport the imposed load.

Several extremal equations have been formulated to provide the additional equation necessary to obtain a determinate solution. These include:

1. Minimum stream power (Chang 1980). Given the discharge and sediment load, the river adjusts its width, depth, and slope (+ velocity) such that γQS is a minimum subject to given constraints, where γ is the specific weight of water. This implies that slope (S) is a minimum as discharge (Q) is prescribed.

2. Minimum unit stream power (Yang, 1976). The river adjusts its width, depth, slope, velocity and roughness to minimize the unit stream power required to transport a given sediment and water discharge. Unit stream power, or stream power per unit weight of water is given by:

$$\frac{Q\gamma LS}{\rho\, gWdL} = VS \qquad (38)$$

where L is the length of the reach, W the width, d the mean depth, ρ the density of water, g the acceleration due to gravity, and V the mean velocity.
3. Minimum energy dissipation rate (Brebner and Wilson, 1967; Yang et al., 1981). The river at equilibrium adjusts to the condition where its rate of energy dissipation is a minimum. For a channel reach of length L, the latter is given by

$$(Q\gamma + Q_s\, \gamma_s)\, LS \qquad (39)$$

where Q and Q_s are the discharge and sediment load and γ and γ_s the specific weights of water and sediment respectively.
4. Maximum friction factor (Davies and Sutherland, 1983). A channel will stabilize when its shape and dimensions correspond to a local maximum of the friction factor.
5. Maximum sediment transport rate (Kirkby, 1977; White et al., 1982). This hypothesis suggests that for a given discharge and slope, channel width adjusts to maximize the sediment transport rate.

In a comprehensive review of these alternative models, Bettess (1987) showed that maximum sediment transport rate was equivalent to the minimum stream power for a given discharge. Equally for sediment concentrations less than 1000 ppm the $\gamma_s\, Q_s$ term in the minimum energy dissipation rate equation is negligible and, therefore, is equivalent to minimum stream power. Effectively this leaves three independent hypotheses:
1. Minimum stream power
   Maximum sediment transport rate
   Minimum energy dissipation rate
2. Minimum unit stream power
3. Maximum friction factor

In order to predict width, depth, and slope (+ velocity) an extremal hypothesis has to be used in conjunction with the continuity equation and chosen flow resistance and sediment transport equations.

Bettess compared the predicted channel widths, for several combinations of extremal hypotheses, friction and transport equations, with empirical width relations for gravel-bed rivers. For fixed sediment diameter, discharge, and sediment load, regime dimensions were determined for minimum stream power (= maximum sediment transport rate), minimum unit stream power, minimum energy dissipation, and maximum friction factor. One set of results was calculated using the White et al. friction equation and the Ackers–White transport equation (White et al., 1982) and another using the Chang–Parker transport function and the Keulegan resistance equation (Griffiths, 1984). These were selected as the former had previously been shown to perform well for a range of conditions (White et al., 1982). While the latter produced curious results

(Griffiths, 1984). The results, based on a bed material size, D_{35}, of 0.01 m and a sediment concentration of 10 ppm are shown in Figure 5.2. The White *et al.* width function is very comparable to the regime equations for minimum stream power and minimum unit stream power and there is no justification for suggesting one is better than the other. The Griffiths approach grossly overestimates width and reflects the inappropriateness of these particular friction and transport equations for gravel-bed rivers. No results were obtained for the maximum friction factor approach as a maximum could not be found.

However, it should be noted that the empirical curves used for comparison purposes (Figure 5.2) ignore sediment transport, a factor which is incorporated in the calculation procedure. Hey and Thorne's regime equations for channel width do include the effect of bedload transport based on an independent estimate of its value obtained from the Parker *et al.* (1982) transport equation. For a sediment concentration of 10 ppm their curve for channels with more than 50 per cent trees and shrubs on the bank corresponds well with the Bettess predictions. The conformance is not as good for channels with less than fifty per cent tree and shrub cover on the bank.

This general approach for determining channel width has been roundly criticized for lacking a physical basis. White *et al.* (1982) freely admit that this is a problem but in the

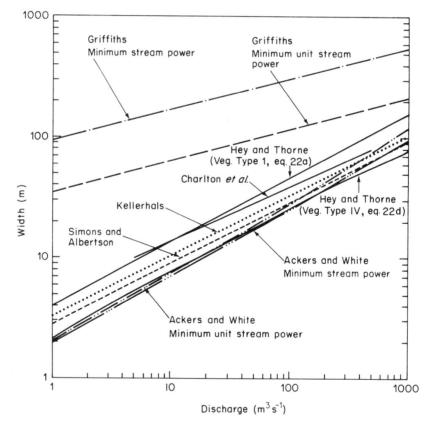

Figure 5.2. Regime widths for gravel-bed rivers (after Bettess, 1987)

absence of suitable physically based models for bank erosion and bar deposition, it affords a practical approach for the determination of channel width. The only alternative is to use some form of empirical width equation (section 5.2.2).

Problems can arise in the application of either minimum stream power or unit stream power. As Bettess points out, it is necessary to ensure that the friction and transport functions used are applicable to the problem in question. Some combinations may not give a maximum or minimum value. In addition it is often difficult to identify the extreme values of stream power, slope or load since the turning points are very flat. Consequently a large range of widths, depths and slopes are near optimal.

One major assumption in the use of these optimization models relates to the relative erodibility of the bed and banks. Regime equations have shown that width is a function of bank material and bank vegetation. If the banks are more resistant to erosion White *et al.* (1982) have argued that the width will be narrower and slope greater than the maximum efficiency solution, while the width will be larger and the slope greater for channels with banks which are more susceptible to erosion than the bed. Empirical evidence, however, suggests that slope adjustment to changes in bank erodibility may not occur, on gravel-bed rivers at least, as slope is independent of bank vegetation (equation 24). Although slope may not respond to the relative erodibility of the bed and bank in the way that White *et al.* (1982) have envisaged, there is no question that the effect of their relative erodibility on channel width needs to be fully investigated for this method to be developed further. Equally consideration has to be given to differences in aggradation between bed and banks. Significantly the same problems have to be investigated in order to develop a rational physically based solution for width adjustment.

Parker (1978) in a mathematical treatment of gravel-bed rivers with stable banks and a mobile bed argued that the shear stress would exceed the critical value for particle entrainment across the full bed width. Using a Keulegan friction law and an excess tractive force bed load function, this required a shear stress twenty percentage above the critical value in the centre of the channel, reducing to approximately threshold values adjacent to the bank. The implication is that this condition occurs during dominant flow conditions, often taken as bankfull discharge. Field evidence from gravel-bed rivers indicates that erosion of composite banks, gravel overlain by fine sandy silty clay, starts when bed material thresholds are exceeded immediately adjacent to the bank. For banks not protected by trees and shrubs this occurred when flows exceeded approximately half the bankfull flow (Hey, 1986). This suggests that for such channels the banks are relatively easily eroded while for channels lined with trees and shrubs the bed and banks could be considered to be equally erodible. Bettess' results, Figure 5.2, appears to support this hypothesis, since predicted widths using minimum unit stream power and minimum energy dissipation rate are comparable to observed values for channels with tree and shrub lined banks. It could, however, simply be a function of the flow resistance and sediment transport equations used in the formulation (*cf.* White *et al.* (1982) and Griffiths (1984), Figure 5.2).

5.2.4 Discussion

Regime and rational equations should be comparable provided the former are derived from a large data set which maximizes the variability within and between the variables, and the latter are based on physically deterministic process equations which have general application.

The two approaches being illustrative of the inductive and deductive approaches to developing morphological models.

5.3 DYNAMIC MODELS OF UNSTABLE CHANNELS

5.3.1 Introduction

Mathematical modelling procedures are now being developed to predict the effect of erosion and deposition on channel morphology. Such dynamic models can be used to study channel change due to natural causes as well as man's activities. As the physical controls on water and sediment movement in rivers are complex and only partially understood, several simplifications are generally made in the modelling procedure in order to produce a tractible problem. Initially most attention was directed towards the development of one-dimensional models for quasisteady gradually varied flow. In these models all the properties of the river, such as discharge, stage height and bed elevation, depend only on distance downstream and time. No account is taken of variations in these values over the width of the channel and the simulation is based on average cross-sectional values. Consequently one-dimensional models cannot directly predict the width and plan shape of a river. More complex two-dimensional models for unsteady gradually varied flow are currently being developed.

In all mathematical models choice has to be exercised in the designation of the governing equations and in the selection of the numerical procedure for solving them. Boundary and initial conditions also need to be specified, which requires further decisions about the characterization of the sediment properties, river geometry, and flow regime. Ultimately some form of sensitivity analysis should be undertaken to assess the models performance.

For a detailed review of mathematical aspects of modelling procedures reference should be made to texts by Cunge *et al.* (1980) and Simons, Li and Associates (1982). In this section attention is focussed on the physical basis of the models and future research requirements.

5.3.2 Governing Equations

The equations governing gradually varied unsteady flow in alluvial channels are (a) the equation of water continuity, (b) the equation of motion for water, and (c) the equation of sediment continuity. The first two are often referred to as the Saint Venant equations and the latter as the Exner equation.

Continuity Equation

For incompressible and homogeneous fluids the continuity equation can be derived by considering the conservation of mass between two channel sections Δx apart.

This yields

$$\frac{\partial Q}{\partial x} + \frac{\partial A}{\partial t} = 0 \tag{40}$$

where discharge Q changes with distance at rate $\partial Q/\partial x$ and flow area A changes with time at rate $\partial A/\partial t$.

If lateral inflow exists

$$\frac{\partial Q}{\partial x} + \frac{\partial A}{\partial t} = q_1 \tag{41}$$

where q_1 is the lateral inflow per unit channel length.

In terms of unit channel width, equation 40 reduces to

$$\frac{\partial q}{\partial x} + \frac{\partial d}{\partial t} = 0 \tag{42}$$

in which q is the discharge/unit width and d the flow depth.

Equation of Motion

Three major forces act on unit volume of water to cause motion, namely hydrodynamic pressure, gravity, and frictional resistance. In the derivation of the equation of motion for the fluid, four important assumptions are made regarding the nature of these forces:

1. Hydrostatic pressure distribution applies as streamline curvature is small;
2. No gravitational forces act parallel to the flow since the channel gradient is small (sine of angle approximates zero);
3. Resistance to flow can be approximated by equations developed for steady uniform flow;
4. Flow is one dimensional.

The resolution of these forces produces the general linear momentum equation which, when expressed in differential form yields

$$S_o - S_f = \frac{\partial(d)}{\partial x} + \frac{1}{gA}\frac{\partial}{\partial x}\left(\frac{Q^2}{A}\right) + \frac{1}{gA}\frac{\partial Q}{\partial t} \tag{43}$$

where S_o is the channel bed slope. S_f the friction slope, g the acceleration due to gravity, and d the flow depth. In this equation S_o represents the gravitational forces and S_f the shear forces, while the three terms of the right hand side represent the pressure gradient, the convective acceleration of the flow and the local acceleration of the flow respectively.

Expressed in terms of unit channel width it reduces to

$$\frac{\partial(ud)}{\partial t} + \frac{\partial(u^2 d)}{\partial x} + gd\frac{\partial(z+d)}{\partial x}\;(+gdS_f = 0 \tag{44}$$

where u is the mean fluid velocity, d the flow depth, and z the bed elevation above some arbitrary datum.

Sediment Continuity Equation

The continuity equation for sediment movement, expressed in terms of unit channel width, is

$$\frac{\partial q_s}{\partial x} + \frac{\partial}{\partial t}(C_s d) + (1-p)\ \rho_s g \frac{\partial z}{\partial t} = 0 \tag{45}$$

where q_s is the bed material transport rate. C_s is the spatial bed material concentration, p the bed porosity, ρ_s the mass density of the bed material, and z the bed elevation above an arbitrary datum.

Solution of Governing Equations

Equations 40 to 45 contain three unknowns: discharge, Q; flow depth, d; and sediment load, Q_s. The remaining variables need to be expressed as functions of these unknowns in order to obtain a solution. This is achieved in the following way:
1. The cross-sectional geometry of the channel is expressed as a function of stage height on the basis of survey information;
2. The mean bed slope S_o is determined from changes in bed elevation: initial conditions specified;
3. The friction slope S_f is a function of flow and channel characteristics and can be related to the unknown variables by a flow resistance function such as Manning's, Chezy's or Darcy–Weisbach's equations;
4. The lateral inflow discharge is prescribed as a function of time and space;
5. Sediment discharge information is available from field data or from available sediment transport theories.

Numerical Methods

Finite difference methods are now generally employed to solve the governing equations. These define a relationship between neighbouring values of dependent variables on a rectangular grid defining the space (x) and time (t) plane. Solutions can be obtained using either an explicit or an implicit scheme. Explicit schemes advance the solution through time by solving for the dependent variables, for example flow depth, velocity and bed elevation, one grid point at a time. In implicit schemes the dependent variables are derived simultaneously for a number of grid points over time.

Explicit schemes are relatively easy to develop and operate. However, there are stability problems and in order to limit the growth of rounding errors small space and time steps are required. This requires large data input and leads to lengthy computational time. In contrast implicit schemes are slightly more complex, requiring the inversion of a matrix, but remain stable over much wider ranges of space and time steps. This means that accurate results can be obtained using coarser step lengths than those required for an explicit scheme. This leads to a saving in computational time.

5.3.3 Model Application

Boundary Conditions

For one-dimensional models which are solved using an implicit scheme three boundary conditions are required when flow is subcritical. With regard to discharge, these are the discharge at the upstream end of the model and the water stage at the downstream boundary for each time step used in the model. This enables backwater curve calculations throughout the reach to determine stage heights and velocities. For the morphological element of the model, information on input sediment load at the head of the study reach is required for each time step. The specification of an appropriate sediment input poses a number of problems since feedback mechanisms resulting from channel instability can cause headward erosion or deposition, thereby altering the input conditions over time. Bettess and White (1981) have investigated several alternative ways of approaching this problem and these include:

1. Known sediment inflow, e.g. downstream from dams;
2. Upstream extension of reach under study, i.e. to ensure upstream boundary conditions will not affect reach under investigation;
3. Inflow determined by conditions immediately downstream of upstream end of model, i.e. sediment inflow calculated on basis of hydraulic and sedimentary conditions at head of study reach;
4. Inflow determined by conditions immediately upstream of upstream end of model, i.e. assumes sediment inflow unaffected by changes within reach being modelled.

Initial Conditions

Choice has to be exercised in the selection of computational points which are used to represent the nature of the river system. It is essential that all the important topographic and hydraulic features are included given the study objectives. More local detail is required to investigate the effect of dredging or weir construction than for large scale instability problems (100 m compared with 0.5–5 km).

For the reach under investigation the following information is required at each computational point:

1. Cross-sectional profiles
2. Longitudinal bed profiles
3. Distance between cross-sections
4. Bed sediment size, surface and subsurface, at each cross-section
5. Channel roughness characteristics
6. Tributary locations

Often, due to deficiencies in the data base, interpolation may be required between surveyed sections to generate the information for each designated computational point.

Time Steps

Mobile bed models are used to simulate water discharges, water surface slopes, sediment transport rates and changes in bed elevations. As water levels can change rapidly, time steps ranging from minutes to hours are normally used for flow routing purposes. In contrast time steps of several days can be employed for sediment routing provided this does not adversely affect the stability or accuracy of the simulation (section *Numerical Methods* above).

Calibration

Ideally historical information relating to the reach in question should be used to compare observed channel changes with those predicted by the model. Where necessary modifications can be made to the flow resistance and sediment transport functions used in the model during the calibration exercise to improve the model's predictive capability. Care should be exercised to ensure that any changes made can be physically justified. If realistic model parameters produce inaccurate results, then the model's construction should be reevaluated.

Once calibrated the model can be operated to predict the response of the channel to any imposed changes. Simulated changes should be judged in the light of geomorphological and river engineering experience.

5.3.4   Evaluation of Dynamic Models

All one-dimensional models are based on the continuity, motion and sediment continuity equations (equations 40–45).

Some, for example HEC-6 (Hydrologic Engineering Centre, 1977), HEC2SR (Simons, Li and Associates Inc., 1980), KUWASER (Simons *et al.*, 1979). WALLINGFORD (Bettess and White, 1981), and ILLUVIAL (Karim and Kennedy, 1982) consider unsteady flows as a series of steady flows which reduces the determination of fluid flow to a simple backwater calculation. Although rivers experience unsteady flow, both seasonally and during flood events, these can be approximated by steady discharges over varying time lengths. Generally the models determine flow depths, velocities and discharges at each computational point independently of sediment transport on the assumption that over the time step considered any change in bed elevation will not significantly affect the flow. Subsequently the sediment continuity equation is solved to determine the change in bed elevation at each computational point over the specified time step. In turn this influences flow calculations over the subsequent time interval.

Other models, Fluvial 11 (Chang and Hill, 1976), UUWSR (Simons *et al.*, 1975), SEDIMENT–4H (Resource Management Assoc., 1982) and MOBED (Krishnappan, 1981) have been developed for unsteady flow conditions. Although this obviates the need to specify time steps during which flows can be regarded as steady, there are two disadvantages. First there is usually a significant increase in computer time. Second, due to lack of knowledge about the effect of non steady flow on flow resistance and sediment transport, it is necessary to use relations developed for steady flows in unsteady flow models. This assumes that during a time step in an unsteady flow model, the friction and sediment transport are similar to those under steady flow conditions. To what extent this negates any advantages accruing from the unsteady flow formulation is open to question.

More recently Chang (1982) has developed an unsteady model, FLUVIAL 14, which, although being one dimensional, enables changes in channel width to be predicted. This is achieved at each time step by minimizing the total stream power subject to limits on width adjustment which are controlled by local sediment removal and transport, the rate of change in stage height and the bank configuration. Surprisingly no knowledge of the velocity distribution in the cross section is required, two-dimensional affect, nor information on the erosional resistance of the banks, particularly the role of vegetation.

Most modellers usually present case studies to show how successful their model simulates channel change. To date only one objective study has been carried out to evaluate several erodible bed models under a range of conditions. This was carried out by a Special Committee under the Chairmanship of Dr. J. F. Kennedy (1983). The numerical simulations produced by six models, HEC2SR, KUWASER, UUWSR, HEC6, FLUVIAL–11 and SEDIMENT–4H were evaluated using data from the San Lorenzo, San Dieguito and Salt rivers. The San Lorenzo being a channelized, stable sand-bed river, the Sand Dieguito an unstable, disturbed sand-bed river and the Salt an unstable gravel-bed river. It was found that none of the movable bed models gave completely satisfactory results. Provided friction factors based on field data were used in the computations, reasonably accurate predictions of flood water surface profiles were obtained. However, where reliance had to be placed on roughness equations, predictions by the different models deviated widely. As the friction factor has a marked effect on river stage, this is a major deficiency of the modelling procedures. Calculations of sediment transport rates, which is a prerequisite for assessing scour and fill activity, also differed very markedly between the models. All failed to adequately model bed armouring and its effect on flow resistance and sediment transport.

5.3.5 Future Developments

There is considerable scope for developing mobile-bed modelling procedures. Essentially it is a question of improving the physical basis of these models as the accuracy of numerical methods is currently an order of magnitude better than the physical uncertainty.

Flow Resistance

All models use a resistance coefficient, usually either Manning or Chezy, to characterize the friction factor. Best results are achieved when their values are obtained from field data for a range of flow conditions. If this information is not available, then estimated values are used based on engineering experience. During erosion and deposition when channel geometry, bed sediment size, and bed forms can be modified this type of approach can lead to inaccuracies. In these circumstances it would be preferable to incorporate into the modelling procedure resistance equations in which the friction coefficient is determined from information on bed sediment size, flow depth and bed form dimensions. Equations are available for predicting the Darcy–Weisbach friction factor for sand-bed rivers (White *et al.*, 1980, 1987) and gravel-bed rivers (Hey, 1979) which are based on sound theoretical principles. These values could be used in the Darcy–Weisbach equation to enable flow depth and mean velocity to be determined at each computational point for each time step.

Although these equations are strictly only applicable to steady uniform flow, they offer a significant improvement on current procedures. Further research is required to determine the effect of unsteady flow on flow resistance, while for two dimensional models the effect of cross sectional shape (Hey, 1979) and plan form needs to be investigated.

Sediment Transport

A reliable sediment transport relation is a prerequisite for accurately modelling erosion and deposition. There are a plethora of different equations each applicable to a particular set of river conditions, but few if any having general application. Often these equations are developed for stable conditions, rather than eroding or depositing ones where sediment sorting and changes in sediment supply affect transport rates. In gravel-bed rivers this leads to bed armouring which can severely affect channel roughness as well as the sediment transport capacity of the flow. In sand-bed channels bed forms can vary significantly as flow conditions change which in turn affects roughness values and transport rates. Research is required on both processes, with a particular need for additional research on sediment routing by size fraction, sediment pulsing and cross sectional variation in transport.

Bank Erosion and Deposition

With one-dimensional models it is possible only to consider values of variables which have been averaged over a cross-section. In order to develop two-dimensional models it will be necessary to determine cross-sectional variations in velocity, shear stress, sediment transport and bed material size particularly as they are affected by secondary flows. Research is progressing on this topic (de Vriend, 1981).

Cross-sectional variations in these parameters, together with the geotechnical properties of the bank material and the bank geometry, have a major influence on bank erosion and deposition (Thorne, 1982; Hey and Tovey, 1987). Equations need to be specified to define both processes if a physically based model is to be developed to define width adjustment. The assumption in all one-dimensional models that width remains constant during erosion and deposition is fallacious. Channel width responds very readily to degradation and aggradation. During degradation width can increase rapidly if the bed becomes armoured or if banks are steepened by toe scour. Equally aggradation can result in an increase in channel width, particularly if braiding occurs.

Meander and Braiding Mechanisms

In one-dimensional models it is not possible to simulate changes in the plan shape of the river. In reality alluvial channels can rapidly change their plan shape in response to erosion and deposition. Erosion, which tends to reduce channel gradients, can increase the river's sinuosity to accommodate this change, while deposition has the opposite effect. Differential bank erosion and deposition is responsible for these changes.

Tributary processes

The Kennedy Committee report (1983) identified the need to improve methods for modelling the effect of sediment load introduced by tributaries.

5.4 CONCLUSIONS

Recent years have witnessed a rapid advance in the development of morphological models for predicting stable channel dimensions and for simulating morphological changes during erosion and deposition.

Empirical models, provided they are based on a well conditions data set, enable the stable three dimensional geometry of alluvial channels to be predicted. Rational models, which are physically deterministic, currently have limited application due to their inability to determine channel width and plan shape.

Dynamic models for predicting scour and fill activity are also limited in their application due to the lack of process-based equations for modelling width and plan shape changes.

Further research is required to investigate both these processes, in particular three-dimensional flow patterns in alluvial channels, and their local effect on sediment transport, bank erosion, and bar deposition. This will best be achieved through collaboration between engineers and geomorphologists.

REFERENCES

Ackers, P. (1964). Experiments on small streams in alluvium. *J. Hydr. Div., Amer. Soc. Civ. Eng.*, **90**, (HY4), 1–37.

Allen, J. R. L. (1977). Changeable rivers: some aspects of their mechanics and sedimentation. In Gregory, K. J. (Eds.), *River Channel Changes*, Wiley, Chichester, 15–45.

Andrews, E. D. (1980). Effective and bankfull discharges of streams in Yampa basin, Colorado and Wyoming. *J. of Hydrology*, **46**, 311–330.

Andrews, E. D. (1984). Bed material entrainment and hydraulic geometry of gravel-bed rivers in Colorado. *Geol. Soc. Amer. Bull.*, **95**, 371–378.

Bathurst, J. C. (1979). Distribution of boundary shear stress in rivers. In Rhodes, D. D. and Williams, G. P. (Eds.), *Adjustment of the Fluvial System*, Kendall/Hunt Publ. Co., Dubuque, 95–116.

Bathurst, J. C., Thorne, C. R., and Hey, R. D. (1979). Secondary flow and shear stress at river bends. *J. Hydr. Div., Amer. Soc. Civ. Eng.*, **105** (HY10), 1277–1295.

Bettess, R. and White, W. R. (1987). Extremal hypothesis applied to river regime. In Thorne, C. R., Bathurst, J. C., and Hey, R. D. (Eds.). *Sediment Transport in Gravel-Bed Rivers*, Wiley, Chichester, 767–789.

Bettess, R. and White, W. R. (1981). Mathematical simulation of sediment movement in streams, *Proc. Inst. Civ. Eng., Pt. II*, **71**, 879–892.

Blench, T. (1969). *Mobile-bed Fluviology*, Univ. Alberta Press, Edmonton, Canada.

Bray, D. I. (1982). Regime equations for gravel-bed rivers. In Hey, R. D., Bathurst, J. C., and Thorne, C. R. (Eds.), *Gravel-bed Rivers*, Wiley, Chichester, 517–542.

Brebner, A. and Wilson, K. C. (1967). Determination of the regime equation from relationships for pressurized flow by use of the principle of minimum energy degradation. *Proc. Inst. Civ. Engs.*, **36**, 47–62.

Bridge, J. S. (1977). Flow, bed topography, grain size and sedimentary structure in open channel bends; a three dimensional model. *Earth Surface Processes*, **2**, 401–416.

Bridge, J. S. and Jarvis, J. (1982). The dynamics of a river bend: a study in flow and sedimentary processes. *Sedimentology*, **29**, 499–542.

Chang, H. (1980). Stable alluvial canal design. *J. Hydr. Div., Amer. Soc. Civ. Eng.*, **106** (HY5), 873–891.

Chang, H. (1982). Mathematical model for erodible channels. *J. Hydr. Div., Amer. Soc. Civ. Eng.*, **108** (HY5), 678–689.

Chang, H. and Hill, J. C. (1976). Minimum stream power for rivers and deltas. *J. Hydr. Div., Amer. Soc. Civ. Eng.*, **103** (HY12), 1375–1389.

Charlton, F. G., Brown, P. M., and Benson, R. W. (1978). The hydraulic geometry of some gravel-bed rivers in Britain. *Report IT 180*, Hydraulics Research Station, Wallingford.

Chitale, S. V. (1966). *Design of alluvial channels*, 6th Cong. Internat. Commission on Irrigation and Drainage, Question 20, 363–395.

Church, M. (1972). Baffin Island sandurs: a study of arctic fluvial processes. *Bull. Geol. Survey Canada No. 216*, 208 pp.

Cunge, J., Holly, F. M., and Verweg, A. (1980). *Practical Aspects of Computational River Hydraulics*, Pitman, London.

Davies, T. R. H. and Sutherland, A. J. (1983). Extremal hypothesis for river behaviour. *Water Resources Research,* **19**(1), 141–148.

de Vriend, H. J. (1981). Velocity redistribution in curved rectangular channels, *J. Fluid Mech.,* **107**, 423–439.

Dietrich, W. E., Smith, J. D., and Dunne, T. (1979). Flow and sediment transport in a sand bedded meander. *J. Geol.,* **87**(3), 305–315.

Engelund, F. (1974). Flow and bed topography in channel bends. *J. Hydr. Div., Amer. Soc. Civ. Eng.,* **100** (HY11), 1631–1648.

Gilbert, G. K. (1914). The transportation of debris by running water. *US Geol. Survey Prof. paper 86,* US Government Printing Office, Washington D.C.

Griffiths, G. A. (1984). Extremal hypotheses for river regime: an illusion of progress, *Water Resources Research,* **20**(1), 113–118.

Henderson, F. M. (1963). Stability of alluvial channels. *Trans. Amer., Soc. Civ. Eng.,* **128**, Part I, No. 3440, 657–686.

Hey, R. D. (1975). Design discharge for natural channels. In Hey, R. D., and Davies, T. D. (Eds.), *Science, Technology and Environmental Management,* Saxon House, UK, 73–88.

Hey, R. D. (1978). Determinate hydraulic geometry of river channels. *J. Hydr. Div., Amer. Soc. Civ. Eng.,* **104** (HY6), 869–885.

Hey, R. D. (1979). Flow resistance in gravel-bed rivers. *J. Hydr. Div., Amer. Soc. Civ. Eng.,* **105** (HY4), 365–379.

Hey, R. D. (1982a). Gravel-bed rivers: process and form. In Hey, R. D., Bathurst, J. C., and Thorne, C. R. (Eds.), *Gravel-bed Rivers,* Wiley, Chichester, 5–13.

Hey, R. D. (1982b), Design equations for mobile gravel-bed rivers. In Hey, R. D., Bathurst, J. C., Thorne, C. R. (Eds.) *Gravel-bed Rivers,* Wiley, Chichester, 553–574.

Hey, R. D. (1986). River response to inter-basin water transfer: Craig Goch Feasibility Study. *J. Hydrology,* **85**, 407–421.

Hey, R. D.(1987). Regime Stability. In *River Engineering—Part I,* Water Practice Manual No. 7, Inst. Water Eng. and Scs., London, 139–147.

Hey, R. D. and Thorne, C. R. (1986). Stable channels with mobile gravel beds. *J. Hydr. Eng., Amer. Soc. Civ. Eng.,* **112**(8), 671–689.

Hey, R. D. and Tovey, N. K. (1987). Mechanics of bank failure. In *Protection of UK River and Canal Banks,* CIRIA, London.

Hydrologic Engineering Centre (1977). *Scour and Deposition in Rivers and Reservoirs,* HEC–6, Corps. of Engineers, Davis, California.

Inglis, C. C. (1946). Meanders and their bearing on river training. *Maritime Paper No. 7,* Institution of Civil Engineers, London.

Inglis, C. C. (1949). The behaviour and control of rivers and canals. *Res. Publ. No. 13,* Central Water Power, Irrigation and Navigation Research Station, Poona, India.

Karim, M. F. and Kennedy, J. F. (1982). I ALLUVIAL: a computer based flow and sediment routing model for alluvial streams and its application to the Missouri river. *IIHR Report No. 250,* Iowa Inst. of Hydraulics Research, Iowa.

Kellerhals, R. (1967). Stable channels with gravel-paved beds. *J. Waterways and Harbours Div., Amer. Soc. Civ. Eng.,* **93** (WW1), 63–83.

Kennedy, J. F. (Chairman) (1983). *An Evaluation of Flood-level Predictions Using Alluvial-River Models,* National Academic Press, Washington D.C.

Kennedy, R. G. (1895). The prevention of silting in irrigation canals. *Proc. Inst. Civ. Engs.,* **119**, 281–290.

Kirkby, M. J. (1977). Maximum sediment efficiency as a criterion for alluvial channels. In Gregory, K. J. (Ed.), *River Channel Changes,* Wiley, Chichester, 429–442.

Krishnappan, B. G. (1981). *Unsteady, Non uniform, Mobile Boundary Flow Model—MOBED Users Manual.* Hydraulics Division, National Water Research Institute, Canada Centre for Inland Waters, Burlington, Ont.

Lacey, G. (1958). Flow in alluvial channels with sandy mobile beds. *Proc. Inst. Civ. Engs.,London,* **9**, 145–164.

Lane, E. W. (1955). Design of stable channels. *Trans. Amer. Soc. Civ. Eng.,* **120**, 2776, 1234–1279.

Laursen, E. M. (1958). Sediment transport mechanics in stable channel design. *Trans. Amer. Soc. Civ. Eng.,* **123**, 2918, 195–206.

Leopold, L. B. and Maddock, T. (1953). Hydraulic geometry of stream channels and some physiographic implications. *US Geol. Surv., Prof. Paper* **252–D**, US Government Printing Office, Washington D.C.

Lindley, E. S. (1919). Regime channels. *Proc. Punjab Engineering Congress.*

Maddock, T. (1970). Indeterminate hydraulics of alluvial channels. *J. Hydr. Div., Amer. Soc. Civ. Eng.,* **96** (HY11), 2309–2323.

Neill, C. R. (1968). A re-examination of the beginnings of movement for coarse granular bed materials. *Rept. No. INT 68*, Hydraulics Research Station, Wallingford, UK

Nixon, M. (1959). A study of the bankfull discharges of rivers in England and Wales. *Proc. Inst. Civ. Eng.,* **12**, 157–174.

Parker, G. (1978). Self-formed straight rivers with equilibrium banks and mobile bed, Part 2, The gravel river. *J. Fluid Mechs.,* **89**(1), 127–146.

Parker, G., Klingeman, P. C., and McClean, D. G. (1982). Bedload and size distribution in paved gravel-bed streams. *J. Hydr. Div., Amer. Soc. Civ. Eng.,* **108** (HY4), 544–571.

Raynov, S., Pechinov, D., Kopaliany, D., and Hey, R. D. (1987). *River Response to Hydraulic Structures*, UNESCO, Paris, 117 pp.

Resource Management Associates (1982). *Flood Routing with Movable Beds*, prepared by Anathurai, R. and Smith, D. J.

Richards, K. (1982). *Rivers*, Methuen, London.

Simons, D. B. and Alberston, M. L. (1963). Uniform water conveyance channels in alluvial material. *Trans. Amer. Soc. Civ. Eng.,* **128**, 1, 3339, 65–167.

Simons, D. B. *et al.* (1975). *A geomorphic study of pools 24, 25 and 26 in the Upper Mississippi and Lower Illinois rivers*, Rept. No. CER74–75DBS–SAS–MAS–YHC–PFL8, Colorado State Univ.

Simons, D. B., Li, R. M., and Brown, G. O. (1979). *Sedimentation Study of the Yazoo River Basin — User Manual for Program KUWASER*, Colorado State Univ.

Simons, Li and Associates Inc. (1980). *Erosion, Sedimentation and Debris Analysis of Boulder Creek, Boulder, Colorado*, prep. for URS Company, Denver, Colorado.

Simons, Li and Associates Inc. (1982). *Engineering Analysis of Fluvial Systems*, Simons, Li and Associate Inc., Fort Collins, USA.

Thorne, C. R. (1982). Processes and mechanisms of river bank erosion. In Hey, R. D., Bathurst, J. C., and Thorne, C. R., (Eds.) *Gravel-bed Rivers*, Wiley, Chichester, 227–259.

White, W. R., Paris, E., and Bettess, R. (1980). The frictional characteristics of alluvial streams: a new approach. *Proc. Inst. Civ. Eng.,* Pt 2, 69, 737–751.

White, W. R., Bettess, R., and Paris, E. (1982). Analytical approach to river regime. *J. Hydr. Div., Amer. Soc. Civ. Eng.,* **108** (HY10), 1179–1193.

White, W. R., Bettess, R. and Wang Shiquang (1987). The frictional characteristics of alluvial streams in the lower and upper regimes. *Proc. Inst. Civ. Eng. Pt. 2,* **83**, 685–700.

Wolman, M. G. and Leopold, L. B. (1957). River flood plains: Some observations on their formation. *US Geol. Survey, Prof. Paper,* **282–C**, US Government Printing Office, Washington D. C.

Wolman, M. G. and Miller, J. P. (1960). Magnitude and frequency of forces in geomorphic processes. *J. of Geol.,* **68**, 54–74.

Yang, C. T. (1976). Minimum unit stream power and fluvial hydraulics. *J. Hydr. Eng., Amer. Soc. Civ. Eng.,* **102** (HY7), 919–934.

Yang, C. T., Song, C. C. S., and Woldenberg, M. J. (1981). Hydraulic geometry and minimum rate of energy dissipation. *Water Resources Research,* **17**(4), 1014–1018.

Zimmerman, C. and Kennedy, J. F. (1978). Transverse bed slopes in curved alluvial streams. *J. Hydr. Div., Amer. Soc. Civ. Eng.,* **104**, 33–48.

Modelling Geomorphological Systems
Edited by M. G. Anderson
©1988 John Wiley & Sons Ltd.

Chapter 6

Flow processes and data provision for channel flow models

JAMES C. BATHURST

Natural Environment Research Council, Water Resource Systems Research Unit, Department of Civil Engineering, University of Newcastle upon Tyne

6.1 INTRODUCTION

In recent decades there has been rapid development of mathematical models for channel flow, based on the ability of computers to solve by numerical techniques the partial differential equations of unsteady fluid flow and wave movement. Powerful flood routing models can now be applied to a wide range of river problems, including rainstorm, snowmelt and dambreak flood predictions. Increasingly, though, channel models are called upon to consider not only the in-channel flow but interactions with overland flow, aquifer flow, overbank flow and the transport of sediment, pollutants and heat. The complexity of the processes involved in these interactions is such that models are inevitably based on approximations and assumptions. In addition the field data needed to support the construction, calibration, and application of such models are frequently unavailable, at least in the quantity or at the spatial and temporal scales required. The accuracy of flow simulations is thus tending to be limited more by difficulties in data provision and determination of basic process equations than by deficiencies in numerical solution techniques.

This chapter examines the degree to which the current understanding of channel flow processes is matched by model capabilities. Derivation of the various flow equations and the numerical techniques used in their solution are not considered since much has already been presented in this area (e.g. Mahmood and Yevjevich, 1975; Cunge *et al.*, 1980). Instead the aim is to describe the problems which arise in application of the equations, especially those associated with flow resistance, overbank flow, stream/aquifer interactions and the transport of sediment, pollutants and heat. The emphasis is thus on the framework in which the equations are used, indicating the simplifying assumptions on which they are based and the modifications needed for particular applications. A further, and important, theme is the contrast between the sophistication of current modelling techniques and the relatively poor availability of field data to support simulations. The final section therefore examines a number of approaches to the problem of data collection.

The chapter begins with a brief review of the basic equations of channel flow.

6.2 EQUATIONS OF CHANNEL FLOW

A complete account of unsteady fluid flow in three dimensions is provided by the equation of mass continuity and the Navier–Stokes force–momentum equations (e.g. Cebeci and Bradshaw, 1977, pp. 26–33). For most flow routing cases the full equations are more complex than is warranted by requirements and data availability and a one-dimensional, gradually-varied flow version is applied. First derived by Saint-Venant, the resulting equations are often quoted in a form similar to

$$\frac{\partial Q}{\partial x} + \frac{\partial A}{\partial t} - q = 0 \tag{1}$$

$$\frac{\partial Q}{\partial t} + \frac{\partial (Q^2/A)}{\partial x} + gA\left(\frac{\partial h}{\partial x} + S_f\right) + L = 0 \tag{2}$$

where t = time; x = longitudinal distance; Q = flow discharge; A = flow cross-sectional area; g = acceleration due to gravity; h = water surface elevation; and q = lateral inflow (positive as inflow). Fread (1985) has given the following equalities for L: $-qu_x$ for surface lateral inflow, where u_x = component of inflow velocity in the x-direction; $-qQ/A$ for bulk lateral outflow; and $-qQ/(2A)$ for seepage lateral outflow. The friction slope S_f is calculated from a resistance formula such as the Darcy–Weisbach equation

$$S_f = \frac{Q^2 f}{8gdA^2} \tag{3}$$

where d = depth; and f = the Darcy–Weisbach resistance coefficient.

In the above, equation 1 expresses conservation of mass while equation 2 relates rate of change of momentum to applied forces. Derivations of the equations are presented by Strelkoff (1969), Liggett (1975) and Cunge et al. (1980) among others.

Assumptions implicit in the equations include the following (Strelkoff, 1969; Yevjevich, 1975, p. 4).

1. Wave depths vary gradually, meaning that the pressure distribution along a vertical should be hydrostatic, or that vertical accelerations should be small: there should be no evidence of rapidly-varied flow such as hydraulic jumps.
2. Longitudinal changes in cross-sectional shape, channel alignment and frictional properties are continuous.
3. Friction losses in unsteady flow are not significantly different from those in steady flow.
4. Variation in the distribution of velocity across the channel does not significantly affect wave movement.

5. Wave movement can be considered to be two-dimensional, with the effects of lateral differences in surface elevations at cross sections being negligible.
6. Average bed slope is small enough that $\Theta \cong \sin\Theta \cong \tan\Theta$ and $\cos\Theta \cong 1$, where Θ = angle between the channel bed and the horizontal.
7. The fluid density is constant.
8. The channel geometry does not change with time.

These assumptions are generally satisfied in most instances of flood routing, although assumptions 3, 7 and 8 may require careful consideration when sediment transport is significant.

The Saint-Venant equations form the basis of most channel flow models. In their full form, though, they are nonlinear and have no known general analytical solution. In order to facilitate solution, therefore, a variety of simplifications has been introduced, in which certain or all the terms in equation 2 are neglected.

Solutions based on equation 1 only, with some more or less empirically derived relationship between storage, inflow and/or outflow at a reach, are known as hydrological routing models. Although limited in their application, they can be solved analytically and have a number of sophisticated versions.

Solutions involving equation 1 and parts of equation 2 are known as simplified hydraulic routing methods. They include the kinematic wave model

$$S_f - S_o = 0 \tag{4}$$

where $\partial h / \partial x = \partial d / \partial x - S_o$; d = depth; and S_o = bed slope; and the diffusion wave model

$$S_f + \frac{\partial h}{\partial x} = 0 \tag{5}$$

The kinematic model should be applied only where the depth-discharge rating curve is single-valued and where backwater effects are insignificant. The diffusion model allows for the attenuation of a flood wave and for backwater effects but should otherwise be limited to slowly or moderately rising flood waves in channels of generally uniform geometry.

Solutions involving the full Saint-Venant equations are known as the complete hydraulic or dynamic wave model. This is the most general model (within the assumptions noted above), allowing for backwater effects and the upstream propagation of waves. It is the most efficient and versatile of the wave routing models but is equally the most complex.

Hydraulic models generally require numerical solution. Summaries of some of the techniques involved, together with more detailed reviews of the characteristics of flow routing models, are given by Price (1974), Amein and Chu (1975), Miller and Cunge (1975), Weinmann and Laurenson (1979), Cunge et al. (1980) and Fread (1985).

6.3   PROBLEMS IN APPLICATION

Application of channel flow models is open to a number of possible types of error (e.g. Cunge and Liggett, 1975; Cunge et al., 1980, pp. 222–225). Their sources include the following:
1. Model simplifications and approximations. The Saint-Venant equations should not be applied to conditions beyond the validity of their basic assumptions. For example, the

choice of a simplified hydraulic model or of a particular equation for calculating the flow resistance coefficient should be appropriate to the prevailing conditions. Similarly, the numerical scheme is only an approximation to the equations themselves. Its accuracy depends, at least in part, on the correct choice of time and distance steps (as discussed in point 4).

2. Lack of accurate and appropriate data. This may arise from inaccurate measurements, badly located gauges and other deficiencies in field expertise. In addition, though, as models have become more sophisticated, their requirements have begun to outstrip the ability of existing field techniques to provide the necessary data, at least on a routine basis (e.g. Dunne, 1983). Few data, for example, are available to support detailed models of stream/aquifer interactions. Scaling is a further problem, with most measurements being made at a local, even point, scale, whereas model parameters are required at the larger scale of the grid or reach network used in the numerical representation of the river and its catchment.

3. Phenomena not taken into account. Many models do not allow directly for certain in-channel processes (e.g. secondary flow), overbank processes (e.g. energy losses in the interaction between overbank and in-channel flows) and processes of lateral inflow (e.g. bank storage effects). Usually the energy losses and other impacts of these phenomena are allowed for by empirical calibration in which the resistance coefficient is adjusted until predicted flow behaviour matches observed flow behaviour as closely as possible. However, the extent to which the coefficient is representative of the flow resistance processes then becomes uncertain.

4. Poor schematization of topographic features. Again this is partly a question of scale. The grid or reach network used to represent the river must have a density of calculation nodes sufficient to account for the significant topographic features in any application. These features could include meanders, pool/riffle sequences and breaks of slope. Similarly, the model time step should be small enough to allow adequate resolution of temporal variations in river conditions. In practice a certain amount of leeway is available since, for example, the effect of misrepresenting channel slope along a reach could be absorbed in the calibration process, based on the resistance coefficient.

Items (1) and (4) of the above are related more to the numerical formulation of a model and are not considered in detail here. The emphasis instead is on data provision and allowance for relevant phenomena. These aspects are discussed in the following sections, progressing from in-channel processes, to overbank processes, to coupled channel and aquifer models and finishing with coupled channel and transport models.

6.4 IN-CHANNEL FLOW

Outside the numerical solution itself, the two most important problems arising in channel flow modelling are data provision and evaluation of the flow resistance coefficient. Two other difficulties examined here include simulation of secondary circulation (which requires an approach different from that based on the Saint-Venant equations) and simulation of interactions between flows in tributary channels and in the main channel.

6.4.1 Data Provision

Construction of a channel model requires that, for each calculation node, data be supplied describing, or enabling to be described (e.g. Miller and Cunge, 1975; Cunge et al., 1980, pp. 225–232): the channel bed slope, S_0; the channel cross-sectional shape as a function of elevation; and the flow resistance characteristics (which are considered in the next section). Models dealing with overbank or floodplain flow also need information on the topography of the river valley, lakes and reservoirs and the storage volume of these features as a function of water surface elevation. Model calibration requires, in addition, spatially and temporally distributed data on flow characteristics (e.g. stage and discharge) for a suitable range of conditions.

Full topographic surveys of river channels and floodplains are time-consuming and expensive. Cunge (1975, p. 735) notes for example that a survey of the Mekong delta waterways involved the installation of 200 staff gauges and 40 recording gauges, together with 2,400 km of precise surveying to establish reference datums for the gauges. In any application there is therefore likely to be considerable incentive to keep this data collection to the minimum compatible with the desired objectives. Two means of achieving this aim which hold promise for the future are remote sensing techniques and geomorphological relationships which allow extrapolation of values from limited field surveys.

Remote sensing techniques can provide, on a cost-effective basis, large amounts of spatially and temporally distributed data, at a variety of spatial scales. For river flows they can indicate the extent of flooding, the movement of pollutant and sediment plumes, and, increasingly accurately, topographic elevations (e.g. Skinner and Ruff, 1973). For flow resistance evaluation for flood plains, they can provide information on vegetation types and land-use patterns, although research is still needed to link, for example, observed land-use patterns to the flow resistance on an other than empirical basis. At present, data collection by remote sensing techniques is in its infancy. However, it is likely to become an increasingly powerful approach in the future, especially for remote areas.

Extrapolation of channel geometry data from limited surveys to the entire river length being modelled can be carried out with a number of geomorphological relationships. For example, Leopold and Maddock (1953) showed that changes in both at-a-site and along-channel or bankfull geometry can be related to water discharge Q as

$$w = aQ^b \tag{6}$$

$$d = cQ^e \tag{7}$$

where w = width; d = depth; and the coefficients and exponents, which must be evaluated empirically for each river, take different values for the at-a-site and along-channel relationships. Thus, once surveys at a few sites along the channel have established the values of the coefficients and exponents, the equations can be used to obtain the channel geometry at all other sites. Similar relationships can also be derived to give, for example, channel characteristics as a function of upstream basin area or channel length. Use of relationships such as these can reduce the effort expended on field surveys and enables data collection to be concentrated on critical reaches of channel and reaches not susceptible to description by the relationships.

6.4.2 Flow Resistance Coefficient

The total resistance to flow is composed of frictional boundary resistance, resistance arising from flow nonuniformities and distortions related to channel geometry and plan form and resistance arising from free surface effects (e.g. Bathurst, 1982). However, the resistance coefficient in equation 3 is usually evaluated as a function of the boundary roughness only, since the friction slope S_f is defined to be the energy gradient needed to overcome the frictional resistance of the boundary roughness. In principle this evaluation can be achieved on the basis of field measurements of bed material and bedform characteristics and one of the many resistance formulae available. In practice, current evaluation techniques are semiempirical at best and a certain amount of calibration is usually necessary to allow for errors in the coefficient. More importantly, though, the coefficient is commonly used as a calibration factor to allow for all the other uncertainties in an application, for example scaling errors and phenomena not accounted for. It is therefore possible for the final value of the coefficient to vary significantly from a value which represents only the frictional resistance of the channel boundary. Nevertheless, it is important that the coefficients should retain a physically realistic value since the predictive capability of the calibrated model may otherwise be endangered (Cunge et al., 1980, p. 381). The following factors must therefore be taken into acount:

1. The value of the coefficient must be appropriate to the length of channel being simulated. For a uniform channel, this presents little difficulty since the resistance is the same at all sections. In natural channels, though, the presence of alternate bars, pools and riffles, meander bends and longitudinal variations in bed material composition cause the resistance to vary spatially along the channel. Generally, model distance steps (perhaps 10–30 channel widths) are much larger than the distances over which channel characteristics change. The resistance coefficient must therefore reflect average reach conditions and should be evaluated at the reach scale, for example by tracer measurements (e.g. Beven et al., 1979).

2. For unsteady flows, it is generally acceptable to assume, as is required for the Saint-Venant equations, that flow resistance can be described by formulae developed for steady flows. As noted by Rouse (1965), though, the problem remains of allowing for the effects that the bed scour and bedform variation associated with unsteady flow can have on flow resistance. These effects are most severe for sand-bed rivers, in which there are significant variations in bedform (e.g. Simons and Richardson, 1966). Solution of the problem requires that the flow model be coupled with a sediment transport model (as discussed in a later section).

3. Flow resistance at a site varies with depth or discharge. Generally the resistance is higher at low flows than at high flows since the various elements of channel roughness (bed material particles and bedforms) have proportionally more effect the shallower the flow. This is reflected in a number of available formulae, based loosely on boundary layer or pipe flow theory with coefficients evaluated empirically. Their general form is illustrated in Hey's (1979) formula for gravel-bed rivers

$$\left(\frac{8}{f}\right)^{1/2} = 5.75 \log\left(\frac{[aR]}{3.5\,D_{84}}\right) \tag{8}$$

where f = the Darcy–Weisbach resistance coefficient; a = a function of channel shape, varying from 11.1 to 13.46; R = hydraulic radius; and D_{84} = bed material size for which 84 per cent of the material is finer. Such formulae, though, on the whole account for bed roughness only and, in particular, do not allow for nonuniform flow effects.

Table 6.1. Characteristic values of flow resistance coefficients for different types of channel

Type of channel	Channel slope (%)	Bed material size D_{50} (mm)	Darcy–Weisbach f	Manning n
Sand-bed	≤0.1	≤2	0.01–0.25	0.01–0.04
Gravel-bed	0.05–0.5	10–100	0.01–1	0.02–0.07
Boulder-bed	0.5–5	≥100	0.05–5	0.03–0.2
Steep pool/fall	≥5	≥50	0.1–100	0.1–5

For a channel reach they are therefore probably most relevant to high flows, when nonuniformities in channel characteristics are least apparent. At low flows, large scale features such as bars and riffles create nonuniform conditions which significantly increase the resistance above that obtained by formulae such as equation 8 (e.g. Parker and Peterson, 1980). Currently there is no generally accepted means of evaluating this extra resistance. However, simulations of low flows are increasingly required for designation of minimum flows needed to maintain water quality and wildlife habitat. Miller and Wenzel (1985) have therefore proposed an approach for low flow modelling in which total energy losses are given as the sum of losses due to bed material roughness (e.g. equation 8) and losses due to nonuniform flow along pool/riffle reaches.

4. Table 6.1 indicates that resistance characteristics vary with channel type. Consequently, because of the empirical basis of resistance formulae, different approaches are required for different channel types. Most experience has been acquired for sand-bed and gravel-bed rivers with single-thread channels. Rather less is known about boulder-bed, steep pool/fall and multi-thread (braided) rivers. Further information is also needed on the resistive effects of such factors as ice cover, secondary circulation at bends and man-made obstructions such as bridge piers. Considerable field and laboratory research into the processes involved will be required to provide the necessary relationships.

6.4.3 Secondary Circulation

Most of this chapter is concerned with modelling unsteady flow along a channel, for which the one-dimensional Saint-Venant equations form a sufficient basis. There are occasions, though, in which it is necessary to investigate the transverse flow structure. In particular the modifications brought about by secondary circulation, developing in the plane normal to that containing the primary flow, can have important consequences for channel and bank erosion, dispersion of pollutants, sorting of sediments, distribution of shear stress and flow resistance. A number of mathematical models have therefore been developed to simulate the patterns of circulation.

Secondary circulation can develop in a variety of locations and by different processes. In rivers, though, it is most evident at bends, where it appears as the familiar cell directing surface water towards the outside of the bend and near-bed water towards the inside. This pattern may also be modified to include a small cell of reverse

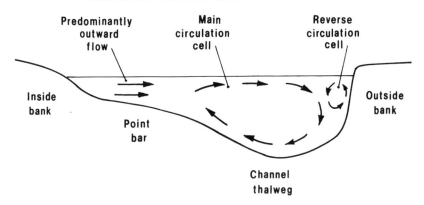

Figure 6.1. Secondary circulation patterns at a river bend cross section

circulation at steep outer banks (Bathurst *et al.*, 1979) and a dominance of outward flow near the inside bank caused by a progressive decrease in depth along the inside bank point bar (Dietrich and Smith, 1983) (Figure 6.1).

Since secondary circulation can be linked to the development of streamwise vorticity, its mathematical description can be based on the equation of vorticity transport (e.g. Perkins, 1970; Cebeci and Bradshaw, 1977, pp. 319–321). This approach has particular advantages for modelling the weak circulation sometimes observed in straight channels. For bend flows, which are of more interest in natural channels, the vorticity approach is less critical and an alternative method based on the equations of mass and momentum continuity is often favoured.

Reviewing the latter approach, Kalkwijk and de Vriend (1980) note that two groups of procedures have been developed.

1 . Procedures based on the integration of the depth-averaged main flow equations. These procedures are relatively simple but have disregarded the convective influence of the secondary flow on the transverse exchange of main flow momentum, leading to significant errors.

2. Procedures based on the integration of the full three-dimensional equations of flow. These procedures are more accurate but are complex and expensive in computer time.

Kalkwijk and de Vriend observe that the second approach is unnecessarily detailed for natural channels which are shallow and have bends of moderate curvature. They therefore propose an intermediate procedure based on depth-averaged computations but with allowance for the convective influence of the secondary flow. The model consists of the equations of continuity of mass, momentum along the channel axis and momentum perpendicular to that axis.

A significant limitation of the model is its requirement that changes of depth across and along the section must be gradual. It is therefore unable to simulate wall effects at vertical banks. De Vriend (1981) indicates that these play an important role in the redistribution of velocity and can lead to local variations in the circulation pattern, such as the reverse cell measured by Bathurst *et al.* (1979). Simulations of these effects are possible only with the more complex three-dimensional models. Further discussion of the influence of bed and bank topography can be found in Smith and McLean (1984).

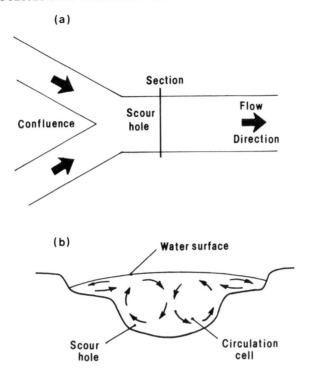

Figure 6.2. Secondary circulation pattern at a channel confluence: (a) plan view of a symmetrical confluence, showing location of section for: (b) circulation pattern in the scour zone downstream of the confluence. Based on Mosley (1976)

Aspects of secondary flow which require further study include the variation in flow strength and pattern which occurs as discharge varies and the decay of secondary circulation downstream of bends. Both these aspects have significant implications for bank erosion in particular.

6.4.4   Tributary Inflows

Simulations of flow along the main river channel can account for inputs from tributaries with the lateral inflow terms in equations 1 and 2. Conservation of both mass and momentum must be satisfied but in many cases it is possible to assume that the inflow has zero momentum, thereby simplifying the calculations. An example of such a case is where the model distance step is large compared with the distance over which confluence momentum effects are significant. An approach based on conservation of mass and momentum which does not make the simplifying assumption is presented by Fread (1973).

Generally, energy losses at a confluence other than from friction are not considered. However, Lin and Soong (1979) found that turbulent mixing losses could be of the same order of magnitude as boundary friction losses. Similarly, a number of studies have shown confluence flows and the associated sediment transport and bed features to be highly three-dimensional (e.g. Mosley, 1976; Best, 1986). Figure 6.2, for example, illustrates the pattern

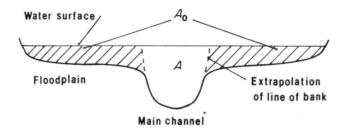

Figure 6.3. Definition diagram for off-channel storage cross-sectional area A_o and main channel flow cross-sectional area A at a section across a channel and its floodplain

of secondary flow responsible for the development of scour holes downstream from confluences. Incorporation of such effects into mathematical flow models has yet to be accomplished.

6.5   OVERBANK FLOW

An extension of in-channel modelling procedures is required to simulate overbank flow brought about by a river spilling over its banks onto the adjacent floodplain. The one-dimensional Saint-Venant equations remain applicable but the general approach must be modified to allow for the relative distribution of flow between floodplain and channel, momentum exchanges between floodplain and channel flows, influence of meandering channels, dambreak floods and evaluation of resistance coefficients for the floodplain.

6.5.1   Distribution of Flow

In one-dimensional flood routing with overbank flow, it is commonly assumed either that flow occurs only in the main channel and that only storage of water occurs on the floodplain, or that the main channel and floodplain can be combined to form a composite channel in which flow conditions are averaged from those in the two components. The first assumption allows the Saint-Venant equations to be used with one modification, incorporating the total flow width or cross-sectional area in the continuity equation. Fread (1985), for example, modifies equation 1 to give

$$\frac{\partial Q}{\partial x} + \frac{\partial (A + A_o)}{\partial t} - q = 0 \tag{9}$$

where A_o = off-channel or overbank storage area (Figure 6.3). The second assumption allows equations 1 and 2 to be used, based on total cross-sectional area and with a flow resistance coefficient evaluated as a weighted average for the floodplain and main channel.

   Each of these approaches is approximate and, strictly, is valid only for certain limited conditions. Bhowmik and Demissie (1982) found from field data that the discharge-carrying capacity of floodplains can vary from a few per cent (when the assumption of storage only applies) to 80 per cent of the total flow (when the combined approach may be appropriate). The capacity depends on several factors, including the nature of the floodplain and the main channel. In particular, though, it increases with the return period of the flood. Thus Bhowmik and Demissie found for their data that, for return periods greater than about forty years, the floodplain and main channel behave as a single unit. This observation has special significance for the modelling of dambreak floods, which, in terms of peak discharge, are usually equivalent to natural floods with extreme return periods.

Allowance for the different flow characteristics of floodplain and main channel can be made within a one-dimensional framework by defining the Saint-Venant equations separately for the two components (Fread, 1976). More accurate simulations can be achieved with two-dimensional (in the horizontal) models (e.g. Cunge *et al.*, 1980, pp. 138–143). However, these are also more complex and, in particular, their calibration requires extensive field measurements of velocity and depth over the floodplain and main channel, data which are rarely available (Vreugdenhill and Wijbenga, 1982).

6.5.2 Momentum Exchange

A common procedure for calculating discharge at cross sections with a significant lateral variation of flow characteristics is to split the cross sections into regions of homogeneous properties. These regions are considered to be separated by vertical interfaces with no flow interaction (i.e. no shear stress) across them. Discharge is then calculated separately for each subsection and the total discharge is obtained by summing the subsection values.

For overbank flow calculations this approach separates the floodplain and main channel flows, neglecting any interaction at the interface between the two. However, a number of flume studies have demonstrated that there can be a significant apparent shear stress at the interface, caused by an intensive momentum exchange from the main channel flow to the adjacent overbank flow. Calculations which assume zero shear stress at the interface are then found to overestimate total discharge (e.g. Myers, 1978; Rajaratnam and Ahmadi, 1981; Wormleaton *et al.*, 1982; Knight and Demetriou, 1983).

Currently there is no allowance for the effect of the momentum exchange in one-dimensional routing models for unsteady flow. Those approaches which have been proposed, both empirical and theoretical, are for steady flow. On the empirical side, relationships have been derived between the apparent shear force at the interface and the ratios of main channel depth to floodplain depth and of floodplain width to main channel width. Generally the shear is greatest when the respective depths and widths are most different, in other words when floodplain depths are low and floodplain widths are high (e.g. Rajaratnam and Ahmadi, 1981; Knight and Demetriou, 1983). Pasche and Rouvé (1985) have carried this approach to practical application, with a one-dimensional model based on a set of equations for the lateral distribution of velocity and incorporating apparent shear stresses at defined interfaces. Constants within the equations have to be determined empirically and further development of this approach will therefore require field data on floodplain roughness.

On the theoretical side, Krishnappan and Lau (1986) have developed a model based on the equations of continuity and three-dimensional turbulent momentum. Although complex, it represents a considerable advance over one-dimensional models in simulating the lateral momentum transfer, secondary circulation and other three-dimensional effects associated with the interaction between floodplain and main channel flows. It has been successfully tested against flume data and the authors note that more data on shear stress

(a)

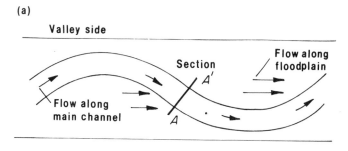

(b)

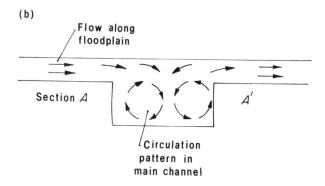

Figure 6.4. Secondary circulation pattern in a meandering channel with floodplain flow: (a) plan view of meandering channel within a floodplain, showing location of section AA' for: (b) circulation pattern in the main channel. Based on Toebes and Sooky (1967)

and velocity distributions would enable the model to be validated for wider ranges of flow conditions.

6.5.3 Meandering Main Channel

Most studies of overbank flow have considered the main channel to be straight. Very few have considered the more natural case of a meandering channel. Then, not only is the interaction between floodplain and main channel flows altered but there is a tendency for the floodplain flow to follow a path down the valley which is shorter than the more circuitous route taken by the flow in the main channel (e.g. Fread, 1976).

The interaction of flows has been studied in the flume only. Toebes and Sooky (1967) found that floodplain flow across a meandering channel produced spiral vortices or secondary currents in the main channel (Figure 6.4). These were strikingly different from and more pronounced than those occurring in a meander channel with inbank flow (Figure 6.1). The exchange of momentum between the floodplain and main channel flows was thus accentuated. Similar evidence is provided by Ghosh and Kar (1975) and Smith (1978). However, there has been no field study of the phenomenon, nor has the interaction been modelled other than by empirical incorporation in the flow resistance coefficient.

Similarly little research has been carried out into one-dimensional models capable of allowing for the different path lengths taken by the floodplain and main channel flows. Tingsanchali and Ackermann (1976) carried out empirical adjustments of roughness coefficients and floodplain geometry in their case study in order to produce the same equivalent distances for the two flows. Fread (1976), in more detail, modified the Saint-Venant equations to account for the floodplain and main channel flows separately. The differences in hydraulic properties and path lengths could then be allowed for within a one-dimensional model that preserves the familiar approach associated with the Saint-Venant equations.

6.5.4 Dambreak Flood

Dambreak flood waves represent an extreme case of overbank flow. They may arise from the failure of a man-made dam, the failure of a natural ice or debris dam or overtopping of a dam by landslide generated waves.

Fread (1984) observes that this type of flood has some important differences from precipitation derived floods, making it difficult to analyse on the basis of common techniques used for the latter. Particularly obvious is the great magnitude of a dambreak flood relative to precipitation derived floods. Thus, Jarrett and Costa (1986, pp. 20–21) describe a dambreak flood in which the discharge approached forty times the estimated 100-year flood. Modelling flows so far above recorded levels means that certain coefficients used in the various flood routing techniques must be obtained by extrapolation and that full calibration of the model cannot be carried out. A further feature of dambreak floods is their very short duration time and particularly the extremely short time between the beginning of rise and the occurrence of the peak. These two features combine to produce acceleration components in the flood wave much larger than those associated with a precipitation derived wave. However, as long as the failure of the dam is slow enough that the resulting wave resembles a rapid precipitation derived wave, it is not inappropriate to use the Saint-Venant equations, albeit the full dynamic wave version (e.g. Fread, 1984). If, on the other hand, failure is instantaneous, an alternative approach is required (Cunge et al., 1980, pp. 357–365).

Evaluation of the resistance coefficient remains a particular problem. Descriptions of flood waves, especially in narrow mountain valleys, refer to trees, mud and boulders being churned about in the flow and to obstructions caused by the buildup of temporary dams (e.g. Jarrett and Costa, 1986, p. 27; Vuichard and Zimmermann, 1986). In the ensuing conditions of high turbulence, alternating supercritical and subcritical flow, unsteady flow and possibly altered flow viscosity, the usual approaches to flow resistance evaluation may not be valid.

6.5.5 Floodplain Flow Resistance

As indicated above there is a lack of data, especially field data, for calibrating and verifying overbank flow models. This refers particularly to the flow resistance coefficient for floodplains with natural vegetation, farmland and man-made structures. A rare study is that of Klaassen and Zwaard (1974) who, on the basis of measurements, related the flow resistance of hedges and orchards to parameters such as the average spacing between hedgerows, the number of trees per unit area and the water depth over the floodplain. A few studies have also investigated the flow resistance of trees (e.g. Li and Shen, 1973) and grass-lined channels (e.g. Kouwen and Li, 1980), while a general analysis is presented by Petryk and Bosmajian (1975). However, considerably more fieldwork is required in this area.

6.6 STREAM/AQUIFER INTERACTION

The foregoing has considered the flow along a channel and its floodplain as an entity, disconnected from other hydrological processes. Often, though, the channel flow model is just one component in a more general hydrological model involving components of overland and subsurface flow. Then the parallel operation of all the components must allow for the exchanges which take place between them.

Two particular exchanges with the channel flow must be considered—overland flow inputs and stream/aquifer interactions. The first is not considered here since it is modelled with the same flow equations as for the channel flow. Also, the overland flow is not affected by the channel flow and can be modelled independently. Aquifer conditions, though, are affected by the channel flow and interaction between the two requires careful representation.

6.6.1 Model Characteristics

On the rising stage of a flood wave, seepage of water into the bank above the phreatic surface can cause a temporary loss of water from the river, bringing about significant attenuation of the wave. On the falling stage, some of the water seeps back to the river augmenting the recession and baseflow. In ephemeral streams (and in floodplain flow) there can be considerable loss of flow by infiltration into the ground, where the phreatic surface is disconnected from the channel. Such disconnection can also result from drawdown of the phreatic surface caused by pumping. These and other phenomena highlight the importance of allowing for stream/aquifer interactions and bank storage effects in channel flow models.

Pinder and Sauer's (1971) general statement of the problem is as follows. 'The time and space distribution of stream elevation must be determined by solving the differential equations which describe one-dimensional open channel flow and include the effect of groundwater flow normal to the channel through its wetted perimeter. This flow through the wetted perimeter of the channel is a function of the hydraulic head in the aquifer as well as the elevation of the stream. To compute the head in the aquifer the mathematical model must simulate transient horizontal two-dimensional unconfined groundwater flow because flow in the aquifer is not necessarily normal to the stream. The interdependence of the open channel flow and the groundwater flow models necessitates the simultaneous solution of the differential equations describing each system. The equation coupling the two systems is Darcy's Law, which describes flow through the wetted perimeter of the channel.'

Work along this line has been undertaken by Zitta and Wiggert (1971) and Cunningham and Sinclair (1979) among others. Freeze (1972), on the other hand, notes the importance of taking into account the unsaturated zone so that water table recessions and the resulting variations in baseflow can be related to the surface hydrological events which are their source. He therefore followed the more complex route of coupling the channel flow equations to a three-dimensional saturated zone/unsaturated zone model. Similarly, where stream and phreatic surface are disconnected, the effect of the intervening unsaturated zone on the exchange flows must be taken into account (e.g. Dillon and Liggett, 1983).

Simulation of the exchange through the channel wetted perimeter must also allow for the layer of relatively low permeability sediment which is sometimes found to line the channel (e.g. van't Woudt *et al.*, 1979). Appropriate versions of the Darcy equation have been developed by Prickett and Lonnquist (1971) and Pinder and Sauer (1971) among others. For a phreatic surface elevation H_p above the bottom level H_b of the lining, the exchange is

$$q = \frac{K(H_p - H_r)}{T} w \qquad (10)$$

where q = exchange discharge per unit length of channel; K = lining conductivity; T = lining thickness; H_r = river surface level; and w = river surface width. If the phreatic surface falls below the level H_b (Figure 6.5), the exchange depends on the pressure head in the river and is given by its maximum value

$$q = \frac{K(H_r - H_b)}{T} w \qquad (11)$$

6.6.2 Model Assumptions

In order to reduce computational requirements, modellers have frequently resorted to simplifying assumptions in their representation of stream/aquifer interactions. Common assumptions examined by Sharp (1977) include:
1. The alluvial aquifer is homogeneous, isotropic and infinite in extent;

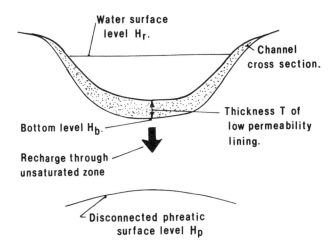

Figure 6.5. Definition diagram for a stream/aquifer system with a low permeability lining and a disconnected phreatic surface. Based on Prickett and Lonnquist (1971)

(a) **ASSUMPTION**.

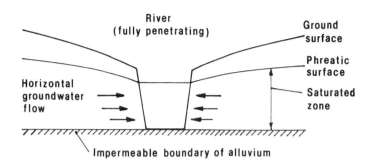

(b) **REALITY**.

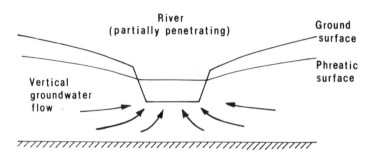

Figure 6.6. Patterns of groundwater flow for stream/aquifer interaction: (a) commonly assumed pattern; (b) pattern likely to be observed in practice

2. The bottom boundary of the alluvium is horizontal and impermeable;
3. The Dupuit–Forchheimer conditions are valid: (a) in any vertical section the groundwater flow is horizontal (Figure 6.6(a)); (b) the velocity is uniform over the depth of flow; and (c) the slope of the free surface, in unconfined flow, is small enough that $\tan \Theta \cong \sin \Theta$, where Θ = angle between the free surface and the horizontal;
4. All flow is saturated;
5. Water from storage is discharged instantaneously upon a reduction in head;
6. Streams are fully penetrating, i.e. through the entire aquifer (Figure 6.6(a));
7. The groundwater system is solely unconfined or confined.

Analysis shows, though, that certain of these assumptions are not generally valid (Sharp, 1977). In particular, accurate simulation of stream/aquifer interaction must allow for the following conditions.

1. Aquifers in shallow alluvial basins are not homogeneous because the hydraulic conductivity varies with depth, except possibly in the top layer containing the river. Groundwater flow velocity is therefore not uniform over the depth of flow.

2. The flow is not always horizontal. In the vicinity of the river, flow is predominantly vertical and the vertical component often predominates near the water table. This is a significant phenomenon which is not easily incorporated into one-dimensional groundwater flow models. One reason for vertical groundwater flow near the river bed is that most major alluvial aquifers possess only partially penetrating streams (Figure 6.6(b)). It is not then realistic to simulate bank storage effects on the basis of semipermeable banks and impermeable beds, particularly as river beds tend to be more permeable than the banks and most of the stream/aquifer exchange therefore takes place through the river bed. In man-made irrigation canals, on the other hand, bank seepage may be more important than bed seepage if the phreatic surface is shallow (Kraatz, 1971).
3. Groundwater systems are not always solely unconfined or confined and alluvial aquifers may be combinations of the two cases.

6.6.3 Data Availability

Further advances in stream/aquifer modelling will clearly require research in the areas identified by Sharp. However, there is a hampering lack of the field data needed for calibrating and verifying models. Few studies, for example, have provided information on bank seepage rates or spatial and temporal variations in phreatic surface level near a river at a sufficiently fine scale. Information is also lacking on bed conductivity values, which are difficult to specify accurately because of the effect of siltation, clogging and organic growth (e.g. Cedergren, 1977, pp. 26–85). The conductivity is particularly sensitive to the quantity, character and size distribution of the finest particle sizes (Cedergren, 1977, p. 38). It is also nonisotropic, with horizontal conductivities typically exceeding vertical conductivities (Cedergren, 1977, p. 40). Van't Woudt and Nicolle (1978), for example, measured conductivities of 500–3000 m/day in the horizontal and 20–80 m/day in the vertical for bed sediment of the Waimakariri River, New Zealand.

6.6.4 Interface Processes

The fine detail of flow interactions at the streambed may not always be appropriate to the design of stream/aquifer models. However, from the very limited information available, derived from laboratory studies, two points are significant.

The first is that a permeable boundary has a flow resistance higher than that of an equivalent impermeable boundary (e.g. Zagni and Smith, 1976; Zippe and Graf, 1983; Ho and Gelhar, 1983). This is thought to be related to an induced stress generated by the seepage disturbances and extra momentum exchange at the permeable boundary (e.g. Nezu, 1977). Seepage into a channel with a sand bed may also cause a small increase in flow resistance by altering the bedform (e.g. Richardson et al., 1985). At present these effects are not allowed for in flow resistance equations. In practice, though, they may be small compared with the effects of other uncertainties.

The second point is that the seepage flow near the bed surface may not always be laminar (e.g. Nezu, 1977; Cedergren, 1977, p. 140). Further research is therefore needed to show whether stream/aquifer exchanges should in some circumstances be based on a turbulent flow formula rather than Darcy's law, which applies essentially to laminar flow.

6.7 TRANSPORT MODELS

Modelling the variation of flow discharge and stage in a channel is important in its own right for such purposes as flood routing and low flow studies. In many cases, though, there is a practical interest in the transport of quantities other than flow momentum, for example sediments, pollutants and heat. The movement of these quantities may also have an effect on the flow field, by affecting the velocity distribution and flow resistance. Their transport is determined by the flow field, so transport models must couple the relevant equations of movement for both water and the quantity in question. This section indicates very briefly the equations involved and certain of the problems arising in coupling.

6.7.1 Sediment Transport

Routing of sediment along rivers is required for a variety of applications, including reservoir sedimentation studies, prediction of the impact of land-use change on sediment yield and prediction of the effects of bedform development on flow resistance. Accurate modelling is hampered by the complexity of the transport process and a realistic model must be able to account for the characteristically nonuniform and unsteady nature of sediment movement.

Bennett and Nordin (1977) indicate that sediment routing models must contain components for routing the flow of water (the Saint-Venant equations), routing the flow of sediment (the equation for conservation of sediment mass) and keeping an account of the local channel-bed elevation and sediment size composition. The general equation for conservation of sediment mass can be written as (e.g. Bennett and Nordin, 1977; Chen, 1979)

$$\frac{\partial Q_s}{\partial x} + (1-p)\frac{\partial(wz)}{\partial t} + \frac{\partial(AC)}{\partial t} - q_s = 0 \tag{12}$$

where Q_s = volumetric sediment discharge; p = bed porosity; w = active bed width in which erosion or deposition occurs; z = local bed elevation, or alternatively the depth of loose sediment in the bed layer; A = flow cross-sectional area; C = mean sediment concentration, Q_s/Q, where Q = flow discharge; and q_s = lateral sediment inflow per unit length of channel. The sediment discharge Q_s is usually related to water discharge through a sediment transport equation. In some cases the bed and suspended sediment loads must be considered separately and two versions of the mass conservation equation are then required (e.g. Bennett and Nordin, 1977). Where the effect of sediment transport on the flow field is to be considered, the sediment and flow equations must be solved simultaneously, allowing for interaction between the two fields. Solution procedures are discussed by Bennett and Nordin (1977), Chen (1979), and Cunge et al. (1980).

Data requirements include the active width of the bed, sediment size distributions and the bed layer thickness. Collection of these data is hampered both by the wide spatial and temporal variations in these parameters and by the relatively unsophisticated measurement techniques currently available. A further problem is selection of the equation linking sediment transport to flow conditions. A wide range of relationships is available but all have an empirical component based on data for particular river sites or laboratory flumes. Great care must therefore be taken in selecting a relationship which is appropriate to the given flow conditions. Guidelines on selection are presented by Vanoni (1975), White et al. (1975) and Bathurst et al. (1987) amongst others.

6.7.2 Pollutant Transport

Pollutant routing is required for such applications as simulating the results of accidental large-scale release of contaminants into rivers and specifying the maximum quantity of pollutant which may be discharged to a river while maintaining an adequate water quality for aquatic life and recreational and water supply purposes. As with sediment routing the transport processes are complex, involving point and non-point inputs, storm and low flow conditions, organic and inorganic substances, interactions between biological and chemical components and the role of the bed as a source or sink for the pollutants. An understanding of the sediment transport processes is required, since pollutants are often in the form of sediments or are adsorbed onto sediments. Water temperature also has an important effect on the various biological and chemical processes.

In similar vein to sediment routing, pollutant routing models must contain components for routing the water, routing the pollutants and modelling the removal or decay of pollutants by biodegration, sedimentation and interaction in the bed (Bedford *et al.*, 1983). The mass conservation equation is applied to each biological or chemical component as

$$\frac{\partial(A\alpha)}{\partial t} + \frac{\partial(Q\alpha)}{\partial x} - \gamma q - S^* = 0 \tag{13}$$

where A = flow cross-sectional area; Q = flow discharge; q = lateral inflow per unit channel length; α = mass concentration of the particular pollutant; γ = concentration of that pollutant in the lateral inflow; and S^* = source/sink term portraying the biochemical interactions. Since the flow routing equations can be solved independently of the pollutant routing equations (but not *vice versa*), the flow equations can be solved first and the output used as input for different runs of the pollutant model. This uncoupling enables solutions and sensitivity tests to be carried out very economically (Bedford *et al.*, 1983). A detailed examination of procedures for simulating pollutant dispersion is given in Cunge *et al.* (1980).

A number of phenomena associated with pollutant transport are poorly understood. These include settling processes, infiltration of pollutants and sediments into the bed and their consolidation in the bed during interstorm periods. However, modelling of solute interaction processes in the bed has been carried out by Bencala (1984). Most water quality models, based on gradient-type diffusion, can also make only empirical allowance for the effect of mixing by secondary circulation at bends. The provision of data for model construction and calibration may similarly be inadequate. Particular difficulties include delineation of the spatial and temporal distribution of each pollutant concentration and designation of representative sampling frequencies and locations (e.g. Sanders, 1979).

6.7.3 Heat Transport

Heat transport models are required for predicting, for example, the effects of discharging heated water from power stations and cool water from reservoirs and of increasing water temperatures by removing shading trees from the river bank. Increases in temperature reduce the solubility of dissolved oxygen, intensify many types of toxicity and affect bedform development and thence flow resistance. Reductions in temperature can similarly affect aquatic life adversely.

A dynamic model, to be used in conjunction with an unsteady streamflow model, has been developed by Bowles *et al.* (1977). This is based on a heat balance for a control volume in a stream, considering nonadvective heat exchange across the stream surface; nonadvective heat exchange across the stream bed; advection of heat associated with stream velocity; and other advective heat fluxes by lateral and tributary inflow, groundwater infiltration and seepage, precipitation and evaporation. The model assumes that there is complete and instantaneous mixing across each section and that there is negligible longitudinal dispersion. However, it does not allow for reservoir effects or major inputs of heated water for which records do not exist. A review of simpler (generally steady flow) models for predicting the effects of reservoir releases is presented by Gosink (1986), while Sayre and Caro-Cordero (1979) consider the modelling of thermal plumes.

The data required by models such as that of Bowles *et al.* (1977) include meteorological as well as hydraulic inputs, while calibration requires that spatial and temporal variations in temperature be measured. Bowles *et al.* (1977) also stress the importance of the heat exchange derived from lateral inflow, for which temperature data are often unavailable. As with sediment and pollutant routing, detailed field studies are needed to provide the data required for supporting model development.

6.8   REQUIREMENTS IN DATA COLLECTION

The further development of mathematical channel flow models will continue to depend on the collection of data for such purposes as calibration, delineation of boundary conditions and elucidation of the basic process equations. For channel models, defined in their widest sense to include bank storage, overbank flow and transport of other quantities, such data requirements are likely to be wideranging and extensive. They will not generally be satisfied by laboratory studies, in which the required parameter ranges cannot be spanned, information is provided at scales smaller than those used in models and difficulties arise in representing field processes at the laboratory scale. There will therefore be an increasing requirement for field data. This will be especially so for flow and transport models based on the equations of physics, since the parameters of such equations can in principle be measured in the field. As has been noted throughout this chapter, though, many aspects of channel modelling are not supported by an adequate data supply. Also, the data required are often of a type or in a form not currently available, at least on a routine basis. This section therefore examines some of the requirements for field data and the areas in which new approaches to data collection may develop.

6.8.1   Data Requirements

Three particular types of data are required:

1. Parameters which fit a model to a particular river system. These describe characteristics which are essentially constant during a simulation. Examples include the saturated conductivity of bed and bank material, floodplain vegetation characteristics (at least on a seasonal basis) and, depending on the type of simulation, bed material size distribution. These parameters need to be measured only once at each location but should otherwise be available on a spatially distributed basis.

2. Input data required to keep a simulation running through time. These include meteorological data, lateral inflows of all quantities (where these are not otherwise simulated) and upstream hydrographs forming boundary data. They should be available at suitable time steps and locations.
3. State variables which describe the various hydrological and other conditions throughout the period of simulation as a function of time and space (except where specified as boundary data). They include river discharge, aquifer phreatic levels, bed seepage rates, pollutant and sediment concentrations and water temperature. These variables provide the basis for model calibration and verification.

Because of the wide range of hydrological processes, each operating at different time and space scales, large amounts of data with different temporal and spatial distributions will be needed to support simulations. Further, a given process, such as seepage or flow resistance, can itself be considered at different scales (large or small) and its mathematical formulation may be different in each case. The data should therefore be appropriate to the scale at which the process itself is represented in the model. Typically, the model scale is relatively large while data have tended to be evaluated at a point or small scale. Measurement techniques and instrumentation appropriate to the model scale must therefore be developed.

Flow data should also be collected at locations and frequencies which ensure that the data are representative of the flow in question and that temporal variability in flow conditions is adequately accounted for. This is particularly so for sediment and pollutant concentrations, which may be nonuniformly distributed at a section and which can vary significantly over short time periods (e.g. Sanders, 1979). Finally, data collection should include as wide a range of conditions as possible in order to provide the extremes which are the best test of model generality, for example low and high flows, large and small rivers, arid and humid environments.

6.8.2   Future Approaches

Requirements for large amounts and new types of data may be expected to stimulate developments in methods of data collection and analysis. Some aspects are as follows:
1. Where large amounts of information are needed, they can be obtained directly, with intensive data collection exercises, or indirectly by extrapolating from a more limited data base. Intensive data collection exercises are likely to be limited to research studies, where time and inclination are favourable. Even so, the requirements in manpower and equipment may still be daunting. For example, measuring the impact of a flood wave on bed and bank seepage requires a large array of piezometers, wells and other equipment, in addition to the current meters and stage recorders for monitoring the wave movement. Measurement of secondary currents at bends requires the deployment of expensive and specialized instruments, while collection of bed material data must allow for the wide spatial variations in sediment characteristics which are typically observed. In many cases techniques are applicable only to relatively small channels and are overwhelmed by the sheer size of large channels. In others there is difficulty even in devising a practical and representative measuring technique, for example for sampling subsurface bed material or monitoring seepage within the bed.

   Given the equipment and manpower needed to collect the data necessary for full tests of channel flow, stream/aquifer interaction, overbank flow and sediment/pollutant transport models, there are clearly advantages in collaboration. These may also coincide with a spread of interests between various organizations. Such collaboration can conveniently be organized around a particular event chosen to provide a suitable range of conditions, for example releases from regulating reservoirs. This author, for example, took part in an exercise based on test releases from the Cow Green Reservoir on the River Tees in England in June 1976. The combination of manpower and equipment provided by four organizations enabled continuous discharge gaugings to be maintained at a number of sites while other measurements covered flow resistance, bank storage, tracer movement and flood wave dynamics.

Where time or money do not permit intensive data collection exercises, for example in engineering applications, the required model data must be obtained by extrapolating from a more limited data base. Two approaches are important here. Firstly, by integrating the data collection with model sensitivity tests, it is possible to indicate which model parameters are the most significant and therefore where the data collection should be concentrated. Secondly, powerful statistical techniques are now available for interpolating between existing data points to provide information at intermediate points. Kriging, for example, developed for the mining industry (e.g. Journel and Huijbregts, 1978), is being increasingly applied in groundwater and rainfall studies and could play an important role in seepage studies. A less sophisticated approach is characterized by the empirical equations of hydraulic geometry (equations 6 and 7).

2. Evaluation of data at the reach scale is likely to require the development of new techniques and instrumentation. This will be important in obtaining synoptic and mean, rather than localized, values of, for example, bed seepage and sediment transport along a river reach. Increased automation of data collection will similarly be required to provide information at the frequencies needed to define the variation in input data and state variables adequately. Considerable potential is offered by remote sensing techniques which can obtain large amounts of data on a spatially and temporally distributed basis and at a variety of scales. However, there are still deficiencies associated with the conversion of the measured electromagnetic radiation characteristics into certain of the physical characteristics required in models.

3. With the large amounts of data used in models, developments in computing are needed to increase automation of data handling, allowing manual treatment to be minimized. Thus, the use of microcomputers in a field environment may assist the recording and initial analysis of data, as well as supporting the integration of model sensitivity tests with data collection. In a related area, presentation of data and output of simulation results should be in a form which can be easily assimilated by the user, primarily a problem of computer graphics.

4. As the data requirements of models become more rigorous, and in order to determine the degree to which uncertainty in the data affects uncertainty in the model output, the accuracy of the data must be specified. Various aspects of this requirement are considered by, for example, Hey and Thorne (1983) for sampling bed material, Ferguson (1986) for estimating sediment and pollutant loads and Herschy (1978) for gauging discharge.

6.9 CONCLUSION

The past two decades have seen the development of powerful channel flow models, capable of integrating main channel flow with overbank flow, aquifer interaction and the transport of sediments, pollutants and heat. Much of the emphasis has been on the numerical solution techniques which have now advanced to a level where they are no longer the principal limiting factor in model application. Their sophistication, though, is not always matched by that of available data provision methodologies, which are not generally capable of supplying the large amounts of data ideally required. Certain types of data are simply unavailable, at least on a routine basis or at the necessary scales. Also, not all models incorporate a sufficient understanding of the relevant field processes. Deficiencies in this area may thus present a significant impediment to model construction, calibration and application. Advances in data provision and the understanding of field processes are therefore needed to match those in numerical technique.

ACKNOWLEDGEMENTS

The author is most grateful to the following for helpful comments on this paper: Dr. K. J. Beven (University of Lancaster), Mr. R. Mackay (NERC Water Resource Systems Research Unit, University of Newcastle upon Tyne) and Dr. P. G. Samuels (Hydraulic Research Ltd., Wallingford).

REFERENCES

Amein, M. and Chu, H.-L. (1975). Implicit numerical modeling of unsteady flows. *Proc. Am. Soc. Civ. Engrs., J. Hydraul. Div.*, **101**(HY6), 717–731.

Bathurst, J. C., Thorne, C. R., and Hey, R. D. (1979). Secondary flow and shear stress at river bends. *Proc. Am. Soc. Civ. Engrs., J. Hydraul. Div.*, **105**(HY10), 1277–1295.

Bathurst, J. C. (1982). Theoretical aspects of flow resistance. In Hey, R. D., Bathurst, J. C., and Thorne, C. R. (Eds.), *Gravel-bed Rivers*, Wiley, Chichester, 83–105.

Bathurst, J. C., Graf, W. H., and Cao, H. H. (1987). Bed load discharge equations for steep mountain rivers. In Thorne, C. R., Bathurst, J. C., and Hey, R. D. (Eds.), *Sediment Transport in Gravel-bed Rivers*, Wiley, Chichester, 453–477.

Bedford, K. W., Sykes, R. M., and Libicki, C. (1983). Dynamic advective water quality model for rivers. *Proc. Am. Soc. Civ. Engrs., J. Env. Engrg.*, **109**(3), 535–554.

Bencala, K. E. (1984). Interactions of solutes and streambed sediment 2. A dynamic analysis of coupled hydrologic and chemical processes that determine solute transport. *Wat. Resour. Res.*, **20**(12), 1804–1814.

Bennett, J. P. and Nordin, C. F. (1977). Simulation of sediment transport and armouring. *Hydrol. Sci. Bull.*, **22**(4), 555–569.

Best, J. L. (1986). The morphology of river channel confluences. *Progress Phys. Geogr.*, **10**(2), 157–174.

Beven, K., Gilman, K., and Newson, M. (1979). Flow and flow routing in upland channel networks. *Hydrol. Sci. Bull.*, **24**(3), 303–325.

Bhowmik, N. G. and Demissie, M. (1982). Carrying capacity of flood plains. *Proc. Am. Soc. Civ. Engrs., J. Hydraul. Div.*, **108**(HY3), 443–452.

Bowles, D. S., Fread, D. L., and Grenney, W. J. (1977). Coupled dynamic streamflow–temperature models. *Proc. Am. Soc. Civ. Engrs., J. Hydraul. Div.*, **103**(HY5), 515–530.

Cebeci, T. and Bradshaw, P. (1977). *Momentum Transfer in Boundary Layers*, McGraw-Hill, New York, 391pp.

Cedergren, H. R. (1977). *Seepage, Drainage and Flow Nets*, 2nd edn., Wiley, New York, 534pp.

Chen, Y. H. (1979). Water and sediment routing in rivers. In Shen, H. W. (Ed.), *Modeling of Rivers*, Wiley, New York, 10.1–10.97.

Cunge, J. A. (1975). Two-dimensional modeling of flood plains. In Mahmood, K. and Yevjevich, V. (Eds.), *Unsteady Flow in Open Channels*, Water Resources Publications, Fort Collins, Colorado, 705–762.

Cunge, J. A. and Liggett, J. A. (1975). Discussion of 'Comparison of four numerical methods for flood routing' by R. K. Price. *Proc. Am. Soc. Civ. Engrs., J. Hydraul. Div.*, 101(HY4), 431–434.

Cunge, J. A., Holly, F. M., and Verwey, A. (1980). *Practical Aspects of Computational River Hydraulics*, Pitman, London, 420pp.

Cunningham, A. B. and Sinclair, P. J. (1979). Application and analysis of a coupled surface and groundwater model. *J. Hydrol.*, 43, 129–148.

de Vriend, H. J. (1981). Velocity redistribution in curved rectangular channels. *J. Fluid Mech.*, 107, 423–439.

Dietrich, W. E. and Smith, J. D. (1983). Influence of the point bar on flow through curved channels. *Wat. Resour. Res.*, 19(5), 1173–1192.

Dillon, P. J. and Liggett, J. A. (1983). An ephemeral stream-aquifer interaction model. *Wat. Resour. Res.*, 19(3), 621–626.

Dunne, T. (1983). Relation of field studies and modeling in the prediction of storm runoff. *J. Hydrol.*, 65, 25–48.

Ferguson, R. I. (1986). River loads underestimated by rating curves. *Wat. Resour. Res.*, 22(1), 74–76.

Fread, D. L. (1973). Technique for implicit dynamic routing in rivers with tributaries. *Wat. Resour. Res.*, 9(4), 918–926.

Fread, D. L. (1976). Flood routing in meandering rivers with flood plains. In *Rivers 76*, Symposium on Inland Waterways for Navigation Flood Control and Water Diversions, *Proc. Am. Soc. Civ. Engrs., Waterways Harbors Coastal Engrg. Div.*, 16–35.

Fread, D. L. (1984). *DAMBRK: the NWS dam-break flood forecasting model*. Report, Hydrologic Research Laboratory, National Weather Service, Silver Spring, Maryland, 60pp.

Fread, D. L. (1985). Channel routing. In Anderson, M. G. and Burt, T. P. (Eds.), *Hydrological Forecasting*, Wiley, Chichester, 437–503.

Freeze, R. A. (1972). Role of subsurface flow in generating surface runoff 1. Base flow contributions to channel flow. *Wat. Resour. Res.*, 8(3), 609–623.

Ghosh, S. N. and Kar, S. K. (1975). River flood plain interaction and distribution of boundary shear stress in a meander channel with flood plain. *Proc. Instn. Civ. Engrs.*, Part 2, 59, 805–811.

Gosink, J. P. (1986). Synopsis of analytic solutions for the temperature distribution in a river downstream from a dam or reservoir. *Wat. Resour. Res.*, 22(6), 979–983.

Herschy, R. W. (1978). Accuracy. In Herschy, R. W. (Ed.), *Hydrometry*, Wiley, New York, 353–397.

Hey, R. D. (1979). Flow resistance in gravel-bed rivers. *Proc. Am. Soc. Civ. Engrs., J. Hydraul. Div.*, 105(HY4), 365–379.

Hey, R. D. and Thorne, C. R. (1983). Accuracy of surface samples from gravel bed material. *Proc. Am. Soc. Civ. Engrs., J. Hydraul. Engrg.*, 109(6), 842–851.

Ho, R. T. and Gelhar, L. W. (1983). Turbulent flow over undular permeable boundaries. *Proc. Am. Soc. Civ. Engrs., J. Hydraul. Engrg.*, 109(5), 741–756.

Jarrett, R. D. and Costa, J. E. (1986). Hydrology, geomorphology, and dam-break modeling of the July 15, 1982, Lawn Lake Dam and Cascade Lake Dam failures, Larimer County, Colorado. *US Geol. Surv., Prof. Pap.*, 1369, Washington DC, 78pp.

Journel, A. G. and Huijbregts, C. J. (1978). *Mining Geostatics*, Academic Press, New York, 600pp.

Kalkwijk, J. P. T. and de Vriend, H. J. (1980). Computation of the flow in shallow river bends. *J. Hydraul. Res.*, 18(4), 327–342.

Klaassen, G. J. and van der Zwaard, J. J. (1974). Roughness coefficients of vegetated flood plains. *J. Hydraul. Res.*, 12(1), 43–63.

Knight, D. W. and Demetriou, J. D. (1983). Flood plain and main channel flow interaction. *Proc. Am. Soc. Civ. Engrs., J. Hydraul. Engrg.*, 109(8), 1073–1092.

Kouwen, N. and Li, R.-M. (1980). Biomechanics of vegetative channel linings. *Proc. Am. Soc. Civ. Engrs., J. Hydraul. Div.*, 106(HY6), 1085–1103.

Kraatz, D. B. (1971). Irrigation canal lining. *Irrig. Drain. Pap. 2*, Food Agric. Organization, Rome, 170pp.

Krishnappan, B. G. and Lau, Y. L. (1986). Turbulence modeling of flood plain flows. *Proc. Am. Soc. Civ. Engrs., J. Hydraul. Engrg.*, 112(4), 251–266.

Leopold, L. B. and Maddock, T. (1953). The hydraulic geometry of stream channels and some physiographic implications. *US Geol. Surv., Prof. Pap.*, 252, Washington DC, 57pp.

Li, R.-M. and Shen, H. W. (1973). Effect of tall vegetations on flow and sediment. *Proc. Am. Soc. Civ. Engrs., J. Hydraul. Div.*, **99**(HY5), 793–814.

Liggett, J. A. (1975). Basic equations of unsteady flow. In Mahmood, K. and Yevjevich, V. (Eds.), *Unsteady Flow in Open Channels*, Water Resources Publications, Fort Collins, Colorado, 29–62.

Lin, J. D. and Soong, H. K. (1979). Junction losses in open channel flows. *Wat. Resour. Res.*, **15**(2), 414–418.

Mahmood, K. and Yevjevich, V. (Eds.) (1975). *Unsteady Flow in Open Channels*, Water Resources Publications, Fort Collins, Colorado, 3 volumes.

Miller, B. A. and Wenzel, H. G. (1985). Analysis and simulation of low flow hydraulics. *Proc. Am. Soc. Civ. Engrs., J. Hydraul. Engrg.*, **111**(12), 1429–1446.

Miller, W. A. and Cunge, J. A. (1975). Simplified equations of unsteady flow. In Mahmood, K. and Yevjevich, V. (Eds.), *Unsteady Flow in Open Channels*, Water Resources Publications, Fort Collins, Colorado, 183–257.

Mosley, M. P. (1976). An experimental study of channel confluences. *J. Geol.*, **84**, 535–562.

Myers, W. R. C. (1978). Momentum transfer in a compound channel. *J. Hydraul. Res.*, **16**(2), 139–150.

Nezu, I. (1977). *Turbulent structure in open-channel flows.* Ph.D. thesis, Kyoto Univ., Kyoto, Japan.

Parker, G. and Peterson, A. W. (1980). Bar resistance of gravel-bed streams. *Proc. Am. Soc. Civ. Engrs., J. Hydraul. Div.*, **106**(HY10), 1559–1575.

Pasche, E. and Rouvé, G. (1985). Overbank flow with vegetatively roughened flood plains. *Proc. Am. Soc. Civ. Engrs., J. Hydraul. Engrg.*, **111**(9), 1262–1278.

Perkins, H. J. (1970). The formation of streamwise vorticity in turbulent flow. *J. Fluid Mech.*, **44**(4), 721–740.

Petryk, S. and Bosmajian, G. (1975). Analysis of flow through vegetation. *Proc. Am. Soc. Civ. Engrs., J. Hydraul. Div.*, **101**(HY7), 871–884.

Pinder, G. F. and Sauer, S. P. (1971). Numerical simulation of flood wave modification due to bank storage effects. *Wat. Resour. Res.*, **7**(1), 63–70.

Price, R. K. (1974). Comparison of four numerical methods for flood routing. *Proc. Am. Soc. Civ. Engrs., J. Hydraul. Div.*, **100**(HY7), 879–899.

Prickett, T. A. and Lonnquist, C. G. (1971). Selected digital computer techniques for groundwater resource evaluation. *Bull. 55*, Illinois State Water Survey, Urbana, Illinois.

Rajaratnam, N. and Ahmadi, R. (1981). Hydraulics of channels with flood plains. *J. Hydraul. Res.*, **19**(1), 43–60.

Richardson, J. R., Abt, S. R., and Richardson, E. V. (1985). Inflow seepage influence on straight alluvial channels. *Proc. Am. Soc. Civ. Engrs., J. Hydraul. Engrg.*, **111**(8), 1133–1147.

Rouse, H. (1965). Critical analysis of open-channel resistance. *Proc. Am. Soc. Civ. Engrs., J. Hydraul. Div.*, **91**(HY4), 1–25.

Sanders, T. G. (1979). Data collection planning and survey of water quality models. In Shen, H. W. (Ed.), *Modeling of Rivers*, Wiley, New York, 16.1–16.57.

Sayre, W. W. and Caro-Cordero, R. (1979). Shore-attached thermal plumes in rivers. In Shen, H. W. (Ed.), *Modeling of Rivers*, Wiley, New York, 15.1–15.44.

Sharp, J. M. (1977). Limitations of bank-storage model assumptions. *J. Hydrol.*, **35**, 31–47.

Simons, D. B. and Richardson, E. V. (1966). Resistance to flow in alluvial channels. *US Geol. Surv., Prof. Pap.*, **422–J**, Washington DC, 61pp.

Skinner, M. M. and Ruff, J. F. (1973). Application of remote sensing to river mechanics. In Shen, H. W. (Ed.), *Environmental Impact on Rivers*, Shen, Fort Collins, Colorado, 16.1–16.22.

Smith, C. D. (1978). Effect of channel meanders on flood stage in valley. *Proc. Am. Soc. Civ. Engrs., J. Hydraul. Div.*, **104**(HY1), 49–58.

Smith, J. D. and McLean, S. R. (1984). A model for flow in meandering streams. *Wat. Resour. Res.*, **20**(9), 1301–1315.

Strelkoff, T. (1969). One-dimensional equations of open-channel flow. *Proc. Am. Soc. Civ. Engrs., J. Hydraul. Div.*, **95**(HY3), 861–876.

Tingsanchali, T. and Ackermann, N. L. (1976). Effects of overbank flow in flood computations. *Proc. Am. Soc. Civ. Engrs., J. Hydraul. Div.*, **102**(HY7), 1013–1025.

Toebes, G. H. and Sooky, A. A. (1967). Hydraulics of meandering rivers with flood plains. *Proc. Am. Soc. Civ. Engrs., J. Waterways Harbors Div.*, **93**(WW2), 213–236.

Vanoni, V. A. (Ed.) (1975). *Sedimentation Engineering*, Am. Soc. Civ. Engrs. Manuals and Reports on Engineering Practice, No. 54, New York, 745pp.

Vreugdenhill, C. B. and Wijbenga, J. H. A. (1982). Computation of flow patterns in rivers. *Proc. Am. Soc. Civ. Engrs., J. Hydraul. Div.*, **108**(HY11), 1296–1310.

Vuichard, D. and Zimmermann, M. (1986). The Langmoche flash-flood, Khumbu Himal, Nepal. *Mountain Res. Dev.*, **6**(1), 90–94.

Weinmann, P. E. and Laurenson, E. M. (1979). Approximate flood routing methods: a review. *Proc. Am. Soc. Civ. Engrs., J. Hydraul. Div.*, **105**(HY12), 1521–1536.

White, W. R., Milli, H., and Crabbe, A. D. (1975). Sediment transport theories: a review. *Proc. Instn. Civ. Engrs.*, **59**, Part 2, 265–292.

Wormleaton, P. R., Allen, J., and Hadjipanos, P. (1982). Discharge assessment in compound channel flow. *Proc. Am. Soc. Civ. Engrs., J. Hydraul. Div.*, **108**(HY9), 975–994.

van't Woudt, B. D. and Nicolle, K. (1978). Flow processes below a gravelly riverbed. *N.Z. J. Hydrol.*, **17**(2), 103–120.

van't Woudt, B. D., Whittaker, J., and Nicolle, K. (1979). Ground water replenishment from riverflow. *Wat. Resour. Bull.*, **15**(4), 1016–1027.

Yevjevich, V. (1975). Introduction. In Mahmood, K. and Yevjevich, V. (Eds.), *Unsteady Flow in Open Channels*, Water Resources Publications, Fort Collins, Colorado, 1–27.

Zagni, F. E. and Smith, K. V. H. (1976). Channel flow over permeable beds of graded spheres. *Proc. Am. Soc. Civ. Engrs., J. Hydraul. Div.*, **97**(HY2), 303–321.

Zippe, H. J. and Graf, W. H. (1983). Turbulent boundary-layer flow over permeable and nonpermeable rough surface. *J. Hydraul. Res.*, **21**(1), 51–65.

Zitta, V. L. and Wiggert, J. M. (1971). Flood routing in channels with bank seepage. *Wat. Resour. Res.*, **7**(5), 1341–1345.

Modelling Geomorphological Systems
Edited by M. G. Anderson
©1988 John Wiley & Sons Ltd.

Chapter 7

Hydrology and sediment models

G. PICKUP

Division of Wildlife and Rangelands Research, CSIRO

7.1 BASIC MODEL TYPES

Mathematical models are numerical representations of conceptual models and provide a quantitative description of the spatial and temporal behaviour of a system. Typically, a mathematical model uses a set of equations to transform information on inputs or driving forces into outputs which describe the state of the system or the rate of removal of material from it. The equations represent the processes involved and may be analytically-based, purely empirical or somewhere in between. Models are invariably simpler than the system behaviour they describe and are usually formulated as computer programs. While a wide range of model types exist, they may be conveniently divided in lumped parameter models, distributed process models and hybrids.

Lumped parameter models are those in which complex system behaviour is averaged or integrated over space and time and represented by a few mathematical functions. Such models are relatively simple and therefore easy to calibrate so that observed and modelled behaviour match as closely as possible. In some lumped parameter models, calibration may be achieved by adjusting parameter values in accordance with known relationships between system characteristics and system behaviour. For example, in runoff routing models, catchment lag may be calculated as a function of catchment area and slope. No such relationships exist for sediment yield so calibration is usually carried out using fitting or optimisation procedures which change parameter values until observed and modelled outputs match. Lumped parameter models must therefore be used on drainage basins for which sediment data already exists. This restricts their usefulness for forecasting in situations where there are no calibration data. Their reliability under conditions different from those for which they are calibrated can also be suspect.

Distributed process models, such as those used for catchment or channel sediment routing, attempt to reproduce the spatial and temporal behaviour of a system by modelling it from physical considerations and in detail. Ideally, the basic relationships in these models are analytically-based mathematical descriptions of physical laws. In practice, because the processes they represent are so complex, they use physical laws based on simplified assumptions, finite difference approximations of continuous processes, and empirical relationships describing aspects of behaviour which we do not understand. They are, however, regarded as being the nearest thing we have to general models which can be used for forecasting.

Hybrid or quasi-analytical models are those which contain both distributed and lumped parameter elements. They typically apply to situations in which only part of the process to be modelled is well understood or described by physically-based mathematical relationships. Other processes may not be understood so well and have to be represented by stochastic processes or relationships which approximate their behaviour without being analytically based. Hybrid models are frequently set up as a series of interlinked compartments which represent spatial variability or spatial behaviour in the modelled system. They may also contain a few parameters which have to be adjusted by trial and error or by automatic search techniques in order to tune the model.

7.2 LUMPED PARAMETER SEDIMENT YIELD MODELS

7.2.1 Sediment Rating Curves

The sediment rating curve is an empirically derived relationship between water discharge and suspended sediment concentration at some point along a stream. Water discharge is obtained either by gauging or from a stage–discharge rating curve. Sediment concentration is measured by collecting samples of the water–sediment mixture which are then analysed for sediment content. A wide variety of samplers and sampling schemes are used to collect sediment data. These schemes range from occasional gulp samples at a single point in a stream using a beer bottle through to complex depth or point integrating studies with specially designed samplers which are intended to give more representative data. Other methods include the use of automatic pumping samplers and continuous monitoring of turbidity, a parameter which can sometimes be related to sediment concentration. An equally wide range of sample processing techniques is available, the result being that it is rarely possible to compare results from different river systems with any degree of certainty. In spite of these problems, the sediment rating curve continues to be the most widely-used method for the determination of sediment yield, largely because, until recently, there was nothing better.

Sediment rating curves are traditionally presented in the form:

$$c = a Q^j$$

in which c is sediment concentration, Q is water discharge, and a and j are empirical coefficients derived by regression. The exponent, j, usually lies between 1.0 and 2.0 with the higher values associated with smaller catchments. The scatter in sediment rating curves may involve several orders of magnitude (see Figure 7.1 for example) and correlation coefficients associated with the regression equation are frequently low. The result is that estimates of sediment yield for short periods based on rating curves are frequently 100 per cent and sometimes up to 1000 per cent in error. Some of these errors arise from using

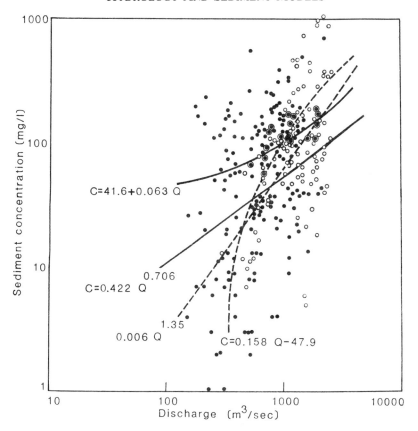

Figure 7.1. Sediment rating curves for the Alice River at Konkonda in Papua New Guinea. While seasonal differences in discharge–sediment relationships concentration occur, the scatter of points is such that virtually any regression equation could be fitted. (From Pickup, Higgins and Warner, 1979 and reproduced by permission of the authors)

mean daily flows instead of discharges for shorter periods when calculating sediment yield. Others arise because both systematic and random variations occur in the relationship between discharge and sediment concentration. These include different relationships in different seasons, hysteresis effects in which the wave of sediment concentration either leads or lags behind the flood wave, and depletion effects whereby the amount of sediment available for transport is progressively reduced during the flood event.

The complex nature of the relationship between sediment concentration and discharge is illustrated by a study of 39 storm events in five catchments of the Wallagaraugh River in New South Wales reported by Olive and Rieger (1984). Seven different types of response were observed (Figure 7.2) including sediment concentration leads, lags, synchronous behaviour and the exhaustion of sediment supply during flood events with multiple peaks. Situations in which the relationship between concentration and discharge was apparently random also occurred regularly. Individual streams did not display the same response throughout the study period which makes modelling difficult because different models are required for different situations on the same stream.

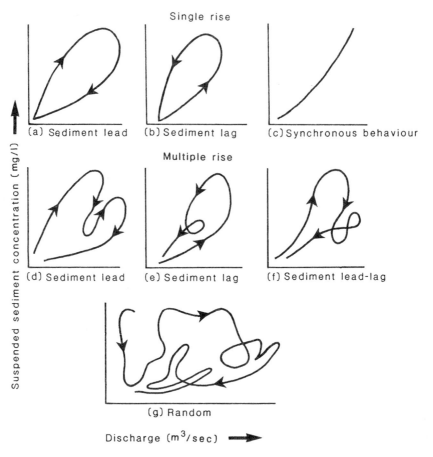

Figure 7.2. Suspended sediment discharge response types during storms in catchments of the Wallagaraugh River, New South Wales. (From Olive and Rieger, 1984 and reproduced by permission of the authors)

Most of the responses described by Olive and Rieger (1984) have been reported elsewhere (e.g. Walling, 1974; Wood, 1977; Beschta, 1981). In spite of this only a few attempts have been made to incorporate them into the sediment rating curve model. Such attempts include the derivation of separate c–Q relationships for different seasons or for the rising and falling stages (e.g. Loughran, 1977, Walling, 1977, Pickup, 1984). They also include a mixing model derived by Walling and Webb (1982) in which it is assumed that suspended sediment generation is essentially limited to storm events. Where sheet and rill erosion are the dominant source of material, sediment will be transported to the stream by surface runoff. The sediment concentrations recorded during a storm event will therefore reflect the mixing of sediment-laden storm runoff with the prevailing baseflow. When baseflow is high, storm runoff concentrations will be considerably diluted whereas little dilution occurs during periods of low baseflow. Baseflow and stormflow can be distinguished using standard hydrograph separation techniques while stormflow sediment concentrations are calculated from:

$$c_s = (c_t Q_t - c_b Q_b)/Q_s$$

in which c and Q are concentration and discharge, and the subscripts, t, b and s refer to total flow, baseflow, and stormflow respectively. The c_b values are obtained by linear interpolation between concentrations before and after the flood event. Since the amount of baseflow dilution varies during a flood event, the mixing model can partially explain the tendency for the graph of sediment concentration to lead the hydrograph. It does not, however, provide an explanation for other types of variation.

The usefulness of the sediment rating curve approach is restricted by the assumptions on which it is based. The idea that a single valued relationship between c and Q can exist implies that sediment concentration depends on the transport capacity of the flow. This may be true when most of the load is made up of suspended channel bed material and it is not supply-restricted. In most streams, a proportion of the suspended load is, in fact, fine material derived from the catchment surface and transported as washload. Washload can usually be carried in quantities well in excess of those delivered. The concentration therefore depends on the rates of erosion within the catchment and rate of delivery to the channel, neither of which have to be related to discharge.

7.2.2 Complex Response Models

The failure of simple functions like the sediment rating curve to describe the behaviour of sediment concentration with discharge over time has led to the development of lumped parameter models with more complex response functions. Such models include the sedimentgraph approach of Rendon-Herrero (1974) in which the sediment discharge equivalent of a unit hydrograph is used. Other, more general methods could be developed using Box–Jenkins (1976) transfer function time series models which are capable of representing a wide range of response functions. The parameters of these models cannot be determined from catchment characteristic or physical considerations. They can therefore only be used in situations for which data already exist.

One of the best complex response models currently available is the dynamic basin (DB) sediment yield model of Moore (1984). This is a lumped parameter scheme developed to resemble the physical processes involved in catchment sediment transport yet still remain simple enough for gradient-based automatic calibration. All functions within the model are smooth and twice-differentiable so threshold situations, in which a process begins to operate only once some critical value is passed, are avoided. This not only makes calibration easier, it also makes physical sense given the variability in sediment characteristics, sediment availability and the intensity of the transport processes which occur at the catchment scale. The DB model is an improvement on time series models which have relatively simple time-invariant response functions and cannot handle the complex behaviour observed in sediment yield data. Its response function varies with the model input and the system state making it both time-variant and capable of handling complex system behaviour.

The DB model uses three relationships to describe the variation in suspended sediment yield from a catchment over time. These are: a sediment availability function, a sediment removal function and a sediment translation function. The availability function describes how the supply of material available for erosion increases with time between events. The removal function determines the detachment of sediment from the available material as a function of the amount available and the storm magnitude. The translation function describes the movement of sediment through the catchment and determines the time distribution of a pulse of eroded material at the outlet. An abbreviated description of these functions drawn from Moore (1984) is presented below.

The sediment availability function in the DB model is based on the assumption that the amount of material which can be eroded builds up in the period between storms but at an exponentially decreasing rate (Figure 7.3). This rate, $R(t)$ is given by:

$$R(t) = \frac{dL}{dt} = R_0 - KL = R_0 \exp[-K(t-t_0)]$$

and the volume of sediment accumulating over a time interval (t_0, t) is calculated as:

$$L(t) = \frac{R_0}{K} \{1 - \exp(-K\Delta t)\}$$

in which R_0 is the initial maximum rate of increase in sediment availability at t_0 when no sediment is available for erosion; K is the availability rate constant and $\Delta t = t - t_0$.

The relationship is capable of reproducing the depletion effect which occurs when a number of storms occur over a short interval but it contains two parameters, K and R_0 which need calibration.

Sediment removal calculations in the DB model are based on a simple lumped model in which the rate of removal varies with the amount of sediment available and what Moore (1984) terms 'direct runoff'. This relationship may be stated as:

$$\frac{dL}{dt} = -k^* q(t)^\beta$$

in which $\frac{dL}{dt}$ is the rate of removal, $q(t)$ is direct runoff at time, t, and k^* and β are coefficients. The relationship between rate of removal and sediment availability is made by allowing k^* to vary with the amount of material available so that $k^* = kL$. The removal equation then becomes:

$$\frac{dL}{dt} = -kLq(t)^\beta$$

The mass of sediment removed in a given interval $(t, t+\Delta t)$ may be obtained by integrating this equation such that:

$$M = L(t) - L(t + \Delta t) = L(t)\{1 - \exp(-kV)\}$$

in which M is the material removed and V is the volume of runoff which is given by:

$$V = \int_t^{t+\Delta t} q(\tau)^\beta d\tau$$

Because the DB model has to be calibrated, model results are dependent on factors such as the time interval used in the calculations. They are also dependent on the method used

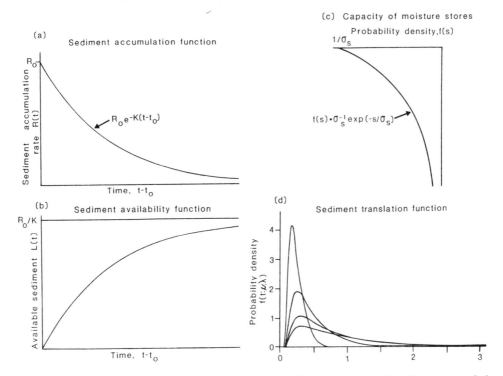

Figure 7.3. Some relationships used in the DB model of Moore (1984). (a) and (b) are the sediment accumulation and availability relationships. (c) is the probability density function of the storage capacities of the population of moisture stores used to represent a basin. (d) is the inverse Gaussian probability density function for various values of the drift parameter, μ with $\lambda = 1$. (Copyright by the American Geophysical Union)

to calculate runoff from rainfall. Runoff within the DB model is generated using the algorithm of Moore and Clarke (1981) in which the catchment is treated as an array of moisture storages. The individual storages vary in capacity but conform with a particular frequency distribution. When rainfall occurs, the area contributing runoff will be restricted to that occupied by storages whose capacity is less than the amount of rain, thus:

$$q(t) = F\{C^*(t)\}\pi_i$$

where π_i is the net rainfall during time increment, i and $F\{C^*(t)\}$ is the distribution function of store capacities, s. This function may be stated as:

$$\text{Prob } [s \leqq C^*(t)]$$

in which $C^*(t)$ is the size of the largest store full of water at time, t. $F\{C^*(t)\}$ is assumed to have an exponential distribution (Figure 7.3) so the proportion of a basin generating direct runoff at time t is:

$$F\{C^*(t)\} = \int_0^{C^*(t)} \sigma_s^{-1} \exp(-s/\sigma_s)\mathrm{d}s = 1 - \exp\{-C^*(t)/\sigma_s\}$$

where σ_s is the mean store depth. The basin runoff rate is then:

$$q(t) = \pi_i\{1 - \exp[-C^*(t)/\sigma_s]\}$$

The model has the advantage that only one parameter, σ_s, is needed to describe the store capacity distribution function. That parameter is also the only one requiring calibration.

The sediment removal function and the runoff model for an exponential distribution of stores may be combined to give:

$$\frac{dL}{dt} = -\pi_i^\beta kL\{1 - \exp[-C^*(t)/\sigma_s]\}^\beta$$

This equation shows that the rate of sediment removal depends on the rainfall intensity, π_i, and the area of direct runoff generation, $A_c(t) = A\{1 - \exp[-C^*(t)/\sigma_s]\}$, where A is total catchment area. Integrating this equation gives:

$$L(t + \Delta t) = L(t)\exp(-kV_i)$$

where the volume of runoff, V_i, in the interval, $t + \Delta t$, is:

$$V_i = \pi_t^\beta \int_t^{t+\Delta t} \{1 - \exp[-C^*(\tau)/\sigma_s]\}^\beta d\tau$$

Moore (1984) assumes a β value of 1 so that:

$$V_i = \pi_i\Delta t + \sigma_s\{\exp[-C^*(t+\Delta t)/\sigma_s] - \exp[-C^*(t)/\sigma_s]\}$$

however, other β values are possible.

The sediment translation function in the DB model describes the temporal distribution of pulse of transport sediment at the basin outlet. This distribution is represented by an inverse Gaussian density function which may be written as:

$$f(t; \mu, \lambda) = \left(\frac{\lambda}{2\pi t^3}\right)^{1/2} \exp\left\{\frac{-\lambda(t-\mu)^2}{2\mu^2 t}\right\} \text{ for } t > 0$$

$$f(t; \mu, \lambda) = 0 \text{ otherwise}$$

This function is controlled by two parameters, μ and λ both of which have to be calibrated and its shape is illustrated in Figure 7.3. There is some physical justification for using the inverse Gaussian density function because it is a solution to the convection–diffusion equation:

$$\tfrac{1}{2}\sigma^2 \frac{\partial^2 c}{\partial x^2} - v\frac{\partial c}{\partial x} = \frac{\partial c}{\partial t}$$

in which c is sediment concentration and v is sediment velocity. This equation is also used in the BR model described in the section on channel sediment routing models. The inverse

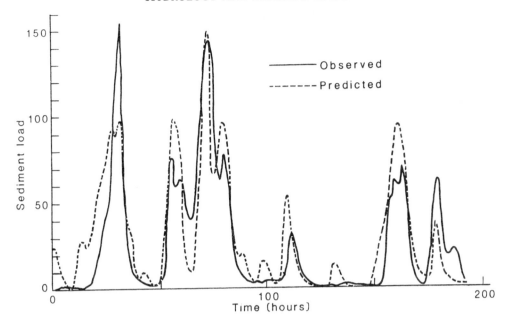

Figure 7.4. Observed and modelled results from the DB model for an 8-day period. (From Moore, 1984, copyright by the American Geophysical Union)

Gaussian density function can also be a solution to the convection–diffusion equation when a sink term is added to allow for the settling out of sediment during transport (Moore and Clarke, 1983).

Some results from the DB model are presented in Figure 7.4. It performs well but not perfectly. Sediment peaks both lead and lag behind hydrograph peaks, sediment concentrations may be overestimated and underestimated for events occurring within a few hours of each other, and there does not seem to be any trend in errors. This leads Moore (1984) to suggest that only a limited level of prediction is possible when the only model input used is rainfall. Better results might be obtained with more detailed information on the spatial distribution of rainfall intensity, runoff, and the behaviour of sediment production zones. While some of these data may become readily available as remote sensing technology improves, they cannot presently be obtained without intensive instrumentation. The cost of this instrumentation cannot be justified outside the research environment so models requiring it are unlikely to be much use for operational purposes. If nothing else, the DB model has realistic input data requirements.

The use of automatic calibration techniques in the DB model also raises questions. During the 1970s attempts were made to relate parameter values of rainfall–runoff models obtained by calibration to physical catchment characteristics to allow their use on ungauged catchments. These attempts met with failure because model performance was dependent on assumed initial conditions, because several minima occurred in the objective function, because some parameters proved insensitive, and because different parameters had identical effects on model output. This experience suggests that it will not be possible to relate calibrated DB model parameters to catchment characteristics either. It is therefore unlikely that the DB model could be extended to ungauged catchments, or indeed to situations outside the calibration range on the calibration catchment. Its value for prediction is therefore limited but that can be said of all the models described in this chapter.

7.3   CATCHMENT SEDIMENT ROUTING MODELS

Catchment sediment routing is a mathematical procedure for calculating erosion rates within a catchment and transporting the eroded sediment to the catchment outlet. The process is described in detail by Lane *et al.* (Chapter 10) so only a brief review of the main types of model and their associated problems is presented here.

A wide variety of catchment sediment routing (CSR) models have been published since the mid-1960s. These models vary in sophistication but are usually claimed to represent or be based on the physical processes of runoff generation, erosion and sediment transport within a basin.

The main aim in developing CSR models has been to predict the impact of land use changes on sediment yield. This can rarely be done using regression equations or calibrated lumped parameter models because complex changes in system behaviour occur which frequently invalidate the simple response functions of such models. It is also not possible to use the prototype approach described in the section on modelling large systems because the database is insufficient for the derivation of enough acceptable prototypes. It has therefore been seen as important for CSR modellers to develop complex physically-based models with measurable parameters for only then can models be transferred between catchments or handle land use changes. The extent to which the resultant models resemble physical processes varies. The best models provide only a partial description of the processes involved in sediment routing. The worst models are little more than exercises in self-delusion on the part of the developers.

Most CSR models use the Universal Soil Loss Equation (USLE) or its variants for the determination of on-site erosion (see Wischmeier and Smith, 1965, 1978, Wischmeier, 1976, and USDA, 1982). The USLE may be stated as:

$$A = R.K.L.S.C.P$$

in which:
  A is the predicted soil loss per unit of area and is normally an estimate of the average annual interrill plus rill erosion from rainstorms for field-sized areas. It does not include erosion from gullies or streambanks, snowmelt erosion, or wind erosion, but can involve eroded sediment which is deposited before it reaches downslope stream channels.
  R is the rainfall and runoff factor for a specific location.
  K is the soil erodibility factor for a specific soil horizon.
  L is the dimensionless slope-length factor expressed as the ratio of soil loss from a given slope length to that from a 72.6-ft. length under the same conditions.
  S is the dimensionless slope-steepness factor expressed as the ratio of soil loss from a given slope steepness to that from a 9-per cent slope under the same conditions.
  C is the dimensionless cover management or cropping management factor, expressed as a ratio of soil loss from the condition of interest to that from tilled continuous fallow.

P is the dimensionless erosion-control practice factor, expressed as a ratio of the soil loss with practices such as contouring, strip cropping, or terracing to that with farming up and down the slope.

The USLE is a statistical rather than a physically-based relationship and its derivation relied heavily on regression analysis. It is based on a very large database consisting of more than fifty years of plot experiments for a wide range of conditions. Its applicability is summarized by Wischmeier (1972) as follows:

'The name "universal" soil-loss equation originated as a means of distinguishing this prediction model from the highly regionalised models that preceded it. None of its factors utilizes a reference point that has direct geographic orientation. In the sense of the intended functions of the equation's six factors, the model should have universal validity. However, its application is limited to states and countries where information is available for local evaluations of the equation's individual factors. There are exceptions to the validity of the EI parameter as a measure of the combined erosive forces of rainfall and runoff. For some situations, a more accurate predictor of runoff–erosion potential needs to be substituted as the value of R. The indicated *nature* of the effects of topographic, cover, and management variables is probably universal, but it has not been shown that the specific ratios for L, S and C, derived on the US mainland, are necessarily accurate on vastly different soils, such as those of volcanic origin for example. Slope effect in situations where gradients appreciably exceed 20 percent is still a serious void in research information.

The relationships, graphs and tables presented for evaluation of the equation's factors cannot be simply transported verbatim to states or countries where the type of rainfall or the soil genesis is vastly different. However, a relatively small amount of well designed local research should enable many countries to adapt the soil-loss equation and basic relationships to their situations.'

The first attempts to develop CSR models from the USLE involved the use of delivery ratios and are summarized by Walling (1983). The delivery ratio is the ratio of sediment delivered at the catchment outlet to gross erosion within the basin. It can therefore be regarded as a time-averaged single parameter equivalent of the sediment routing component of a CSR model. A number of attempts have been made to derive generalized relationships between the value of the sediment delivery ratio and catchment characteristics such as basin area, basin length, number of gullies present and so on. None have proved general for a number of reasons, the major one being that gross erosion is estimated rather than measured so it can be very much in error. It is also likely that delivery ratios vary greatly through time.

Later CSR models still use the USLE or similar equations but contain explicit attempts to describe drainage basin hydrology, interrill, rill and channel sediment transport. The Stanford Sediment Model (Negev, 1967) was the first serious attempt to combine all these processes in a computer model (Figure 7.5) and established the basic approach followed by many authors since then (e.g., Walker and Fleming, 1979). It consists of a rainfall–runoff model which is used to determine the amount of overland flow, a set of relationships which calculate soil splash, entrainment of surface material and the rate of sediment transport by overland flow. There are also relatively primitive routines to calculate transport within the channel. Most of the equations in the Stanford Sediment Model are simple exponential relationships with no physical basis. The parameters have to be determined by guesswork or trial and error making it difficult to apply. It is also a lumped parameter model so the process descriptions

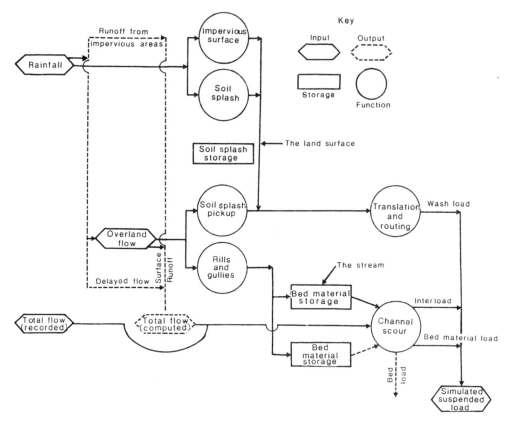

Figure 7.5. Flow chart for the Stanford Sediment Model. (From Negev, 1967 and reproduced by permission of the author)

are averages for the whole catchment. The Stanford Sediment Model is therefore no more physically-based than the DB model described in the previous section.

Recent developments in CSR modelling have taken a distributed approach to both the catchment hydrology and the sediment transport phase (see Foster, 1982 and Rose, in press for substantive reviews). The basis of most models is a rainfall–infiltration–runoff algorithm coupled with a hydraulic model which describes flow on the watershed surface. The kinematic wave equations provide the most commonly used hydraulic model and may be structured for a cascade of planar surface elements or for more complex structures such as converging cone sections (Woolhiser *et al.*, 1971; Smith, 1976). Sediment transport relationships usually involve bedload equations and erosion and deposition are determined using the sediment continuity equation. These relationships and their problems are discussed below in the section on channel sediment routing models. Some distributed CSR models have been generalized to the stage where they are close to being operational tools (USDA, 1980; Borah *et al.*, 1981) and CREAMS, the most versatile model currently available, has even been converted to run on 16 bit personal computers under the MS–DOS operating system.

Distributed CSR models still have their problems. There are basic uncertainties in our understanding of the processes of shallow overland flow, its interaction with rainfall, different slope forms and vegetation. The hydraulic models in use may be physically sound in some cases but quite unrelated to conditions in many catchments especially when they use simplified representations of complex topographic structures such as a cascade of planes. CSR models still involve a certain degree of spatial lumping which may not be consistent with observed behaviour. For example, as Walling (1983) states, storm runoff in humid areas is produced from only a small proportion of the basin and this contributing area varies with antecedent moisture conditions. Sediment is eroded from this runoff contributing area rather than the whole catchment or planar element yet CSR models treat that element as a unit.

CSR models are difficult, if not impossible, to apply to large complex catchments because of the large amount of data required, the difficulty of synchronizing processes in different subcatchments, and the need to represent sediment storage behaviour. They also ignore mass movement processes such as soil creep and landsliding which may be of considerable importance, particularly in mountain areas. They may therefore be no more than complex lumped parameter models and not represent the actual processes of sediment movement in a catchment at all. These deficiencies have led to the sediment budget approach in which attempts are made to quantify actual rates of sediment movement over the whole area (Dietrich and Dunne, 1978; Lehre, 1981).

Sediment budget models are constructed by identifying each sediment source and calculating its yield over a specific period of years. Only a small proportion of sediment moves directly from a source area to the catchment outlet so allowance must be made for storage. This involves identifying storages such as alluvial fans and floodplains and determining the residence time of material in them from observed storage volumes and outflows. Sediment budgets rely heavily on approximate estimates of rates of process based on short-term sediment discharge measurements, stratigraphic analysis, and dating. They have also used plot experiments to derive erosion rate equations in key areas such as sediment yield from logging roads (Reid *et al.*, 1981). The sediment budget approach is not an operational tool for forecasting sediment yield as yet but it does focus attention on the need to model storage behaviour since it is the rate of release from storage that usually determines the rate of transport in many parts of a system.

7.4 CHANNEL SEDIMENT ROUTING MODELS

Channel sediment routing is the mathematical procedure by which inflowing sediment load is passed along the channel to the catchment outlet during a sequence of flow events. It is normally carried out when information is required on the effects of a change in sediment input, water discharge or channel geometry, all of which affect the rate of sediment transport and storage. Since these effects are usually complex, it is necessary to carry out routing or accounting calculations for a whole series of channel reaches to fully assess what the overall impact will be.

Sediment routing requires a model describing the behaviour of the channel and a set of input data describing the system and the events to which it is subjected. The input data usually consists of:

—A sequence of flow events representing a hydrograph or flow duration curve;
—A series of cross-sections and roughness coefficients which describe the channel system;
—The size distribution and amount of the inflowing sediment load;
—The size distribution and amount of sediment available for transport in the channel bed and banks.

The channel behaviour model varies in sophistication depending on the problem. In most cases it consists of:

—The conservation of mass and momentum equations for the flow and a flow resistance equation;
—The conservation of mass equation for the sediment;
—A relationship between flow conditions and the rate of bed-material transport;
—A relationship between the rate of erosion or deposition and the size distribution of sediments in the bed;
—A relationship describing changes in channel geometry with erosion or deposition.

Sediment routing models based on transport equations usually do not handle washload. It is therefore either ignored or specified as an input variable obtained from either a sediment rating curve or a catchment erosion model. Once it enters a model, it is usually routed straight through to the downstream end of the system.

A wide range of sediment routing models have been published and some are available as computer programs. In the sections which follow, a number of different approaches to the sediment routing problem are discussed. These include analytical solutions, finite difference approaches using the known discharge and sequential routing methods, and the moving wave or dispersion approach.

7.4.1 Analytical Models

It is not possible to solve the routing equations in the channel behaviour model analytically except in a few highly idealized cases. Sediment routing procedures therefore usually involve the use of numerical models which solve the partial differential equations by finite difference methods. The analytical models do, however, provide a convenient starting point and when all the uncertainties and problems involved in sediment routing are taken into account, sometimes produce answers which are as good as or better than those obtained from more complex numerical models.

The basic equations describing one dimensional unsteady flow over a moveable bed in a wide channel are usually stated as:

Equation of motion of water:

$$\frac{\partial u}{\partial t} + u \frac{\partial u}{\partial x} + g \frac{\partial h}{\partial x} + g \frac{\partial z}{\partial x} = -g S_f$$

Continuity equation for water:

$$\frac{\partial h}{\partial t} + u \frac{\partial h}{\partial x} + h \frac{\partial u}{\partial x} = 0$$

Flow resistance:

$$u = c\sqrt{hs_f}$$

Sediment transport:

$$q_s = au^b$$

Continuity equation for sediment:

$$\frac{\partial z}{\partial t} + \frac{\partial q_s}{\partial x} = 0$$

where:
   u is velocity
   t is time
   x is distance
   g is the gravity constant
   h is flow depth
   z is bed elevation
   s_f is the friction slope
   c is the Chezy coefficient
   q_s is the sediment transport rate by volume and
   a and b are empirical coefficients.

Solution of these equations requires simplifications. In the parabolic model of de Vries (1973), it is assumed that water movement is steady and uniform and the Froude number is less than 0.5. By omitting $\dfrac{\partial h}{\partial t}$ and $\dfrac{\partial u}{\partial t}$ from the analysis, the equations can be reduced to:

$$\frac{\partial z}{\partial t} - K\frac{\partial^2 z}{\partial x^2} = 0$$

and K can be approximated by:

$$K \cong \frac{1}{3}\, b\, \frac{q_s}{S_o}$$

where S_o is the original slope of the bed. The hyperbolic model solution is more complex and does not assume uniform flow during the entire process. The basic equation describing it is:

$$\frac{\partial z}{\partial t} - K\frac{\partial^2 z}{\partial x^2} - \frac{K}{c}\frac{\partial^2 z}{\partial x \partial t} = 0$$

in which c denotes the celerity of a small disturbance at the bed. There is no simple solution to this equation and further details are available in de Vries (1973).

An example of the use and limitations of the parabolic model is presented by Soni *et al.* (1980). Given boundary conditions:

$$z(x,0) = 0 \text{ for all } x$$
$$z(0,t) = z_o \text{ for all } t > 0 \text{ and}$$
$$z(x,t) = 0 \text{ for } x \to \infty \text{ for } t > 0$$

the parabolic model may be solved as:

$$\frac{z}{z_o} = 1 - erf(\eta) = erfc(\eta)$$

in which $\eta = x/2\sqrt{(Kt)}$. The change in bed elevation which results from a change in sediment input, can be seen from the above solution to decrease asymptomatically downstream. To calculate the value of z_o, the reach length over which erosion or deposition occurs must be known. Soni *et al.* (1980) assume that the change ends at $z/z_o \cong 0.01$. It is then possible to evaluate z_o at time, t for a given change in sediment input, by equating the volume of sediment entering the system Δq_s to the area under the transient bed profile. Thus:

$$(\Delta q_s)t = \int_0^\ell z\,dx$$

and

$$z_o = \frac{\Delta q_s t}{\int_0^\ell (1 - erf(\eta))\,dx}$$

in which ℓ is the reach length over which change takes place.

The pattern of change which occurs in the bed over time for situations in which there is an increase in the sediment input from a point source at the upstream end of the system is summarized in Figure 7.6 and broadly conforms with what is observed in nature. Tests of the model against changes in flumes, however, indicate that while the basic form of the model is correct, the observed value of the aggradation coefficient, K, does not accord with the theoretical value. This is not surprising given the simplifying assumptions used but it does suggest that K values might be derived empirically in the early stages of erosion or deposition and used to predict subsequent changes in the bed profile.

Analytical models of the type described here have many restrictions. They do not handle sediment mixtures, bed armouring or downstream variations in channel cross-section. They are not formulated for variable discharge. The assumptions of uniform quasisteady flow during the transient phase are unrealistic and so on. The trend has therefore been to use numerical models in which these factors can be explicitly evaluated.

7.4.2   Finite Difference Approaches—the HEC–6 Model

Two numerical model approaches to sediment routing are in common use: the sequential routing method and the known discharge method (Ponce *et al.*, 1979). In the sequential

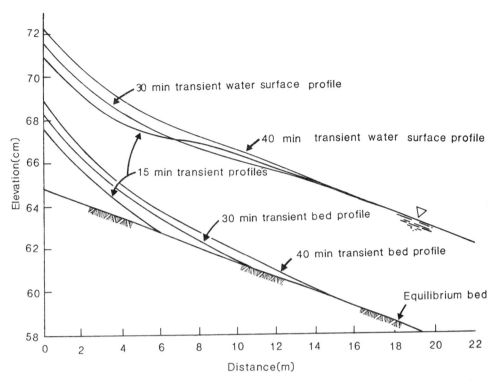

Figure 7.6. Changes in bed profile determined using the parabolic model for an increase in sediment load from a point source at the upstream end of a channel. (From Soni *et al.*, 1980 and reproduced by permission of *Journal of the Hydraulics Division*, American Society of Civil Engineers)

routing method, unsteady flow is assumed and the two flow equations are first solved to provide information on flow conditions along a reach over time. The flow conditions are then used with a bedload equation to provide data on transport rates which allows calculation of sediment inflow, outflow and storage using the continuity equation. In the known discharge method, unsteady flow is represented by a sequence of constant discharges. The routing problem is then reduced to a simple relationship describing water continuity and two differential equations, namely the sediment continuity equation and the equation of motion. These equations may be solved in an uncoupled mode in which step-backwater analysis or some other method is used to compute the flow profile after which erosion and deposition are determined using the sediment continuity equation. Alternatively, a coupled mode may be used in which the equations of motion and sediment continuity are solved simultaneously. The HEC-6 model, described here, uses the known discharge approach and solves the routing equations in an uncoupled mode. The HEC-6 computer program is available commercially making it the most widely used sediment routing model currently available.

HEC-6 is a large and comprehensive sediment modelling program consisting of more than 7 000 lines of FORTRAN code. The flow profile computations handle a wide range of hydraulic and channel conditions and are based on the HEC-2 program (Hydrologic Engineering Centre, 1982). Several routines are offered for the sediment transport calculations including Toffaletti's (1969) modification of the Einstein (1950) procedure, Laursen's (1958) equation and a user-specified function based on the depth–slope product. Bed armouring is allowed for using the method of Gessler (1970). Several detailed descriptions of the model are available in the literature so only a brief outline drawn from the users manual (Hydrologic Engineering Centre, 1977) and the paper by Thomas (1982) is presented here.

The basic flow equations in HEC–6 are:

Equation of motion for water:

$$\frac{\partial h}{\partial x} + \frac{\partial(\alpha u^2/2g)}{\partial x} = S_f$$

Continuity equation for water:

$$Q = uA + Q_1$$

Flow resistance:

$$u = 1.486 \frac{R^{2/3} S_f^{1/2}}{n}$$

in which:
  α is the velocity distribution coefficient
  A is channel area
  Q is discharge
  Q_1 is tributary inflow and
  R is hydraulic radius
  n is a roughness coefficient.

These equations describe a steady, non-uniform flow situation and are solved using the standard step method. The value of n may vary with discharge and it is possible to specify a stage–discharge relationship at any point to represent weirs and other fixed boundary structures. It is also possible to carry out flow computations for complex cross-sections including areas of overbank flow. The model can therefore handle a much wider and more realistic set of hydraulic conditions and relationships than can be used in an analytical model.

The sediment transport relationships in HEC–6 may be stated as:

Continuity equation for sediment:

$$\frac{\partial q_s}{\partial x} - W_b \frac{\partial z}{\partial t} = q_{s\ell}$$

Sediment transport:

$$G = fn(u, h, S_f, W_b, D_{eff}, T, D_{si}, P_i)$$

in which

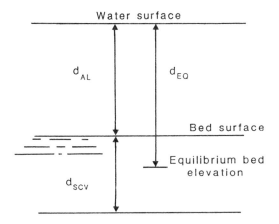

Figure 7.7. Definition diagram for equilibrium depth and active layer thickness in the HEC-6 model. (From Thomas, 1982 and reproduced by permission of John Wiley and Sons)

W_b is bed width

q_{si} is the lateral inflow of sediment from tributaries (by volume)

G is the sediment transport rate by weight

h is effective flow depth

D_{eff} is effective grain size of sediment in size class, i

T is water temperature

D_{si} is the geometric mean of a grain size class interval, i

P_i is the percentage of a particular grain size class in the bed surface layer.

These relationships allow for erosion and deposition in cross-sections of differing widths. They also allow a wide range of factors to affect the transport rate including bed armouring and sediment size grading.

HEC–6 models the interaction between the flow and the bed in a more complex manner than many earlier models. The bed surface layer consists of a control volume (Figure 7.7). The control volume is made up of three zones: the armour layer, the active layer, and the inactive layer.

The armour layer is one grain diameter in thickness and occurs at the bed surface. The active layer includes the armour layer and extends into the sediment control volume by depth, d_{EQ} which is the minimum water depth required for a particular grain size to be immobile. This depth is calculated by combining Manning's, Strickler's, and Einstein's equations:

$$u = \frac{1.486\, R^{2/3} S_f^{1/2}}{n}$$

$$n = \frac{D^{1/6}}{29.3}$$

$$\psi = \frac{\gamma_s - \gamma_f}{\gamma_f}\, \frac{D}{h S_f}$$

in which:

  R is hydraulic radius

  D is grain size and

  γ_f and γ_s are fluid and sediment density and

  ψ is from Einstein's bedload function.

For a condition of no transport, $\psi = 30$. Given a sediment density of 2.65 and replacing R with the depth of flow, h, these equations and constants may be combined to give:

$$d_{EQ} = (q/10.21D^{1/3})^{6/7}$$

where q is discharge per unit width. The active layer thickness is then given by $(d_{EQ} - h)$ and the grain size which determines d_{EQ} is the smallest size which is stable in the armour layer.

Discharge, flow depth, and particle size all change from reach to reach and with time. Each change in one or more of these variables changes the active layer thickness resulting in the exchange of particles between the active and inactive layers. Previous armour layers are maintained but once an armour layer is destroyed, its particles are treated as being uniformly mixed through the active layer in that particular reach.

The extent to which an armour layer develops depends on the flow characteristics and the particle sizes present in the bed. For a given size, the depth of bed material, d_{se} which must be removed to develop a surface layer one grain diameter thick is given by

$$d_{se} = \frac{2}{3}\frac{SAE.D}{P_i}$$

where SAE is the proportion of the bed area exposed to potential scour.

The program designates the sediment between the bed layer and the equilibrium depth as active and the sediment below equilibrium depth as inactive. Only active zone material is liable to scour. When all material is removed from the active zone, the bed is armoured for that flow condition. With a sediment mixture, the rate of armouring is proportional to the volume of material removed and the surface area exposed for scour is given by:

$$SAE = \frac{VOL_A}{VOL_{se}}$$

in which:

  VOL_A is the volume of sediment remaining in the active zone and

  VOL_{se} is the total volume of the active zone.

The stability of the armour layer under a particular flow condition is determined from a relationship showing the probability that grains will stay as function of the ratio of critical to actual tractive force (Figure 7.8). The relationship is based on the work of Gessler (1970) and the tractive forces, τ_0 and τ_c, are calculated as:

$$\tau_0 = \gamma_f h S_f$$
$$\tau_c = 0.047(\gamma_s - \gamma_f)D$$

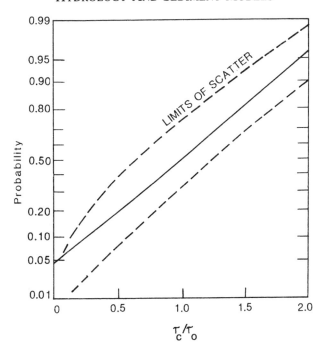

Figure 7.8. Function showing probability that a grain on the bed will not move. (From Thomas, 1982 and reproduced by permission of John Wiley and Sons)

The probabilities are weighted by the proportion of each grain size present in the bed, P_i and are then used to calculate a bed stability coefficient, BSF:

$$BSF = \frac{\Sigma \, p_i^2 P_i D_i}{\Sigma \, p_i P_i D_i}$$

Gessler (1970) recommends a BSF value of 0.65 for a stable armour layer and this is used in the program to indicate when the armour begins to break up.

The ability to handle the armouring process and situations where there is an undersupply of sediment load is a major improvement over earlier sediment routing models and a real necessity in a generalized model. Another advantage of HEC-6 is its attempt to model the behaviour of sediment mixtures. To deal with this situation, a sediment transport potential is calculated for each size class in the bed, irrespective of the amount present. The actual transport rate is then calculated according to the proportion of each size class present, thus:

$$q_s = \sum_{i=1}^{n} q_{si} P_i$$

in which q_{si} is the transport potential in size class i and P_i is the proportion by weight size class i in the active layer.

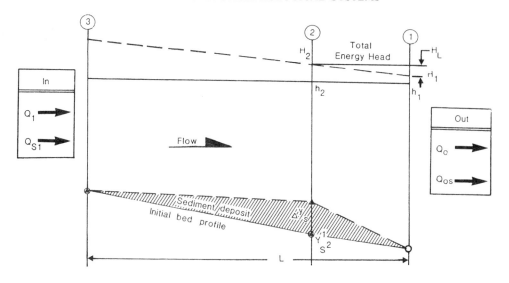

Figure 7.9. Finite difference scheme for bed elevation calculations used in HEC-6. (From Thomas, 1982 and reproduced by permission of John Wiley and Sons)

The effects of armouring on the transport relationship depend on the equation used. In the Einstein procedure, the hiding factor adjusts transport capacity to allow for armouring. In other transport equations, a correction factor is calculated based on the bed area exposed to scour and the transport capacity in excess of that required to transport delivered load. The correction factor is essentially based on guesswork.

HEC-6 begins by calculating the flow profile. A steady subcritical discharge is assumed for both main channel and tributary inflows so computations are carried out in the upstream direction. A stage–discharge rating table must be specified for the downstream boundary of the system. The flow profile is determined by solving the gradually varied flow equation and the standard step method is used. The results of the hydraulic computations are stored for the sediment transport calculations which begin at the upstream end of the system. Sediment inflows are determined from a set of user-specified sediment discharge rating tables which describe the input of each grain size used in the model. Alternatively, they may be calculated from the flow characteristics and the bed composition using a transport equation. The sediment transport calculations proceed downstream reach by reach: armour layer stability is tested; equilibrium depth, equilibrium bed elevation, and active layer thickness are determined; sediment is exchanged between the active and inactive layers and the effective size of the bed material is calculated. The transport potential can now be determined from the selected transport equation and the actual transport rate calculated by solving the sediment continuity equation. This is done using an explicit finite difference form (Figure 7.9) such that:

$$y_s^t = y_s^{t-1} + \frac{\Delta}{w_b}(\frac{\overline{Q}_{sI} - \overline{Q}_{sO}}{0.5L} - q_{st})$$

in which

y_s^t is the bed elevation at timestep t;

y_s^{t-1} is the bed elevation at time step $t-1$;

$\overline{Q}_{sI}$ is the sediment inflow volume averaged over timestep, Δt;

$\overline{Q}_{sO}$ is the time averaged sediment outflow;

L is reach length;

q_{sl} is lateral sediment inflow volume; and

W_b is bed width.

A variable bed gradation during the time step, Δ is used in evaluating this equation. This is done by calculating Q_{sO} over a number of subintervals and then integrating over the time step.

While HEC–6 is a widely used model, only a small number of tests have been performed to see how accurate it is. Figure 7.10(a) shows the result of one such test carried out in a flume with a sand bed containing a step. The model reproduces the downstream migration of the step and gives a fairly accurate representation of the water surface profile. It does not, however, reproduce the detailed pattern of local variations in the profile. A second test is shown in Figure 7.10(b) which illustrates measured and modelled changes in bed elevation just upstream of a dam removal site on the Clearwater River in Idaho. Once again, the general pattern is reproduced but the detailed results are of only limited accuracy.

While large general models like HEC–6 appear attractive, they have their limitations, some of which are not immediately obvious. These include the amount of data required, the difficulties of approximating a continuous process with a finite difference scheme, and a tendency for system behaviour to be determined artificially by the time and distance steps.

It is virtually impossible to collect enough information to run HEC–6 properly. Few rivers have sediment discharge rating curves for individual particle sizes and the costs of collecting such data are very substantial. It is also rare to have data on the depth of bed-material along a reach making it necessary to carry out a drilling program or seismic profiling. Another costly issue is the need to survey cross-sections which are spaced closely enough for an adequate description of channel geometry. If modelling is to be accurate, each pool and riffle, each contraction and expansion must form part of the input. HEC–6 can, of course, be run with less data by using default parameters in the model or by incorporating estimates of them. If this is done, a sensitivity analysis should be carried out to determine how and to what extent errors in parameter estimates might affect model output. Typically, such analyses show that some parameters have little effect on model performance. They also show that changing a parameter may sometimes have quite unexpected effects on model output.

The problems of using finite difference schemes to approximate a continuous process like erosion and deposition are well known. Models have a tendency to become unstable, particularly when long reaches or large time steps are used in the computations. The modeller may then find himself in the difficult position of trying to force water to flow uphill. Such problems can sometimes be overcome by the use of smaller time and distance steps but only at the expense of increased data collection costs and the need for more computer time. It is, however, much better to use a finite difference scheme formulated to be stable. The implementation of these schemes is not always simple and discussions of the problem may be found in Cunge and Perdreau (1973) and Ponce et al. (1979).

Another problem of the finite difference scheme is the tendency for model behaviour to be governed by the time and distance steps chosen. The scheme in HEC–6 effectively imposes a velocity of sediment movement on the system. This can be tolerated if we are only interested in the long-term results of change. It can also be tolerated when there is an even and continuous supply of sediment and enough sediment is stored in the bed to accommodate potential scour in every reach. If, however, the area of interest is short-term changes, such as those which occur as a result of a discontinuous sediment input or during the passage of a slug of material through the system, the problem becomes more serious. Modelling results are then sensitive to the time and distance steps used and the downstream passage of sediment can be controlled by the numerical solution scheme used rather than by hydraulic conditions and sediment properties. Since very little is known about sediment velocities, getting the time and distance steps correct is very much a matter of luck. The result is that HEC–6 does not perform well when modelling transient conditions.

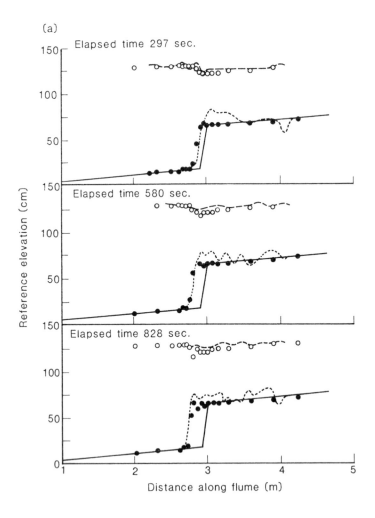

Figure 7.10(a). (*Caption opposite*)

7.4.3 Finite Difference Approaches — the Sequential Routing Method

Sequential routing models incorporate unsteady flow and in some cases use finite difference schemes formulated to avoid the problem with sediment velocity described above. Consequently, they offer a more dynamic approach to the modelling of erosion and deposition and have greater potential for handling transient conditions. A number of sequential routing models have been published (see, for example, Bennett and Nordin, 1977). Here we concentrate on the approach described by Alonso et al. (1981) since the FORTRAN code for the model is available. This approach is also described in Borah et al. (1982a, b) and is hereafter referred to as the SR model.

The basic flow equations in the SR model are:

$$\frac{\partial Q}{\partial t} + \frac{\partial \beta Q u}{\partial x} + gA\frac{\partial h}{\partial x} = gA(S_o - S_f + \frac{q_\ell u_\ell}{gA})$$

Continuity equation for water:

$$\frac{\partial Q}{\partial x} + \frac{\partial A}{\partial t} = q_\ell$$

Flow resistance:

$$u = C\sqrt{RS_f}$$

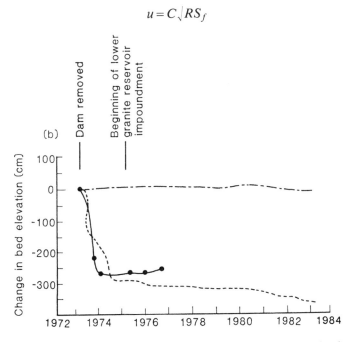

Figure 7.10. Tests of the HEC–6 model in a flume and a river. (a) shows changes over time in a sand bed flume (from Thomas and Prasuhn, 1977 and reproduced by permission of Journal of the Hydraulics Division, American Society of Civil Engineers). (b) shows observed and modelled results of the removal of a dam and subsequent erosion of accumulated sediments (from Thomas, 1982 and reproduced by permission of John Wiley and Sons)

in which:

Q is discharge

t is time

β is the momentum coefficient

u is velocity

g is the acceleration of gravity

A is cross-sectional area

h is flow depth

x is longitudinal distance

S_o is bed slope

S_f is friction slope

q_ℓ is lateral inflow

u_ℓ is the velocity component of lateral inflow in the main flow direction

C is the Chezy coefficient and

R is hydraulic radius.

Complete solution of these equations is difficult so a diffusion or kinematic wave approximation is commonly used. The SR model uses the kinematic wave approximation in which energy gradients due to local and convective acceleration are negligible in comparison with gravitational and frictional effects. The momentum equation then becomes:

$$S_o \cong S_f$$

which is reasonable for relatively steep streams. A number of algorithms for solving the kinematic wave approximation exist. The one used here is the shock fitting technique of Borah *et al.* (1980).

The sediment continuity equation in the SR model is stated as:

$$\frac{\partial Ac}{\partial t} + \frac{\partial Q_s}{\partial x} + (1-\lambda) W_b \frac{\partial z}{\partial t} + \frac{\partial r}{\partial t} = q_{s\ell}$$

in which:

c is sediment concentration

Q_s is the sediment volume flux

X is bed porosity

W_b is bed width

z is bed elevation

r is the net sediment volume flux through the suspended load-bed load interface, and

$q_{s\ell}$ is the lateral influx of sediment.

This formulation is standard except for the term $\dfrac{\partial r}{\partial t}$ which represents the time rate of net local exchange between the suspended load and bed load zones and is given by:

$$\frac{\partial r}{\partial t} = k_0 A \{ (C-c)r - k_1(R-r)c \}$$

in which:

   k_0 and k_1 are constants

   C is the sediment concentration at transport capacity and

   R is the net sediment exchange between the bed and suspended load zones at transport capacity.

This equation may be solved in the following manner (Alonso *et al.*, 1981). Very near equilibrium the right hand side becomes very small, i.e.:

$$k_0 A\{(C-c)r - k_1(R-r)c\} \cong 0$$

At the same time, C deviates from its equilibrium value so that from the previous equation:

$$\frac{\partial r}{\partial t} = \frac{k_1 RC}{\{C + (k_1 - 1)c\}^2} \frac{\partial c}{\partial t}$$

By approximating Q_s with Auc over a short time increment, the sediment continuity equation may then be written as:

$$\frac{\partial c}{\partial t} + v_s \frac{\partial c}{\partial x} = \frac{q_{st}}{A(1 + f_s)}$$

in which v_s represents the celerity of a sediment wave and is given by:

$$v_s = \frac{u}{(1 + f_s)}$$

and:

$$f_s = \frac{k_1 RC}{A\{C + (k_1 - 1)c\}^2}$$

$$q_{st} = q_{s\ell} - (1 - \lambda) W_b \frac{\partial z}{\partial t}$$

The term k_1 seems to be of the order 10^{-3} so f_s is approximated by:

$$f_s \cong \frac{CEL}{AC(1 - 0.999 c/C)^2}$$

where CEL is a calibration parameter used to control sediment velocity through the model so the arrival times of observed and modelled sediment peaks match each other.

    The sediment continuity equation described above is solved by the method of characteristics. Details of the solution and the grid scheme used are available in Alonso *et al.* (1981). An important feature of the model and the solution is the inclusion of sediment velocity. This makes it possible to reproduce the observed tendency for sediment waves to lag behind flood waves particularly when a significant amount of transport as bed load occurs.

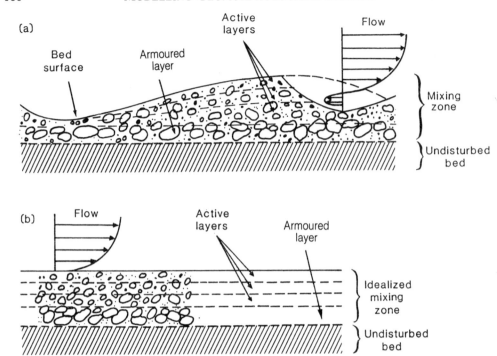

Figure 7.11. Schematic representation of bed processes used in the SR model. (From Borah *et al.*, 1982a and reproduced by permission of the *Journal of the Hydraulics Division*, American Society of Civil Engineers)

The SR model offers a choice of sediment transport equations including those of Yang (1973) for sands, a duBoys type relationship (Graf, 1971) for fine gravel, and the Meyer-Peter and Muller (1948) formula for medium and coarse gravel. It is also capable of handling sediment mixtures. The mixture algorithm is similar to that of Higgins (1979) and uses the idea of residual transport capacity which may be defined as the ability of the stream to carry additional load of a particular size fraction over and above the size fractions already present in the flow. For a flow carrying a particular concentration, c_1 of uniform material of size, D_1, the residual transport capacity for that size, C_{r1}, is given by:

$$C_{r1} = C_1 - c_1 = C_1 - C_1(c_1/C_1)$$

where C_1 is the transport capacity for material of size, D_1. For two sizes of material, D_1 and D_2 present in the flow, the residual transport capacity for size, D_1 is:

$$C_{r1} = C_1 - C_1(c_1/C_1) - C_1(c_2/C_2) = C_1(1 - \sum_{i=1}^{2}(c_i/C_i))$$

The expression may be generalized for any size fraction, D_i and number of load fractions present, $c_1, c_2 \ldots c_n$ as:

$$C_{ri} = C_i\{1 - \sum_{j=1}^{n}(c_j/C_j)\} \text{ for } i = 1, 2 \ldots n$$

The term within the brackets, { } represents the proportion of the potential transport capacity, C_i, taken up by all size fractions in transit. C_{ri} indicates the potential erosion or deposition in a particular size fraction.

The interaction between the flow and the bed is modelled in a more complex way than in HEC–6 although there are similarities. As bed forms move downstream, they mix the bed material down to a particular depth, below which the bed material remains undisturbed (Figure 7.11). The depth of bed affected by bed forms is regarded as a mixing zone. Where a range of particle sizes is present, the slowly moving coarse material tends to accumulate at the base of the mixing zone. When flow conditions are such that the coarse material becomes immobile, finer materials may be washed out leaving an armour coat. During the scouring of this finer material, exchange between the bed and the flow occurs in a thin layer at the bed surface known as the active layer. Several of these layers may be removed during scour. Alternatively, a number of active layers will be deposited during aggradation.

In modelling the behaviour of the bed, the mixing zone is treated as a band of constant thickness made up of several layers, the uppermost of which is the active layer. When all material in the active layer moves, its thickness, ALT, is given by:

$$ALT = \frac{D_n}{P_n(1-\lambda_n)}$$

in which D_n, P_n and λ_n are the size, fraction and porosity of the coarsest material present. When the active layer is an armour coat, its thickness may be considerably less because particles finer than D_n may make up the armour layer. In the case of full or partial armouring, ALT is determined as:

$$ALT = \frac{1}{\sum_{i=\ell}^{n} P_i} \cdot \frac{D_\ell}{1-\lambda_\ell}$$

where D_ℓ is the smallest fraction the flow cannot transport. If deposition occurs, the new material is added to the bed and a revised active layer thickness is calculated for the new sediment mixture.

Over time, the size composition of the active layer varies and must be kept track of. This is done using an accounting procedure based on an entrainment frequency matrix. The matrix is time invariant and is derived in the following manner. Assume that the bed layer contains three particle sizes, $D_1 < D_2 < D_3$ and the flow has sufficient transport capacity to erode all three fractions. The D_1 particles at the bed surface are the first to be removed. The next to be entrained are the D_2 particles at the bed surface followed by the D_1 grains hidden beneath them. Finally the D_3 grains shift followed by the D_2 and D_1 grains hidden beneath them. This sequence can be generalized in the form of an entrainment frequency matrix, $\underset{\sim}{F}$.

$$
F = [F_{ij}] =
\begin{matrix}
1 & 0 & 0 & 0 & - & 0 & - & 0 \\
1 & 1 & 0 & 0 & & & & 0 \\
2 & 1 & 1 & 0 & & & & 0 \\
4 & 2 & 1 & 1 & & & & 0 \\
8 & 4 & 2 & 1 & & & & 0 \\
16 & 8 & 4 & 2 & & & & 0 \\
 & & & \cdot & & & & \\
 & & & \cdot & & & & \\
2^{i-2} & 2^{i-3} & 2^{i-4} & 2^{i-5} & - & 2^{i-j-1} & - & 1
\end{matrix}
$$

in which $j = 1,2 \ldots N$, where N is the number of size fractions. The elements of any column in this matrix express the number of times the D_j fraction is depleted due to the entrainment of fractions $D_j, D_{j+1} \ldots D_n$. The lower off-diagonal elements in each row indicate the number of times the D_j fraction becomes available once D_i is removed from the bed surface.

The amount of eroded material corresponding to the frequencies F_{ij} depends on the volume of each fraction present in the active layer. The equation to obtain the amount of material available for entrainment is:

$$
v_{ij} = \frac{F_{ij} P_i}{\sum\limits_{r=j}^{N} F_{rj} P_r} \cdot V_j, \qquad i,j = 1,2 \ldots N
$$

where:

$$
P_i = \frac{100 \, V_i}{\sum\limits_{k=1}^{N} V_k} \qquad i = 1,2 \ldots n
$$

in which:

v_{ij} is the proportion of volume V_j which becomes available for entrainment when the D_i fraction is eroded

V_i is the volume occupied by fraction D_i

P_i is the percentage of active layer material in fraction D_i.

These equations define a volume entrainment matrix $v = [v_{ij}]$. The diagonals of this matrix represent the volumes of individual fractions on the bed surface. The off-diagonal row elements contain the volumes in each fraction exposed by the erosion of larger sizes. During erosion, the diagonal elements are depleted first, beginning with the finest fractions. After this the off-diagonal elements are scoured, once again starting with the finest fractions. When the residual transport capacity of the flow is not able to remove a whole row, equal amounts are removed from each element of the row. The calculated volumes removed from the matrix v may be added to the flow or reduced by a factor which corrects for different levels of detachability. This factor is a model calibration parameter and is obtained by fitting model sediment yield volumes to observed data.

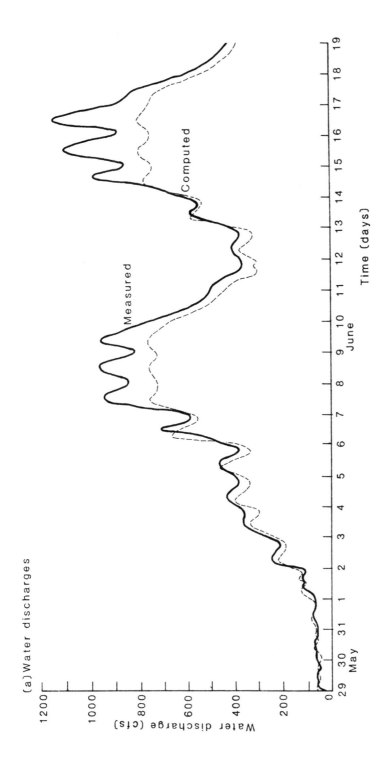

(a) Water discharges

Water discharge (cfs)

Time (days)

Measured

Computed

May 29 30 31 1 2 3 4 5 6 7 8 9 10 11 12 13 14 15 16 17 18 19
June

Figure 7.12a.

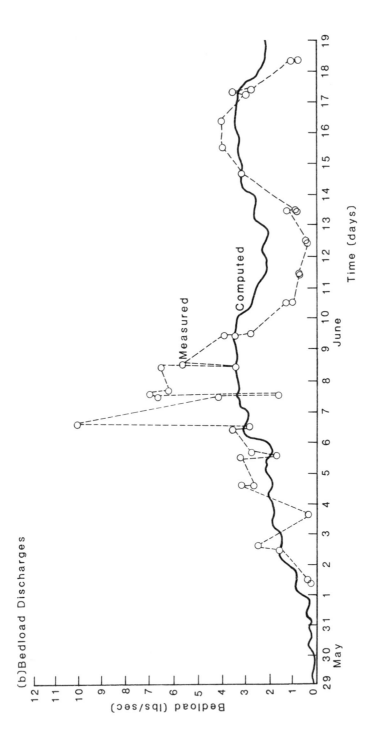

Figure 7.12b.

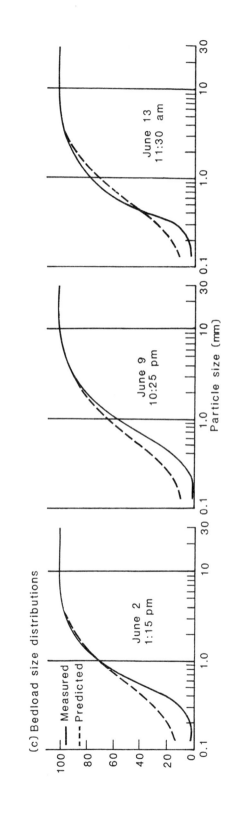

Figure 7.12. Observed and predicted water discharge, sediment discharge and sediment load size at the lower end of the East Fork River study reach as predicted by the SR model. (From Borah *et al.*, 1982b and reproduced by permission of the *Journal of the Hydraulics Division*, American Society of Civil Engineers)

When deposition occurs, material which settles out is added to the active layer starting with the largest size fraction in the flow and continuing with progressively finer fractions. Whether material actually reaches the bed or not depends on its average fall velocity. The actual deposition of a size fraction, G_{di}, during the time interval, Δt is calculated as:

$$G_{di} = \begin{cases} C_{ri} & \text{if } \beta \geq 1 \\ \beta C_{ri} & \text{if } \beta < 1 \end{cases}$$

where:

$$\beta = 2 w_i \Delta t / h$$

in which

C_{ri} is the residual transport capacity for size fraction, i
w_i is the fall velocity of the size fraction
h is the mean flow depth.

Tests of the performance of the SR model have been published by Alonso *et al.* (1981) and Borah *et al.* (1982b). The most detailed of these tests involved an attempt to simulate behaviour of fourteen reaches of the East Fork River, a small sand and gravel-bed stream in Wyoming, over a 22 day period in 1975. A 30 minute timestep was used in the computations which involved 10 separate particle sizes. Some results of these tests for section B–17 at the lower end of the system are shown in Figure 7.12.

Broadly speaking, the SR model performs fairly well for a sediment model. Observed and predicted flows are roughly in accordance. The discrepancies arise at high flow because overbank flow was not taken into account. There could also be errors because of the use of a constant roughness coefficient in the hydraulic calculations. The measured and predicted bedloads are virtually identical over the full 22 day period but short-term fluctuations are modelled very poorly. This occurs with virtually every run with a sediment model and the problem is discussed in the next section. The observed and predicted bedload size distributions show some differences but, again, are quite reasonable for a sediment model given the approximate nature of the bed behaviour model.

7.4.4 The Braided River Model—A Sediment Flow Approach

HEC–6 and the SR model are good examples of the detailed approach to sediment modelling in which an attempt is made to simulate as many of the processes involved in sediment routing as possible. There are, however, situations in which the physical processes are not fully understood or cannot be simulated with the level of precision required. Under these circumstances it may be necessary to use a 'grey box' scheme or to replace some parts of the model with a lumped parameter structure.

The braided river (BR) model (Pickup and Higgins, 1979; Pickup *et al.*, 1983) is typical of this approach and consists of a braided river hydraulics algorithm, a sediment transport relationship and a sediment flow model. Its structure was derived from a consideration of the physical processes but because many of the parameters needed for a conventional sediment routing approach could not be obtained, some of the detailed internal behaviour

Figure 7.13. The Kawerong River below the Bougainville Copper Mine waste dump. The BR Model suits this situation in which a rapidly changing braided river occupies the whole valley floor. (Photo: G. Pickup)

is represented by stochastic processes. There is also a need for the model to be calibrated. The BR model differs from the conventional approach in another way. HEC-6 and the SR model allow the flow of water and the channel characteristics to drive the sediment flow. The BR model, by contrast, models the sediment flow and allows it to control deposition and the rate of sediment discharge within the limits imposed by transport capacity. The sediment flow model is based on physical considerations but uses a lumped parameter scheme to describe system behaviour.

The BR model originated from the need to predict the pattern of movement and deposition of mining waste in the Kawerong River of Papua New Guinea. This river occupies a steep narrow valley with a gradient of 7–2 per cent and in the period 1968–1976 received 226 million tonnes of sediment. Much of this material was fine and passed through the river as washload producing sediment concentrations of 100,000–300,000 ppm. About 46 million tonnes of the total was made up of relatively coarse material eroded from waste rock dumps. This material included gravel, cobbles and boulders and entered the Kawerong River as a series of slugs. These slugs passed downstream as waves filling the Kawerong valley

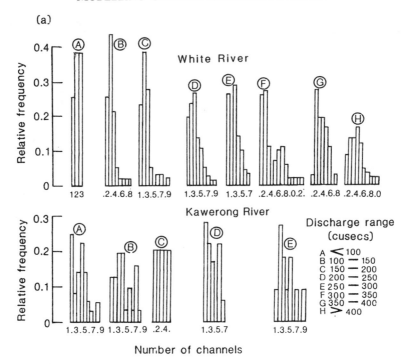

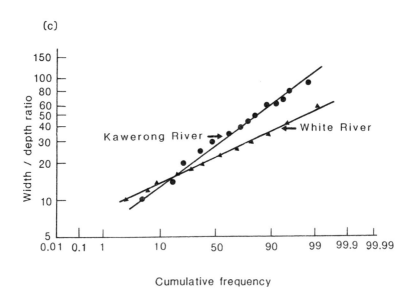

Figure 7.14. (*Caption opposite*)

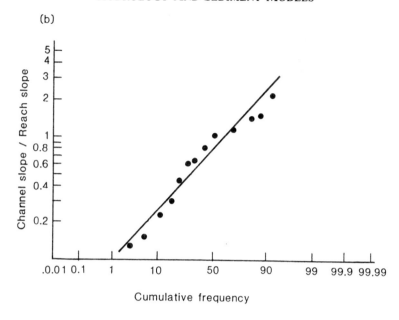

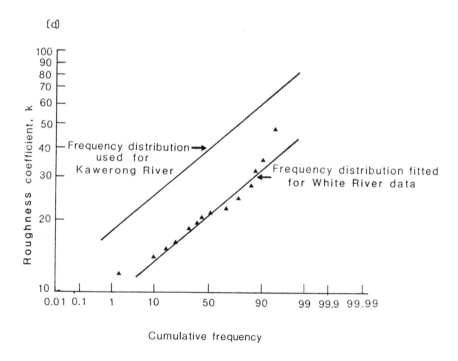

Figure 7.14. Data and functions used in the braided river hydraulics algorithm of the BR model. (From Pickup, 1979 and reproduced by permission of Elsevier Scientific Publishing Co.)

with either temporary or long-term deposits (Figure 7.13) and changing its channel from a narrow, boulder strewn gully to a braided river occupying the whole width of the valley. The braided channels change form and shift position very rapidly making measurement extremely difficult and flow velocities of up to $3 \, m^{-1}$ are not uncommon.

The Kawerong cannot be modelled with the approach used in HEC–6 and the SR model because it represents a different type of behaviour. The original channel has been drowned in sediment and as the slugs of input material pass through the system, they appear to generate their own geometry. Changes in the river are therefore supply-controlled rather than affected by the transport capacity imposed by a pre-existing channel geometry. There is also the problem of modelling the hydraulics and sediment transport processes of a rapidly changing braided river.

The solutions to the flow equations used in HEC–6 and the SR model apply to single channel rivers or to simple branching situations such as mid-channel islands. They cannot be used in the complex, rapidly changing situations found in active braiding rivers. The only method which seems to be available to produce estimates of hydraulic parameters for sediment transport calculations in such systems is that of Pickup and Higgins (1979). In this method, a series of simulation procedures based on empirical data from the Kawerong and White Rivers is used to produce a set of representative values for each reach and time increment used in modelling. The first step in the method is to select the number of channels in a reach from the observed frequency distribution (Figure 7.14). The flow is then proportioned out among the channels according to the relationships:

$$q_n = r_n Q \qquad \text{for } n = 1$$

$$q_n = r_n \left(Q - \sum_{j=1}^{n=1} q_j \right) \text{ for } n < m$$

$$q_n = Q - \sum_{j=1}^{m-1} q_j \qquad \text{for } n = m$$

where Q is total discharge, r_n is a random number ($0 \geq n \leq 1$), q_n is the discharge in channel, n, and m is the total number of channels in the cross-section. Slope, width–depth ratio, and the channel slope/reach slope ratio for each channel are then generated from empirical log normal frequency distributions and the results substituted into the Strickler equation to produce the remaining hydraulic parameters required by most transport equations (Figure 7.14). A number of transport relationships were tried with the braided hydraulics algorithm (Pickup and Higgins, 1979). The best was found to be the Meyer-Peter and Müller (1948) equation which predicted the observed transport in eight reaches of the Kawerong River over a two year period to within -31 per cent to $+39$ per cent with a mean absolute error of 16 per cent.

The braided hydraulics algorithm is not suitable for short-term simulation and only produces representative parameters when it is run for a significant period of time such as a month. It can therefore only produce average transport rates. This could be a problem if a standard sediment routing approach using small time steps is required. When it is used with the sediment flow approach on the Kawerong River, in fact, it produces better results than the normal approach.

The sediment flow model was derived from the observation that when a slug of material entered the system from the waste rock dumps, it passes downstream as a travelling wave of deposition. Similarly, if sediment input is reduced, existing deposits in the upper part of the river are scoured and shifted downstream. This suggests that the sediment in the Kawerong can be treated as a slowly moving system and that the deposits are basically sediment in transit although the movement is intermittent. This conceptual approach can be used because the Kawerong River is braided and channel migration makes it possible to erode or deposit material on every part of the valley floor with equal frequency.

Although sediment may move through the system, wave shape is not constant. Instead, because not all sediment moves at the same velocity, dispersion occurs. There are four main sources of dispersion in the Kawerong:

1. Sediment particles of a given size travel in a series of steps with intervening rest periods. Neither step length nor rest period duration are constant.
2. The range of different particle sizes present produces a range of sediment velocities.
3. Only sediment at the bed surface moves; material which is buried cannot move until it is exposed to the flow by erosion of material above.
4. Only those sections of the valley occupied by active channels contain moving sediment; other deposits are stationary until channel shifting or the development of new channels exposes them to flow.

The behaviour of a moving wave of material can be described by the one-dimensional diffusion equation (Taylor, 1954). This equation may be applied to bed material movement when formulated as:

$$\frac{\partial s}{\partial t} + v_s \frac{\partial s}{\partial x} = k \frac{\partial^2 s}{\partial x^2}$$

in which:

s is the bed material moving through the system at a given point expressed as volume per unit time

t is time

v_s is sediment velocity

x is distance and

k is the dispersion coefficient

Given a series of step inputs of sediment at the top end of the system, $g_1, g_2, g_3 \ldots, g_n$ and without allowing for material already in the system at $t = 0$, the equation for the volume of sediment moving down the river at point, X, and time, t, becomes:

$$s(X,t) = g_1 \left(erf\left[\tfrac{1}{2} X\{k(t-t_1)\}^{-\frac{1}{2}} \right] - erf\left[\tfrac{1}{2} X(kt)^{-\frac{1}{2}} \right] \right)$$

$$+ g_2 \left(erf\left[\tfrac{1}{2} X\{k(t-t_2)\}^{-\frac{1}{2}} \right] - erf\left[\tfrac{1}{2} X\{k(t-t_1)\}^{-\frac{1}{2}} \right] \right)$$

$$+ g_3 \left(erf\left[\tfrac{1}{2} X\{k(t-t_3)\}^{-\frac{1}{2}} \right] - erf\left[\tfrac{1}{2} X\{k(t-t_2)\}^{-\frac{1}{2}} \right] \right) + \ldots$$

in which $X = x - v_s \Delta t$. The term, $s(X,t)$ is the volume of sediment per unit time moving through the system. It is converted to the area of sediment occupying each valley cross-section by the relationship:

$$A_s(X,t)=s(X,t)/v_s(X,t)$$

where v_s is the sediment velocity.

The sediment flow model described above contains two parameters for which no method of estimation is available. These are the sediment velocity, v_s and the dispersion coefficient, k. In the absence of any precedent, v_s was determined from the relationship

$$v_s=c\log_{10}Q_s$$

in which c is a calibration parameter and Q_s is the monthly sediment transport capacity for a particular reach obtained from the braided river hydraulics algorithm assuming a constant sediment mixture. Little is known about the likely behaviour of k although it should depend on the particle size distribution of the sediment input and the rate of transport. It was therefore treated as a constant for the sake of simplicity.

The BR model describes the behaviour of only that part of the load which moves completely through the reach in the time span covered by modelling. In the upper cross-sections of the Kawerong, bed behaviour oscillated about a stationary mean value and all material could be considered to be moving with the wave. In the lower cross-sections deposition occurred which was permanent at the timescale of the model. Wave movement was superimposed on a rising bed level trend. Q_s values from the braided hydraulics algorithm were used to separate the moving from the permanently deposited components of material in each reach. If input was greater than reach transport capacity some of it would be deposited and held in storage. This material was released in future months when input was less than reach transport capacity.

Changes in slope occur as a wave of deposition moves down the system or where there is a long-term buildup of material. In theory these changes should affect the sediment transport calculations. In practice, because the Kawerong River is so steep, the changes in slope which occur in response to deposition are very small so the error introduced by ignoring them is negligible compared with other sources of inaccuracy. A constant slope was therefore assumed for each reach, considerably simplifying the calculations.

Calibration of the BR model was accomplished by using an automatic search algorithm and by setting it up in the form:

$$A_s(CALC)=fn[c,k,g(1),g(2),\ldots,g(n),Q_s(1),Q_s(2),\ldots,Q_s(n)]$$

in which the calibration parameters are c and k. These may be varied and when the model is run with a particular c and k value, an objective function or goodness of fit measure can be calculated. This usually has the form:

$$F=\sum|A_s(CALC)-A_s(OBS)|^j$$

where $A_s(CALC)$ and $A_s(OBS)$ are the calculated and observed values of deposition area and j is an exponent, usually with a value of 1 or 2.

The behaviour of the sediment flow model is illustrated in Figure 7.15 which shows how a single slug input of material moves through a uniform system at a constant discharge for various values of c and k. During the input period, Δt, the area of deposition remains

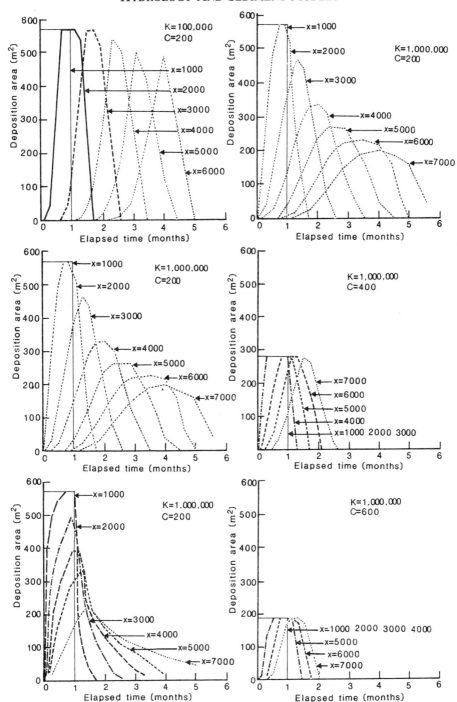

Figure 7.15. Effects of changing parameters on output from the BR model; x denotes distance in metres from the sediment input point at the upstream end of the system. (From Pickup *et al.*, 1983 and reproduced by permission of Elsevier Scientific Publishing Co.)

constant for those reaches where $v_s \Delta t > X$. Further downstream the pattern depends on the dispersion coefficient. For small values of k, the rectangular slug takes on a triangular shape in which height decreases and the base becomes more broad. For larger values of k, the transition from a rectangular to a triangular pattern occurs quickly and the pattern then takes on the shape of a normal distribution. For even larger values of k the transition from a rectangular to a triangular slug occurs quickly and a long downstream tail develops. The slug rapidly flattens and widens with time and distance. Changing the velocity parameter, c, has three effects. Firstly, the area of deposition is modified irrespective of time or distance. This occurs because rapidly moving sediment occupies a smaller area than slow moving material. The second effect is the obvious one where the sediment velocity determines the location of the moving wave of material in the system. The third effect occurs because dispersion increases with time. If the sediment moves through the system slowly, more dispersion will occur by the time it reaches a given point than if the sediment travelled more quickly.

Figure 7.15 indicates that the dispersion model can reproduce a wide range of patterns in the movement of sediment. It also indicates that a change in c can produce similar effects to a change in k. Therefore, when the model is calibrated, its parameters are not independent of each other and there may be more than one 'optimal' set of parameter values.

Because the dispersion model requires calibration, its performance was assessed using a split record approach in which half of the input data were used to fit model parameters and the other half to test the accuracy of the results. The dispersion model was also compared with a conventional model based on the sediment continuity equation but incorporating the effects of sediment velocity. Details of this second model are given in Pickup *et al.* (1983). The results are shown in Figure 7.16 which reproduces the basic pattern along the river in which deposition increases downstream. It also reproduces the basic pattern through time in which deposition gradually increases in all sections. The quality of the results varies from one period to another. Months 6–20 show fluctuating scour and fill in the upper reaches when, in fact, the amount of deposition remained fairly steady. Further downstream, the model shows extensive scour but in reality, deposition occurred. These errors are due to deficiencies in input data. After month 20, model performance improves. Not only is the general pattern of scour and fill reproduced but so also are the individual waves of sediment moving down the system. Only cross-section 6 is poorly modelled. The other sections show good results.

The sediment continuity model produces results which are much poorer than the BR model. While the overall volume of sediment transported is of the correct order of magnitude, the model does not predict the basic pattern of deposition through time. It also fails to predict downstream variations in deposition and on many sections shows fluctuations which did not occur.

7.5 GENERAL PROBLEMS OF SEDIMENT ROUTING MODELS

7.5.1 Relationships between Flow Conditions and Bed-material Load

The rate of transport in a catchment or channel sediment routing model is obtained from sediment balance calculations which compare the transport capacity of the flow with the sediment supply from upstream and the amount of material available for entrainment on the bed. The sediment transport capacity is determined from transport equations which creates a number of problems.

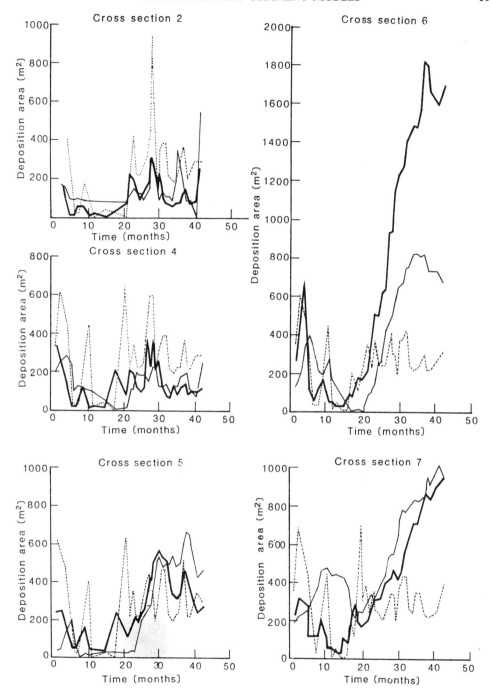

Figure 7.16. Observed and calculated deposition in cross-section of the Kawerong River using the BR model and a standard sediment continuity model. (From Pickup *et al.*, 1983 and reproduced by permission of Elsevier Scientific Publishing Co.)

The first difficulty in using a transport equation is to select the correct one. Most have been derived for a limited range of flow and sediment load conditions and have only been tested extensively in flumes. When such equations are applied in the field, they frequently do not work and errors of several orders of magnitude are common. The errors arise for many reasons, the most common of which are:
— The equation does not apply to the hydraulic conditions or sediment load for which it is being used
— The equation cannot handle a particular sediment mixture
— Spatial and temporal variations in sediment supply.

The problem of selecting the correct equation can be partly overcome by using only those which have been tested in situations similar to the one being modelled. Examples of such tests spanning a wide range of conditions include those presented by Vanoni *et al.* (1961), Higgins (1979), Bagnold (1977) and Alonso *et al.* (1981). It cannot, however, be fully overcome because, as Klingeman and Emmett (1982) suggest, there appear to be two bed sediment transport regimes: one for full bed mobility and the other when varying degrees of armour or other sediment supply limiting factors occur. The full-mobility conditions can be described reasonably well by transport equations as long as the correct flow and sediment parameters are known. The supply-limited condition is poorly defined and transport equations do not handle it well.

The errors introduced by using transport equations with sediment mixtures are largely unknown but probably vary with the equation, especially if it contains an empirical component. Transport equations handle mixtures in two ways. In the simpler equations, a representative or effective grain size such as d_{50} is used. This may be viable when all the sediment is in motion but when flow is closer to the threshold condition, not all particle sizes can be mobilized so d_{50} for the load is not the same as d_{50} for the bed (White and Day, 1982). In the more sophisticated procedures, sediment mixtures are allowed for by calculating a transport rate for each size fraction, multiplying it by the percentage of that fraction present in the bed-material and summing the result. This can produce excessively high sediment discharges because the transport rate of the finer fractions may be supply limited. The characteristics of a sediment mixture can also affect the threshold at which sediment is entrained. Andrews (1983) has shown that the critical dimensionless shear stress, τ_i, required to entrain a given particle size, d_i is equal to

$$\tau_i = 0.0834 \, (d_i / d_{50})^{-0.872}$$

for bed particles between 0.3 and 4.2 times the median diameter of the subsurface bed material. For particles greater than 4.2 times d_{50}, τ_i approaches a constant value of 0.02. Many transport equations use the Shields value of 0.06 for τ_i which can cause large proportional errors at low transport rates.

Sediment transport equations are based on the assumption that, at the time scale involved, there is a unique overall relationship between flow conditions and the transport rate. This is frequently not the case. Instead, both the transport rate and the transport relationship vary in time and space. This situation can be handled to some extent in a sediment flow approach like the BR model. It is more difficult to incorporate into conventional sediment routing schemes like HEC–6 and the BR model.

Temporal variability occurs at a number of scales and has been reported from humid environments-dominated by snowmelt floods (Emmett *et al.*, 1982) through to the arid zone where flash floods are the norm (Lekach and Schick, 1983). Temporal variability may result from both intrinsic or extrinsic causes.

The most common forms of intrinsic variability are those associated with the passage of bed forms. Emmett *et al.* (1982) report that, over the short term, up to fourfold variations in the transport rate may occur as a bedform moves through a cross-section. Since bedform changes occur at a wide range of temporal scales, they are likely to produce a wide range of temporal variability in sediment transport rates. Another common form of intrinsic variability is the loop rating curve in which sediment concentrations are higher at a given flow on the rising stage of a flood than during the falling stage. This is probably due to the gradual depletion of available sediment which produces an increasingly supply-limited condition over time. A reverse form of loop rating may also occur. Reid *et al.* (1985) suggests that long periods between floods allow the gravel framework to bed down and increase particle interlock. The next flood may then produce more bedload transport on the falling stage after rising stage flows have loosened the gravel and washed out the fines. There are also random as well as systematic forms of temporal variability. These result from factors such as bank collapse, local variations in shear stress and so on.

Extrinsic causes of temporal variability are also common. In many rivers discrete input events produce waves of sediment which travel downstream gradually dispersing as they move. Such behaviour is common in mountainous regions affected by landslides (Mosley, 1978) and in rivers receiving large volumes of mining waste as in the Kawerong River (see also Gilbert, 1917). Extrinsic variations in sediment input which involve very large volumes of material generate their own channel geometry. They are consequently difficult to model using the range of hydraulic conditions and the transport relationships previously existing in the river.

Spatial and temporal variations in the transport relationship and the transport rate may or may not be the same thing. When a wave of sediment travels downstream, they are closely related and it may be possible to predict the behaviour at one cross-section from that of its upstream neighbour so spatial and temporal behaviour are closely related. Usually, however, both sediment supply and variations in transport capacity due to longitudinal changes in channel geometry determine system behaviour. In some reaches where the time-averaged transport capacity is less than or equal the supply rate, distinct pulses or waves of material form and gradually move downstream. In other reaches where transport capacity is consistently higher than the sediment supply, the waves are destroyed, little change in morphology occurs and the sediment passes quickly through to the next reach in which short- or medium-term storage occurs.

7.5.2 Relationships between Channel Form and Erosion and Deposition

When the sediment supply and transport capacity of a system are not in equilibrium, the deficiency or excess of load is met by erosion or deposition at the channel bed and banks. This erosion or deposition changes the channel geometry, flow resistance and slope which, in turn, affects the transport capacity of system. It is therefore important to incorporate a relationship between the processes of scour and fill and the changes in channel form which occur in sediment routing models. The precise form of this relationship governs how sediment transport discontinuities travel through the system and how they are preserved or destroyed (see Pickup, 1976).

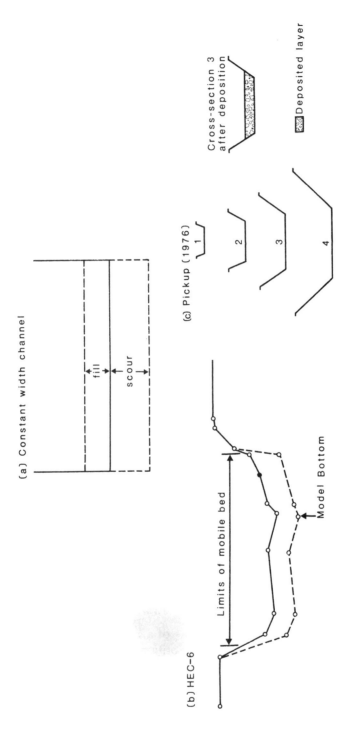

Figure 7.17. Schemes used to represent changes in channel cross-section used in sediment routing models

Most published sediment routing models are one-dimensional and ignore changes in channel form by assuming constant width. In these formulations only the bed level is free to vary (Figure 7.17) making solutions of the sediment routing equations relatively simple. The constant width assumption is acceptable in wide channels which are not susceptible to bank erosion. Elsewhere it is highly dubious, particularly if large changes in discharge or sediment load are involved or if substantial erosion or deposition occurs. Under these circumstances, one-dimensional models are likely to be inaccurate, particularly where the particle size distribution of the load is such that its transport rate is sensitive to changes in channel shape.

Only a few attempts have been made to incorporate the changes in channel shape which occur with erosion and deposition into sediment routing models. Most of these are restricted to narrow channels. Pickup (1976) used an empirical model derived from Crawford's Creek (Figure 7.17) in which the channel becomes wider and the banks more gentle as the bed incises. A more general approach based on a rectangular cross-section is used for rill erosion in the CREAMS model (Foster *et al.*, 1980).

The simple models of channel behaviour described above are only suitable for small streams with cohesive banks in which erosion or relatively minor deposition occurs. Where the banks are non-cohesive, quite different channel forms may result which none of the published sediment routing models are capable of handling.

Perhaps the biggest deficiency of the channel behaviour relationships used in modelling is their complete inability to describe what occurs when a very large increase in sediment load or change in discharge imposes a completely new channel geometry on the system. Changes of this magnitude are not uncommon when very large floods occur or when major slugs of sediment load enter the system and are typical of the type of problem faced in operational modelling. Clearly, more work is required in this area.

7.5.3 Problems of Routing Models

Most of the models described in previous sections are attempts to develop integrated mathematical descriptions of the behaviour of large systems by combining detailed submodels of their parts. Thus, a sediment routing model is essentially a cascade of flow and sediment storage and transport models for a series of short reaches. This approach is computationally extravagant but is usually justified on the grounds that the modelling is physically based and the parameters used have physical meaning. Physically-based models also have the advantage that their parameters can be objectively measured and that a change in system behaviour can be specified in terms of physically realistic changes in parameter values. They are also not system-specific and may be used generally. Whitehead (1977) is particularly critical of the large general model approach used in sediment routing. He considers that the type of model chosen for a particular situation should reflect the nature of the problem and that there is little to be gained from building large multipurpose models which are to be transported from one location to another. Such models tend to be inefficient because they require parameters to describe processes which may not be relevant to the problem in hand and therefore provide the modeller with additional hidden modes of behaviour via ill-defined mechanisms which may be misleading if the model is used for forecasting. Other arguments against using the physically-based model are that different processes are modelled with different degrees of precision and the model is only as good as its weakest link. Given the potential errors involved in using transport equations, the stochastic nature of many of the processes involved, and the fact that some elements of sediment routing models are little more than guesswork, they are not physically-based at all. Modelling the full internal structure is therefore an exercise in pseudoprecision and such models can never be general.

Physically-based analytical models are an objective to strive for rather than an established reality in hydrology and sediment modelling. Sediment routing models contain imperfect descriptions of processes and because they have a cascade structure, there is a tendency for these errors to compound with each successive step in the computations. Many of the internal sequential operations which transform input to output are unverified or unverifiable and are usually ignored. For example, many sediment routing models recalculate the particle size distribution of the bed sediment for each reach and time step before determining the transport rate and passing load downstream. Few attempts are made to test whether the computed bed sizes and the transport rates in individual reaches are accurate. Where such tests have been made, there are, in fact, substantial errors (see, for example, Alonso *et al.*, 1981). What usually happens when a model is tested is that parameter values are 'fiddled' so that model output matches observed behaviour as closely as possible. This calibration procedure is used in the SR model, for example, to control the flood wave, the sediment velocity and the erodibility of bed material. Calibration is a legitimate practice in operational situations. It should, however, be recognized that calibration tends to optimize only a few aspects of model output. When other outputs are examined model performance may be considerably worse than the optimized outputs. Higgins *et al.* (1987) illustrate some of these problems by showing that while sediment routing models produce acceptable estimates of total transport, associated estimates of the amount and location of deposition can be orders of magnitude in error. It should also be recognized that sediment routing models which have been calibrated are no longer analytical and certainly not physically based. Instead, they are hybrid or quasianalytical and are no more general than the lumped parameter models described earlier.

7.6 MODELLING LARGE SYSTEMS

Modelling the distributed behaviour of large systems with analytical or quasianalytical models poses special problems. The obvious procedure is to model subcatchments separately using lumped parameter sediment yield models or catchment sediment routing models. The resultant discharges and sediment yields then enter a channel routing model as a set of tributary inflows. This procedure is complex, computationally expensive and prone to considerable error. Individual models build up total system behaviour from the behaviour of a system's constituent parts so errors have a tendency to compound. It is also very difficult to synchronize the temporal behaviour of different parts of the system correctly. For example, because water and sediment travel at different speeds, a flood wave may overtake the pulse of sediment it transports from upstream source areas, producing new erosion further downstream as new material is entrained. If the behaviour of different parts of the system is not properly synchronized, model results are likely to be inaccurate. A further problem is the huge amount of data required to run a model of a large system. Few organizations have the resources to collect such data and the cost can rarely be justified. An alternative approach is to use a sediment budget but these models also require a great deal of data, some of which may be impossible to obtain.

The difficulties of modelling large systems using conventional methods suggests that alternative approaches might be better. In the remainder of this chapter one such approach based on erosion cell behaviour, spatial process models and satellite data is described. The method applies to flat arid and semiarid lands used for grazing and was developed for speed, cheapness and ease of operation. The models used describe behaviour in each part of the system but use a single stochastic process to determine that behaviour. They can therefore be classed as lumped parameter models. All the computer programs necessary to carry out this type of modelling are available commercially in the erosion module of the Micro-BRIAN image processing package.

7.6.1 The Erosion Cell

The erosion cell concept (Pickup, 1985) provides the basis of the spatial process models used and was developed to describe the complete erosional and depositional structures which occur at a variety of spatial scales in the flat alluvial lands of central Australia. These lands consist of a mosaic of areas of soil erosion and deposition interspersed with areas of apparent stability. The pattern appears to result from erosion by wind and sheetflow. Channel erosion is relatively minor although it can be important locally. Only small depths of erosion and deposition occur (cm rather than m) but because gradients are so low, they spread over large areas.

Most of the active parts of this landscape may be divided into what Schumm (1977) has termed sediment production zones, sediment transfer zones and sinks. The production zone consists of the eroded area itself and sheds both water and sediment whenever rainfall of sufficient magnitude occurs. The transfer zone is an area which occurs downstream of or below the production zone. Most of the sediment from the production zone passes through the transfer zone but because temporary and intermittent storage of material occurs, this may take a considerable time. This is especially true under arid and semiarid conditions where erosion-producing events such as major storms are both intermittent and short-lived. Many erosional events may be required before sediment crosses the transfer zone. The sink occurs downstream of the transfer zone and is an area of sediment accumulation. The rate of accumulation in a sink may be highly variable depending on the rate of delivery, shifts in the main area of deposition and the position within the sink.

Taken together, the production zone, transfer zone and sink zone make up an erosion cell. Erosion cell structures are primarily the result of runoff but they can be reinforced by wind erosion. Wind velocities close to the ground are higher over eroded bare areas than they are over vegetated ground. There is therefore a tendency for eroded material to be deposited in stable areas and sinks which usually have a good vegetation cover. Furthermore, as wind erosion strips a surface, the amount and velocity of runoff increases. The result is the gradual imposition of a pattern of erosion increasingly dominated by the activity of running water. Erosion cell behaviour seems to be the three-dimensional equivalent of the principles derived from arroyos and discontinuous gully systems by Schumm and Hadley (1957). An idealized set of erosion cells is shown in Figure 7.18.

Erosion cells may be fully developed, partly developed, or latent, depending on the climatic and geomorphic history of the landscape. They may also exist at a variety of spatial scales, with the smaller cells embedded in the larger ones forming an erosion cell mosaic. Not all cells have the three zones present. For example, one cell may capture the production

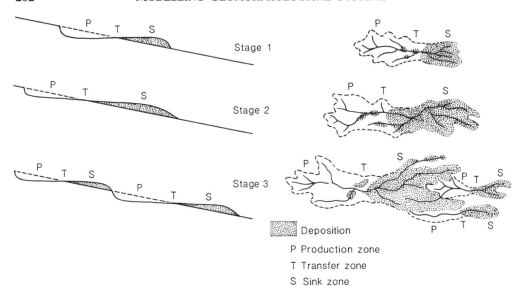

Figure 7.18. Cross-section and plan views showing idealized development of erosion cell structures in a flat landscape. (From Pickup, 1985 and reproduced by permission of *The Australian Rangeland Journal*)

zone of another leaving the transfer zone and sink isolated. In small cells the transfer zone may not develop at all. Where a production zone drains into a large river, all sediment may be removed and there will be no transfer zone or sink.

The rate of activity within an erosion cell mosaic may intensify due to short-term climatic change or more intense grazing pressure which removes vegetation, breaks up soil aggregates and lowers the erosion threshold of the landscape. The initial effect of more intense erosional activity is to enhance the erosion processes which are already in operation (Pickup, 1985). This tends to reinforce existing trends in landscape development, i.e. production areas extend and generate more sediment which results in increased deposition in associated sinks. Transfer zones experience more rapid and more intense variation as an increased volume of sediment passes through. As erosional activity increases, production zones will become larger and may extend both upslope, laterally and downslope into the transfer zone. The smaller sinks may lose their ability to trap sediment and begin to erode. As erosion proceeds, production zones will extend even further and the smaller sinks can no longer be sustained; deposition will then be concentrated only in the largest sinks. The landscape is then dominated by large areas of bare ground separated by smaller areas with very dense vegetation cover which receive most of the runoff, nutrients and transported seed generated elsewhere in the landscape. Alternatively, if the production zones drain into a large river with a high sediment transport capacity, local sinks may not develop and the material removed from the production zones may be completely lost to the local area.

Besides intensifying the existing structure of erosion cells in an area, increased erosion may activate previously stable areas and cause new cells to develop on them. Some of these cells may be latent in that the basic structure is present but the cell was only active under past climatic conditions. Other cells may be completely new but these occupy only a very small part of the landscape.

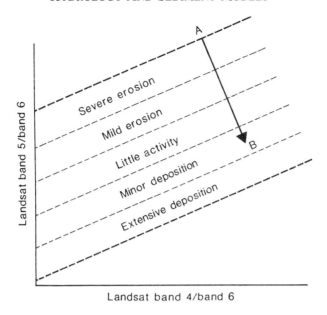

Figure 7.19. Schematic diagram of the arid land soil stability index (A–B). (From Pickup and Chewings, 1986b and reproduced by permission of *The Australian Rangeland Journal*)

The behaviour of erosion cell mosaics is too complex to be handled by conventional catchment or sediment routing models. The patterns of flow and sediment transport within them vary from event to event and are governed as much by the presence or absence of vegetation cover as by factors such as gradient. There are, however, a number of properties of cell behaviour which can be exploited for modelling purposes. These are: the tendency for erosion to enhance existing cell structures; the tendency for both erosion and deposition to spread outwards from existing source zones; and the very limited amount of new cell development which occurs. In cases such as this, where a high degree of temporal and spatial autocorrelation exists, there is considerable potential for extrapolation.

7.6.2   Measurement of Erosion and Deposition from LANDSAT

A prerequisite for the modelling of erosion cell behaviour is the ability to map erosion and deposition at the landscape level and express it as a single variable. The loss or gain of soil material affects both the cover and greenness of vegetation. The effect can be so great that cover and greenness can sometimes be used in arid or semiarid lands as an approximate measure of the extent of erosion and deposition. Cover and greenness can be detected and measured using data produced by the Multispectral Scanner carried on the Landsat series of earth resources satellites. This instrument measures reflected light in the visible green (Band 4, 0.4–0.6 μ), visible red (Band 5, 0.6–0.7 μ) and near infrared (Band 6, 0.7–0.8 μ and Band 7, 0.8–1.1 μ) regions of the spectrum.

Pickup and Nelson (1984) have developed a land stability index from Multispectral Scanner data which exploits the cover-greenness effect and at the same time filters out confounding factors such as differences in the colour and brightness of both soil and vegetation. The index is empirically based and was derived from observations of more than 700 sites in central Australia. It works by taking the ratios of the values in LANDSAT Bands 4 and 6 and 5 and 6 for each pixel. When the ratios are plotted against each other, they occupy a space between two parallel lines (Figure 7.19). Values close to the upper line are usually associated with areas of severe erosion. Erosion intensity decreases away from the upper line and points in the mid-range tend to be areas of little activity. The lower part of the data space is usually associated with deposition which increases in intensity towards the lower line. An approximate measure of erosion or deposition intensity can thus be obtained by taking the perpendicular distance of a point from the upper line. Small values of this parameter indicate erosion, medium range values suggest little activity and large values indicate deposition.

The land stability index depends on the way in which the 4/6–5/6 data space partitions out bare surfaces, dry vegetation and green vegetation. Bare surfaces all plot along the upper line with colour and brightness determining the actual position. Dry vegetation plots below the upper line, usually in the middle of the 4/6–5/6 parallelogram, but as the proportion of green vegetation increases, there is a shift towards the lower limit of the data space.

This relates to erosion and deposition in the following manner. Severe erosion kills vegetation, removes topsoil and leaves a bare surface. Less severe erosion usually produces patches of bare ground interspersed with islands of surviving topsoil still occupied by vegetation. Areas in the initial stages of erosion do not have large patches of bare ground but they do have a relatively poor nutrient and moisture status. They therefore have a tendency to be dry and have a reduced vegetation cover. Deposition areas, on the other hand, are sinks for water, soil, nutrients, and seed from surrounding areas. This produces vigorous vegetation growth, good ground cover and a predominance of the larger long-lived species such as trees. They consequently display a much stronger green signal than other areas which causes them to plot close to the lower limit of the 4/6–5/6 data space.

The index works well on central Australian landscapes where there is a strong relationship between vegetation cover and erosion status. It should be applied elsewhere with caution.

7.6.3   The Statistical Characteristics and Behaviour of Erosion Cell Mosaics

Erosion cell mosaics in arid and semiarid grazing lands appear to go through a distinct series of states depending on the processes operating on them. The changes between states do not occur continuously. Instead, the mosaic structure may remain relatively unchanged for long periods until it is disturbed. Change may then be rapid. The principal agents of disturbance are grazing, which lowers the erosion threshold of the landscape, and major rains occurring after a dry period when the landscape is bare and unprotected. In the periods between such episodic upheavals, the landscape may be regarded as being fairly stable even though there is evidence of past erosion and deposition.

The state of an erosion cell mosaic may be described by the location, mean, frequency distribution and spatial autocorrelation function (*sacf*) of the land stability index values in an area (Pickup and Chewings, 1986a). A one-sided structure is used here to calculate the *sacf* such that the autocovariance function, c is given by:

$$c = f|I|^2$$

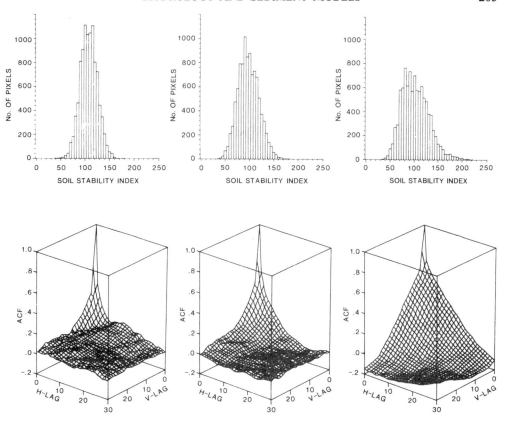

Figure 7.20. Frequency distributions and spatial autocorrelation functions of soil stability index values for stable and eroding sites. Small values of the stability index indicate erosion while larger values indicate deposition. The degree of erosion increases from left to right

in which I is the Fourier transform of soil stability index values and f denotes an inverse Fourier transform. The autocorrelation coefficient at lag, g, is then given by:

$$r_g = c_g / c_0$$

The changes which occur in the mean, frequency distribution and *sacf* as a landscape becomes more eroded are illustrated in Figure 7.20. In the relatively stable landscape, the mean is high, the frequency distribution has a relatively narrow range and the *sacf* declines rapidly with increasing spatial lag. This indicates that only a small part of the landscape is occupied by areas of extreme erosion and deposition and that there is great spatial diversity. Large erosion cell structures have therefore not developed and much of the landscape is both stable and resilient. In the partially degraded landscape, the mean is slightly smaller, the frequency distribution has a wider range, and the *sacf* decreases less rapidly with lag. This suggests a shift towards more erosion *and* deposition and greater spatial uniformity implying expansion of the zones within the various erosion cells. In the degraded landscape these trends continue. The frequency distribution spreads further indicating that

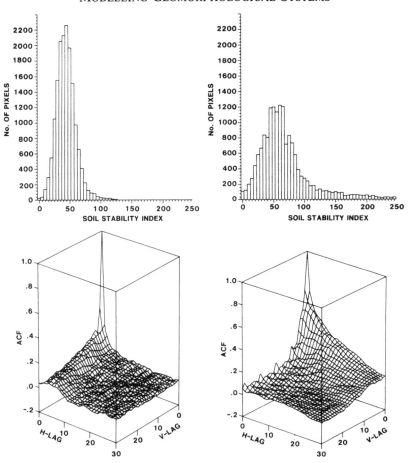

Figure 7.21. Changes in frequency distribution and spatial autocorrelation function of soil stability index values as an area becomes more eroded as a result of grazing. The graphs on the left show site conditions in 1972 while those on the right are for 1984

most of the landscape is made up of areas of erosion or deposition. The *sacf* also increases at each spatial lag greater than zero. The erosion cell structures are therefore becoming increasingly large.

Pickup and Chewings (in press) have proposed the sequence of changes which occur as a landscape moves towards a more eroded state in response to grazing. The frequency distribution and *sacf* on the left hand side of Figure 7.21 show the condition of an area at the end of a period of heavy grazing. The frequency distribution has a narrow range and mean is low, indicating a large amount of bare ground but little deposition. The *sacf* declines quickly with increasing lag suggesting a high degree of spatial diversity and little cell development. This landscape is poised for change as the frequency distribution and *sacf* for the same area twelve years later show. Now there is a higher degree of spatial uniformity resulting from cell development and areas of deposition have expanded as the sediment has moved. During these changes the location of the cells does not change.

7.6.4   Random Field Models

The high degree of temporal and spatial autocorrelation which occurs in erosion cell behaviour suggests that it may be treated as a spatial stochastic process. Behaviour is therefore modelled as a mathematical abstraction of an empirical process whose behaviour is governed by probability laws. A particular area may therefore contain a range of states of erosion and deposition which conform to a particular probability distribution. These states also exhibit a degree of geometric and statistical dependency so that each point is related to its neighbours. The dependency produces spatial patterns and may be regarded as the statistical analogue of the continuity of mass and dispersion equations which describe erosion and deposition in quasianalytical models.

The stochastic process used here is the two dimensional autoregressive random field model defined over a finite lattice (Kashyap, 1981). The random field model expresses each observation in an area as a weighted sum of the surrounding observations plus a noise component. The state of the system at a particular location is given by:

$$Y(s) = \sum_{r \epsilon N} \theta_r Y(s+r) + \sqrt{\rho} \omega(s)$$

where:

Y is the erosion index value at location, s

s refers to grid coordinates with $\{s=(i,j)\epsilon\Omega\}$, $\Omega=(i,j), 1 \leq i,j \leq M$ for a square matrix.

r indicates the coordinates of a member of the neighbourhood set, N, around location, s, with $\{r=(k,1)\epsilon N\}$

θ_r are the weighting parameters

ρ is the variance of the noise series and

$\omega(s)$ is a sequence of random variates with zero mean and unit variance.

The system for labelling the neighbourhood set, N used here is:

$$r=(0,-1)$$

$$r=(-1,-1) \quad X \quad X \quad X \quad r=(1,-1)$$

$$r=(-1,0) \quad X \quad . \quad X \quad r=(1,0)$$

$$r=(-1,1) \quad X \quad X \quad X \quad r=(1,1)$$

$$r=(0,1)$$

where . indicates the location of value $Y(s)$ at coordinates $(0,0)$ and x indicates its neighbours at coordinates $(s+r)$. Any collection of neighbours may be used depending on the spatial interactions present in the data.

The random field model may be simultaneous or conditional Markov depending on whether the noise series, ω, is independent or exhibits various types of dependency. Here we use an independent noise series so the model is simultaneous autoregressive (SAR). Such a model may involve first or higher order autoregressive components in any direction depending on the neighbourhood set chosen. The SAR model structure is used because it resembles a diffusion process. There is, however, probably no particular operational advantage in using simultaneous structure over a conditional Markov one. It may also be possible to obtain equally good results using a moving average structure. All that is required is that the chosen model should be capable of rescaling the original data and acting as a smoothing operator in a particular range of the two-dimensional frequency domain. Short-term soil erosion modelling only requires that the high frequency spatial variability be smoothed so SAR or other such short memory models are appropriate. If the lower spatial frequencies change, which might be the case over a long period of time, a structure such as the long correlation model (Kashyap and Lapsa, 1984) would be more appropriate.

Procedures for fitting SAR models are presented by Kashyap and Chellappa (1983) who have developed a finite lattice model representation for which an explicit log likelihood function may be obtained. They describe an iterative procedure for obtaining maximum likelihood estimates of model parameters which is relatively quick. Tests with synthetic data have shown that the method produces estimates of parameter values close to the true ones (Chellappa, 1980) even when the noise series is non-Gaussian.

Model fitting involves the estimation of θ and ρ. These are obtained as the limits of θ_t and ρ_t where:

$$\theta_{t+1} = (R - \frac{1}{\rho_t}S)^{-1}(V - \frac{1}{\rho_t}U) \quad t=0,1,2 \ldots$$

$$\rho_t = \frac{1}{M^2}\sum_{\Omega}\{y(s) - \theta_t^T z(s)\}^2 \quad t=0,1,2,3 \ldots$$

in which:

$$z(s) = \text{column } \{y(s+r), r \epsilon N\}$$

$$S \quad = \sum_{s\epsilon\Omega} z(s)z^T(s) \qquad\qquad m \times m \text{ matrix}$$

$$U \quad = \sum_{s\epsilon\Omega} z(s)y(s) \qquad\qquad m \times 1 \text{ vector}$$

$$R \quad = \sum_{s\epsilon\Omega} (S_s S_s^T - C_s C_s^T) \qquad m \times m \text{ matrix}$$

$$V \quad = \sum_{s\epsilon\Omega} C_s \qquad\qquad\qquad m \times 1 \text{ vector}$$

$$C_s = \text{column } \{\cos \frac{2\pi}{M}((s-1)^T r), r \epsilon N\}$$

$$S_s = \text{column } \{\sin \frac{2\pi}{M}((s-1)^T r), r \epsilon N\}$$

and m is the dimension of θ. The initial estimate of θ is calculated as:

$$\theta_0 = S^{-1}U$$

Choosing the most appropriate model for a particular body of data involves selecting the most appropriate neighbourhood set. The test statistic used to choose between alternative models is derived from Bayesian decision theory and is described by Chellappa (1980). The most appropriate model is the one with the minimum C value where:

$$c = \{ -\sum_{\Omega} \ln(1 - 2\theta^T C_s + \theta^T R\theta) + M^2 \ln \rho + m \ln(M^2) \}$$

where m is the number of model parameters and θ and ρ are parameter estimates.

If a model structure is known, a synthetic data set can be generated from it. The generating algorithm is used in erosion forecasting (see below) and involves the following steps (Chellappa and Kashyap, 1981):
1. Generate a series of independent random variables, $\{\omega(s), s\epsilon\Omega\}$, with zero mean, unit variance and specified probability density;
2. Determine f_{ij}^* which is the two dimensional discrete Fourier transform of $\omega(s)$;
3. Calculate μ_{ij} where:

$$\mu_{ij} = (1 - \theta^T \psi_{ij})$$

$$\psi_{ij} = \text{column } \{\exp \sqrt{-1} \ \frac{2\pi}{M}(k(i-1) + l(j-1)\} \text{ with } k, l \epsilon N$$

4. Determine x_{ij} as:

$$x_{ij} = \frac{\sqrt{\rho}}{M^2} \ f_{ij}^*/\mu_{ij}$$

5. Find the inverse Fourier transform of x_{ij} and then add the mean to produce $Y(s)$.

It may also be useful to use inverse filtering to remove a known model from a data set to obtain its noise series. A new model may then be applied and the data set regenerated. This approach makes it possible to see the effect of applying different models to the same area. The algorithm for obtaining a noise series for a given data set is to subtract the mean and carry out steps 2 and 3. The variable x_{ij} in step 4 is then calculated as:

$$x_{ij} = \frac{1}{\sqrt{\rho}M^2} \ f_{ij}^* \mu_{ij}$$

and its inverse Fourier transform is the noise series.

7.6.5 Application to Erosion Modelling

A model structure which appears to describe erosion cell mosaic behaviour well is the first order, eight neighbourhood version (Pickup and Chewings, 1986a):

$$Y_{ij} = \theta_1 y_{i-1,j} + \theta_2 y_{i+1,j} + \theta_3 y_{i,j-1} + \theta_4 y_{i,j+1}$$

$$+ \theta_5 y_{i-1,j-1} + \theta_6 y_{i-1,j+1} + \theta_7 y_{i+1,j+1}$$

$$+ \theta_8 y_{i+1,j-1} + \sqrt{\rho}\, w_{ij} + \bar{Y}$$

where i and j are grid coordinates and y values are deviations about the mean. The model is similar to a discretized version of a two-dimensional diffusion process (Angel and Jain, 1978) and it is capable of representing three of the four types of change needed to model different states of an eroding or recovering system. Changes in the overall system state are represented by shifts in the mean. Changes in the frequency distribution of system states and the spread or break up of patches within the mosaic are handled by the θ and ρ parameters. The $\omega(s)$ series contains information on the geographical location of erosion and deposition but is not a true white noise process as used by Kashyap and Chellappa (1983). Instead, it is assumed to remain unchanged over time and its values are dependent on location. It is hereafter termed the underlying pattern series.

The use of an $\omega(s)$ series which is not true white noise violates some of the assumptions on which the model fitting method is based. In operational terms, this does not appear to be a problem. Numerical tests indicate that even if weak spatial structures survive at low spatial frequencies in the $\omega(s)$ series, they have minimal effect on estimates of the θ and ρ parameters (Pickup and Chewings, 1986a). The use of an underlying pattern series rather than a white noise process can also be justified on the grounds that it represents the physical process more closely. Landscape behaviour has a substantial memory component and the basic landscape structure on which erosion cells develop does not change over short periods which would be the case with a true white noise process.

The modelling procedure relies on the high degree of temporal and spatial autocorrelation present in erosion cell behaviour and uses the landscape to forecast itself. The underlying pattern series can be interpreted as representing the latent pattern of erosion and deposition in an area while the mean, variance and θ parameters describe the extent of erosion. As the landscape degrades, a frequency distribution of erosional and depositional states is imposed, scaling the underlying pattern series. This scaling is represented by the ρ parameter. An autocorrelation structure is also imposed as an erosion cell mosaic develops by intensifying very weak features in the underlying pattern series. This structure is controlled by the θ parameters.

The steps involved in modelling the way in which a landscape changes as erosion proceeds may be stated as follows. Firstly, assume that a landscape, $\underline{Y}_1$, in a particular state of erosion can be related to an underlying noise series, $\underline{w}$ by a linear filter, A_1 such that:

$$\underline{Y}_1 = A_1 \underline{w}$$

Similarly, assume that the same landscape in a more eroded state, $\underline{Y}_2$ is related to $\underline{w}$ by

$$\underline{Y}_2 = A_2 \underline{w}$$

Then, given estimates of A_1 and A_2, and assuming that $\underline{w}$ does not change over time, the more eroded state may be forecast as:

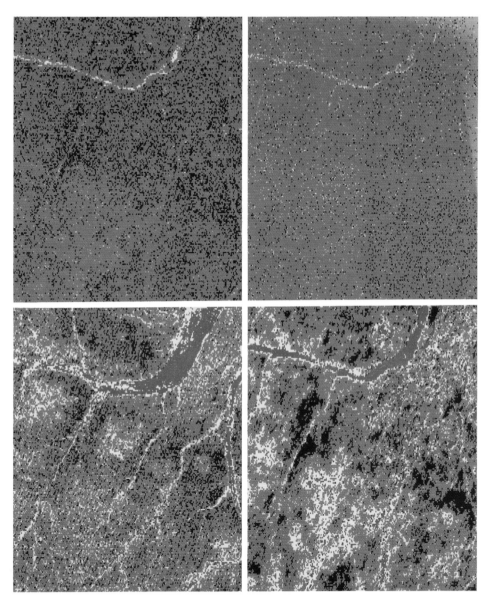

Figure 7.22. Results of erosion forecasting on a 256 × 256 pixel soil stability index image for Murloocoppie in South Australia's arid zone. The images are: upper left – stability index values for 1972; upper right – noise series values; lower left – stability index values for 1984; lower right – forecasted 1984 values. The colours indicate stability index values and may be interpreted as: blue – eroded bare ground; green – slightly eroded to stable; yellow – stable to minor deposition; red – substantial deposition

$$\hat{Y}_2 = A_2 \hat{w} = A_2 A_1^{-1} Y_1$$

where ^ indicates an estimate. The linear filters, A_1 and A_2 are SAR models obtained using the fitting procedures described above. The estimate of A_1 is derived by fitting an appropriate model to Y and $\hat{w}$ is calculated by inverse filtering. The A_2 model is an unknown but can be obtained by using landscapes similar to Y_1 but which are in a more eroded condition as prototypes. Once such a prototype is identified, A_2 is estimated by the normal SAR model fitting procedure.

This approach may appear simple but tests show that it is capable of producing accurate forecasts of change for large areas with very complex patterns of erosion and deposition. It seems to work because the underlying pattern series contains most of the information on the likely pattern of soil movement. Also, because the model parameters used to generate a more eroded condition are derived from a similar or neighbouring area, they contain the averaged response of the landscape bypassing the need for detailed modelling. An example of the use of SAR models in describing changes in the land stability index over time is presented in Figure 7.22 which shows a series of 256×256 pixel density sliced images for an area near Murloocoppie in the arid zone of South Australia. The 1972 image indicates large areas of bare ground resulting from dry conditions, grazing and erosion. Very heavy rains occurred between 1972 and 1984 causing extensive erosion, deposition, and vegetation recovery within the area. The results are apparent in the 1984 image which shows the development of major sinks and areas of regrowth as well as source zones which continue to erode. The modelling process is illustrated in the images on the right hand side of Figure 7.22. After fitting an SAR model to the 1972 and carrying out an inverse filtering operation, an underlying pattern series image is derived. This image shows little structure although lineations are weakly preserved. The range of the data is also compressed. When the underlying pattern series is filtered using a suitable prototype (in this case from a neighbouring area), the forecasted image results. This image is not an exact reproduction of the observed 1984 condition but it correctly identifies the main areas of change and its direction.

The use of prototypes is analogous to model calibration but takes the process a step further. Instead of using a whole data set in which the system may pass through a number of states to calibrate a single set of model parameters, individual system states are explicitly recognized and a model is derived for each. This process is made possible by the existence of a large body of commercially-available LANDSAT data going back to 1972 for many areas. The skill, of course, lies in recognizing and selecting data representative of each system state. The modeller should also be sure the prototypes chosen for a particular system are truly representative.

7.7 CONCLUSIONS

A wide range of techniques is available for the modelling of erosion, deposition and sediment yield. These vary in sophistication from simple regression equations through complex response models requiring sophisticated calibration techniques to simulation models which attempt to reproduce the physical processes involved. Some models are sufficiently general to be used without calibration data but the results are often of limited value. Most models require some degree of calibration which restricts their use as operational tools for forecasting. It also suggests that the simplest models might be the best ones as long as no major changes in system behaviour or morphology occur. Where such changes do take place, it is rarely possible to model them from physical principles and there may be no calibration data for the system in question. It may therefore become necessary to use an already modified system as a prototype to determine the effects of change on another.

REFERENCES

Alonso, C. V., Borah, D. K., and Prasad, S. N. (1981). *Stream Channel Stability, Appendix J. Numerical Model for Routing Graded Sediments in Alluvial Channels*, U.S. Dept. Agriculture Sedimentation Laboratory, Oxford, Mississippi.

Alonso, C. V., Neibling, W. H., and Foster, G. R. (1981). Estimating sediment transport capacity in watershed modelling. *Trans. A.S.A.E.*, **5**, 1211–1220 and 1226.

Angel, E. and Jain, A. K., (1978). Frame to frame restoration of diffusion images. *IEEE Trans. on Automatic Control*, **AC23**, 850–855.

Andrews, E. D. (1983). Entrainment of gravel from naturally sorted riverbed material. *Geol. Soc. Am. Bull.*, **94**, 1225–1231.

Bagnold, R. A. (1977). Bedload transport by natural rivers. *Water Resour. Res.*, **13**, 303–312.

Bennett, J. P. and Nordin, C. F. (1977). Simulation of sediment transport and armouring. *Hydrol. Sci. Bull.*, **XXII**, 555–569.

Beschta, R. L. (1981). Patterns of sediment and organic matter transport in Oregon Coast streams. *I.A.S.H. Pub.*, **132**, 179–185.

Borah, D. K., Alonso, C. V., and Prasad, S. N. (1981). *Stream Channel Stability. Appendix I. Single Event Model for Routing Water and Sediment on Small Catchments*, US Dept. Agric. Sedimentation Laboratory, Oxford, Mississippi.

Borah, D. K., Alonso, C. V., and Prasad, S. N. (1982a). Routing graded sediments in streams: formulations. *Procs. Am. Soc. Civ. Engrs. J. Hydraul. Div.*, **108**, 1486–1503.

Borah, D. K., Alonso, C. V., and Prasad, S. N. (1982b). Routing graded sediments in streams: applications. *Procs. Am. Soc. Civ. Engrs. J. Hydraul. Div.*, **108**, 1504–1517.

Borah, D. K., Prasad, S. N., and Alonso, C. V. (1980). Kinematic wave routing incorporating shock fitting. *Water Resour. Res.* **16**, 529–541.

Box, G. E. and Jenkins, G. M. (1976). *Time Series Analysis: Forecasting and Control*, Holden-Day, San Francisco.

Chellappa, R. (1980). Fitting random field models to images. *University of Maryland Computer Science Centre Report, TR-928*.

Chellappa, R. and Kashyap, R. L. (1981). Synthetic generation and estimation in random field models of images. *Procs. IEEE Computer Soc. Conference on Pattern Recognition and Image Processing*, 577–582.

Cunge, J. A. and Perdreau, N. (1973). Mobile bed fluvial mathematical models. *La Houille Blanche*, **7**, 562–580.

Dietrich, W. E. and Dunne, T. (1978). Sediment budget for a small catchment in mountainous terrain. *Zeit. Geomorph. N. F. Suppl. Bd.*, **29**, 191–206.

Einstein, H. A. (1950). The bedload function for sediment transportation in open channel flows. *Soil Conservation Service Tech. Bull.*, **1026**, U.S. Dept. Agriculture.

Emmett, W. W., Myrick, R. M., and Meade, R. H. (1982). Field data describing the movement and storage of sediment in the East Fork River, Wyoming. Part III. River hydraulics and sediment transport, 1980. *U.S. Geol. Survey Open File Report*, 82–359.

Foster, G. R. (1982). Modelling the erosion process. In Haan, C. T. (Ed.), *Hydrologic Modelling of Small Watersheds. Am. Soc. Agric. Engrs. Monograph* **5**, 297–379.

Foster, G. R., Lane, L. J., Nowlin, J. D., Laflen, J. M., and Young, R. A. (1980). Chapter 3. A Model to estimate sediment yield from field sized areas: development of model. In *CREAMS—A Field Scale Model for Chemicals, Runoff and Erosion from Agricultural Management Systems. U.S.D.A. Conservation Research Report*, **26**.

Gessler, J. (1970). Self-stabilising tendencies of alluvial channels. *Procs. Am. Soc. Civ. Engrs. J. Waterways and Harbours Div.*, **96**, 239–248.

Gilbert, G. K. (1917). Hydraulic mining debris in the Sierra Nevada. *US Geol. Survey Professional Ppr.*, **105**.

Graf, W. H. (1971). *Hydraulics of Sediment Transport*, McGraw-Hill, New York, 124–126.

Higgins, R. J. (1979). Sediment transport in a river with a high induced load. *Civ. Engg. Trans. Inst. Engrs. Australia*, **CE21**, 111–117.

Higgins, R. J., Pickup, G. and Cloke, P. (1987). Estimating the transport and deposition of mining waste at Ok Tedi. In Thorne, C. R., Bathurst, J. C. and Hey, R. D. (Eds.). *Sediment Transport in Gravel-Bed Rivers*. John Wiley and Sons, Chichester, 949–976.

Hydrologic Engineering Centre (1977). *HEC-6, Scour and Deposition in Rivers and Reservoirs, Users Manual*, U.S. Army Corps of Engineers.

Hydrologic Engineering Centre (1982). *HEC-2, Water Surface Profiles, Users Manual*, US Army Corps of Engineers.

Kashyap, R. L. (1981). Analysis and synthesis of image patterns by spatial interaction models. *Progress in Pattern Recognition*, **1**, 149–186.

Kashyap, R. L. and Chellappa, R. (1983). Estimation and choice of neighbours in spatial interaction models of images. *IEEE Trans. on Information Theory*, **29**, 60–72.

Kashyap, R. L. and Lapsa, P. M. (1984). Synthesis and estimation of random fields using long-correlation models. *IEEE Trans. on Pattern Analysis and Machine Intelligence*, **6**, 801–809.

Klingeman, P. C. and Emmett, W. W. (1982). Gravel bedload transport processes. In Hey, R. D., Bathurst, J. C., and Thorne, C. R. (Eds.), *Gravel Bed Rivers*, John Wiley and Sons, Chichester, 141–168.

Laursen, E. M. (1958). The total sediment load of streams. *Procs. Am. Soc. Civ. Engrs. J. Hydraul. Div.*, **84**, (1530)1–6.

Lehre, A. K. (1981). Sediment budget of a small California Coast Range Drainage Basin. *I.A.S.H. Pub.*, **132**, 123–140.

Lekach, J. C. and Schick, A. P. (1983). Evidence for transport of bedload in waves: analysis of fluvial sediment samples in a small upland stream channel. *Catena*, **10**, 267–279.

Loughran, R. J. (1977). Sediment transport from a rural catchment in N.S.W. *J. Hydrol.*, **34**, 357–377.

Meyer-Peter, E. and Müller, R. (1948). Formulas for bedload transport. *Int. Assoc. Hydraul. Struct. Res.*, 2nd Mtg., Stockholm, 39–64.

Moore, R. J. (1984). A dynamic model of basin sediment yield. *Water Resour. Res.* **20**, 89–103.

Moore, R. J. and Clarke, R. T. (1981). A distribution function approach to rainfall–runoff modelling. *Water Resour. Res.* **17**, 1367–1382.

Moore, R. J. and Clarke, R. T. (1983). A distribution function approach to modelling basin sediment yield. *J. Hydrol.*, **65**, 239–257.

Mosley, M. P. (1978). Bed material transport in the Tamaki River near Dannevirke, North Island, New Zealand. *N.Z. J. Sci.*, **21**, 619–626.

Negev, M. (1967). A sediment model on a digital computer. *Stanford University, Dept. of Civil Engineering Tech. Rep.*, **76**.

Olive, L. J. and Rieger, W. A. (1984). Sediment erosion and transport modelling in Australia. *Drainage Basin Erosion and Sedimentation: a conference on Erosion, Transportation and Sedimentation in Australian Drainage Basins*, Univ. of Newcastle and Soil Conservation Service of N.S.W., 81–93.

Pickup, G. (1976). Simulation modelling of river channel erosion. In Gregory, K. J. (Ed.), *River Channel Changes*, John Wiley and Sons, Chichester, 47–60.

Pickup, G. (1984). Geomorphology of tropical rivers. I. Landforms, hydrology and sedimentation in the Fly and lower Purari, Papua New Guinea. *Catena Supp.*, **5**, 1–17.

Pickup, G. (1985). The erosion cell — a geomorphic approach to landscape classification in range assessment. *Australian Rangelands J.*, **7**, 114–121.

Pickup, G. and Chewings, V. H. (1986a). Random field modelling of spatial variations in erosion and deposition in flat alluvial landscapes in arid central Australia. *Ecological Modelling*, **33**, 269–296.

Pickup, G. and Chewings, V. H. (1986b). Mapping and forecasting soil erosion patterns from Landsat on a microcomputer-based image processing facility. *Autralian Rangelands J.*, **8**, 57–62.

Pickup, G. and Chewings, V. H. (in press). Forecasting patterns of soil erosion in arid lands from Landsat MSS data. *Int. J. Remote Sensing.*

Pickup, G. and Higgins, R. J. (1979). Estimating sediment transport in a braided gravel channel—the Kawerong River, Bougainville, Papua New Guinea. *J. Hydrol.*, **40**, 283–297.

Pickup, G., Higgins, R. J., and Grant, I. (1983). Modelling transport and deposition as a moving wave—the transfer and deposition of mining waste. *J. Hydrol.*, **60**, 281–301.

Pickup, G., Higgins, R. J., and Warner, R. F. (1979). *Impact of Waste Rock from the Proposed Ok Tedi Mine on Sedimentation Processes in the Fly River and its Tributaries*, Dept. of Minerals and Energy and Office of Environment and Conservation, Papua New Guinea.

Pickup, G. and Nelson, D. J. (1984). Use of Landsat radiance parameters to distinguish soil erosion, stability and deposition in arid central Australia. *Remote Sensing Env.*, **16**, 195–209.

Ponce, V. M., Garcia, J. L., and Simons, D. B. (1979). Modelling alluvial channel bed transients. *J. Hydraul. Div. Procs. Am. Soc. Civ. Engrs.* **105**, 245–256.

Reid, J., Frostick, L. E., and Layman, J. T. (1985). The incidence and nature of bedload transport during flood flows in coarse grained alluvial channels. *Earth Surface Processes and Landforms*, **10**, 33–44.

Reid, L. M., Dunne, T., and Cederholm, C. J. (1981). Application of sediment budget studies to logging road impact. *J. Hydrol. (N.Z.)*, **20**, 49–62.

Rendon-Herrero, O. (1974). Estimation of washload produced on certain small watersheds. *Procs. Am. Soc. Civ. Engrs. J. Hydraul. Div.*, **100**, 835–848.

Rose, C. W. (in press). Developments in soil erosion and deposition models. *Advances in Soil Science*, **2**.

Schumm, S. A. (1977). *The Fluvial System*, Wiley–Interscience, New York.

Schumm, S. A. and Hadley, R. F. (1957). Arroyos and the semi-arid cycle of erosion. *Am. J. Sci.*, **255**, 161–174.

Smith, R. E. (1976). Simulating erosion dynamics with a deterministic distributed watershed model. *Procs. 3rd Federal Inter-Agency Sediment Conf., Denver, Colorado*, 163–173.

Soni, J. P., Garde, R. J., and Ranga Raju, K. G. (1980). Aggradation in streams due to overloading. *Procs. Am. Soc. Civ. Engrs. J. Hydraul. Div.*, **106**, 117–132.

Taylor, G. I. (1954). The dispersion of matter in turbulent flow through a pipe. *Proc. R. Soc. London Ser. A*, **223**, 446–468.

Thomas, W. A. (1982). Mathematical modelling of sediment movement. In Hey, R. D., Bathurst, J. C., and Thorne, C. R. (Eds.), *Gravel Bed Rivers*, John Wiley and Sons, Chichester, 487–508.

Thomas, W. A. and Prasuhn, A. L. (1977). Mathematical modelling of scour and deposition. *Procs. Am. Soc. Civ. Engrs. J. Hydraul. Div.*, **103**, 851–864.

Toffaleti, F. B. (1969). Definitive computations of sand discharge in rivers. *Procs. Am. Soc. Civ. Engrs. J. Hydraul. Div.*, **95**, 225–248.

U.S.D.A. (1980). CREAMS—A field scale model for chemicals, runoff and erosion from agricultural management systems. *USDA Conservation Research Rep.*, **26**.

U.S.D.A. (1982). Proceedings of the workshop on estimating erosion and sediment yield in rangelands. *Agricultural Research Service, ARM–W–26.*

Vanoni, V. A., Brooks, N. H., and Kennedy, J. F. (1961). Lecture notes on sediment transportation and channel stability. *W.M. Keck Laboratory of Hydraulics and Water Resources, California Institute of Technology, Report*, **KH–R–1**.

Vries, M. de (1973). Riverbed variations—aggradation and degradation. *Int. Assoc. Hydraul. Research Int. Seminar on Hydraulics of Alluvial Streams*, New Delhi, 1–10.

Walker, R. A. and Fleming, G. (1979). The Strathclyde Sediment Model I. User Guide. *University of Strathclyde, Dept. of Civil Engineering Report*, **HHCD–79–10**.

Walling, D. E. (1974). Suspended sediment and solute yields from a small catchment prior to urbanisation. In *Fluvial Processes in Instrumented Watersheds. Inst. Brit. Geogrs. Spec. Pub.* **6**, 169–192.

Walling, D. E. (1977). Assessing the accuracy of suspended sediment rating curves for a small basin. *Water Resour. Res.*, **13**, 531–538.

Walling, D. E. (1983). The sediment delivery problem. *J. Hydrol.*, **65**, 209–237.

Walling, D. E. and Webb, B. E. (1982). Sediment availability and the prediction of storm period sediment yields. *I.A.S.H. Pub.*, **137**, 327–337.

White, W. R. and Day, T. J. (1982). Transport of graded gravel bed material. In Hey, R. D., Bathurst, J. C., and Thorne, C. R. (Eds.), *Gravel Bed Rivers*, John Wiley and Sons, Chichester, 181–210.

Whitehead, P. G. (1977). Mathematical modelling and the planning of environmental studies. *Centre for Resource and Environmental Studies Report*, **AS/R15**, Australian National University.

Wischmeier, W. H. (1972). Upslope erosion analysis. In *Environmental Impact on Rivers*, Water Resources Publications, Fort Collins, Colo.

Wischmeier, W. H. (1976). Use and misuse of the universal soil loss equation. *J. Soil and Water Cons.*, **31**, 5–9.

Wischmeier, W. H. and Smith, D. D. (1965). Predicting rainfall–erosion losses from cropland east of the Rocky Moutains. *U.S.D.A. Agric. Handbook*, **262**.

Wischmeier, W. H. and Smith, D. D. (1978). Predicting rainfall erosion losses—a guide to conservation planning. *U.S.D.A. Agric. Handbook*, **537**.

Wood, P. A. (1977). Controls of variation in suspended sediment concentration in the River Rother, West Sussex, England. *Sedimentology*, **24**, 437–445.

Woolhiser, D. A., Holland, M. E., Smith, G. L., and Smith, R. E. (1971). Experimental investigation of converging overland flow. *Trans. ASAE*, **14**, 684–687.

Yang, C. T. (1973). Incipient motion and sediment transport. *Procs. A.S.C.E. J. Hydraul: Div.*, **99**, 1679–1704.

Modelling Geomorphological Systems
Edited by M. G. Anderson
©1988 John Wiley & Sons Ltd.

Chapter 8

Models of sediment transport in natural streams

PAMELA S. NADEN

School of Geography, University of Leeds

8.1 INTRODUCTION

Sediment transport in natural streams is one of the most complicated modelling problems in geomorphology. The basic difficulty is that of describing the mechanics of a two-phase flow of water and sediment. This problem is exacerbated by the variety of conditions that must be described — from virtually no motion with only spasmodic individual grain movements through to a moving carpet of increasing sediment concentration almost verging on a debris flow. Coupled with this, is the wide range of sediment sizes (from colloidal to boulder-sized material) and sediment mixtures which must be accommodated. This means that even conceptually it is no easy matter to model sediment transport, as both the physical processes involved and their controlling parameters will vary with grain size and sediment concentration. Indeed, Simons and Sentürk (1977) stated that

> 'The mechanics of sediment transport is so complex that it is extremely unlikely that a full understanding will ever be obtained. A universal sediment transport equation is not and may never be available.'

Progress has also been hampered by the problem of measuring sediment transport, especially in the case of large grain sizes, in large quantities and in large rivers. In fact, much of the evidence for sediment transport equations is based upon sand-sized material, and flume experiments which have not always been correctly scaled to a field prototype. Consequently, our knowledge of sediment transport rates and thresholds is rather limited and often unrepresentative of the range of field conditions. Recent advances in the field of continuous monitoring (Leopold and Emmett, 1976; Reid *et al.*, 1980; Reid *et al.*, 1984) have begun to redress this balance, but reliable sediment transport data, backed up by relevant fluid flow parameters, are still extremely rare. Thus, it is not surprising that even a qualitative description of the sediment transport process is hard to come by, let alone reliable predictive equations. Largely because of this, modelling work is also in its infancy and applications are limited to relatively simple field problems.

In modelling sediment transport, it is necessary to calculate the impact of the flow intensity on sediment behaviour, subject to particle characteristics and availability. This simple statement introduces the three key questions which will recur throughout this chapter — the problem of providing a suitable measure of flow intensity, a relevant particle or bed description, and the different strategies used to deal with sediment availability.

Three areas of sediment transport modelling are discussed — the initiation of sediment motion, the transport of suspended sediment and the movement of bedload. In each case, a brief review of current modelling work is presented with its implications for future field investigations.

8.2 INITIATION OF SEDIMENT MOVEMENT

Fundamental to a knowledge of sediment transport is the flow threshold associated with the onset of sediment motion. Although this need not always be modelled explicitly, it does underlie many of the models and equations used to predict sediment transport. It also represents one of the easier modelling problems in the fluvial sediment transport field and it is, therefore, worth discussing in some detail.

8.2.1 Critical Velocity or Critical Shear Stress

Critical conditions for motion are generally modelled either in terms of a critical velocity (Forchheimer, 1914; Hjulström, 1936; Mavis and Laushey, 1948; Neill, 1968; Helley, 1969) or a critical shear stress (Schoklitsch, 1934; Shields, 1936; White, 1940; Leliavsky, 1955; Baker and Ritter, 1975; Costa, 1983). The problem with critical velocity predictions is that a consistent reference height within the flow has not always been used, and this has led to spurious comparisons between calculated thresholds. Other, purely empirical, power functions based upon a measure of critical shear stress, such as those quoted by Baker and Ritter (1975) and Costa (1983), have little general applicability.

Of those critical shear stress conditions having some physical basis, the most commonly-used criterion is that derived by Shields (1936). In this simple model, shown in Figure 8.1 as modified by Miller *et al.* (1977) using the data of White (1970), the threshold of motion is defined in terms of the dimensionless bed shear stress ($\tau_c/(\rho_s - \rho)gD$ where τ_c is critical shear stress, ρ_s is density of sediment, ρ is density of water, g is acceleration due to gravity, and D is particle diameter). This is plotted against the grain Reynolds number ($Re_* = u_*/Dv = f\{D/\delta\}$ where u_* is bed shear velocity, v is kinematic viscosity of water, δ is depth of laminar sublayer) which is a measure of the type of flow boundary. Under smooth boundary conditions ($Re_* < 2$), the particles are enclosed in laminar film and their motion is independent of turbulence. As Reynolds numbers increase, there is a zone of transition as the laminar film begins to break up, while for rough boundary conditions ($Re_* > 400$), the critical dimensionless bed shear stress becomes constant. This constant is referred to as the Shields' parameter and has been variously quoted as 0.06 (Shields, 1936), 0.03 (Neill, 1967, 1968) and 0.045 (Miller *et al.*, 1977). Most natural channel flows

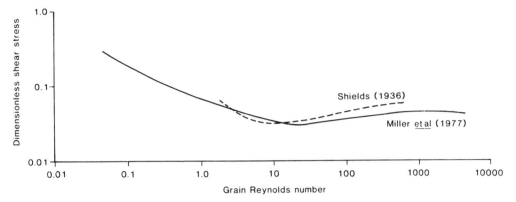

Figure 8.1. Shields' curve as modified and extended by Miller *et al.* (1977). (Reproduced by permission of Blackwell Scientific Publications Limited)

lie in this range of intermediate to large Reynolds numbers and it is, therefore, this area of the curve to which most of the subsequent discussion refers.

Accumulated field evidence reveals a wide scatter of values about the Shields' curve (Church, 1978). Indeed, Williams (1983) has put forward upper and lower limits of 0.24 and 0.01 for the Shields' parameter. A variation of this order of magnitude in the prediction of the onset of incipient motion is clearly unworkable and much recent work has sought to explain and to quantify deviations away from the Shields' curve.

The main reasons for the scatter can be divided into two categories—those factors relating to the sediment characteristics and those concerned with the flow. Details of empirical studies of each of these factors, and their sources in the literature, are given in Table 8.1.

8.2.2 Sediment Characteristics

With regard to the sediment, the most fundamental, although not independent, controls are the protrusion of individual grains into the flow and the pivoting angle between the

Table 8.1. Factors influencing the scatter about the Shields' curve

A. *Sedimentary*	
Grain size distribution	Andrews (1983)
Grain shape	Helley (1969); Li and Komar (1986)
Packing of grains (% free to move)	Church (1978)
Protrusion into the flow	Fenton and Abbott (1977)
Pivoting angles	Helley (1969); Li and Komar (1986)
Imbrication	Komar and Li (1986)
Cluster Bedforms	Brayshaw *et al.* (1983)
B. *Flow*	
Use of average shear stress	Bridge and Jarvis (1976)
Use of time-averaged shear stress	Cheetham (1979)
Spatial distribution of drag force	Brayshaw *et al.* (1983)
Relative roughness	Neill (1968); Baker and Ritter (1975)
Narrow versus wide streams	Carling (1983)

grain to be moved and the bed grains. These two variables are themselves a function of grain size distribution, grain shape, degree of imbrication, and presence of gravel bedforms and structures.

Fenton and Abbott (1977) conducted a series of experiments using uniform spheres in order to establish the effect of relative protrusion (p/D where p is protrusion and D is grain diameter) of individual grains on the threshold of motion. Their results indicated that the Shields' parameter varied from 0.01 for grains resting on the bed surface to 0.1 for coplanar grains. This effect is no doubt partly reflected in the data of Andrews (1983). Using data from rivers with different grain size mixtures, Andrews demonstrates that thresholds for a chosen grain size may vary over an order of magnitude depending on where that grain size falls within the sediment size distribution.

In addition, Li and Komar (1986) have recently presented data on the value of the pivoting angle as influenced by the ratio of the diameter of the grain to be moved and the bed grains, grain shape, and imbrication. In a companion paper (Komar and Li, 1986), these results are applied to the Shields' threshold. They suggest that, in mixed grain sizes, the Shields' parameter should decrease with increasing grain size. This is again in line with the findings of Andrews (1983). Higher thresholds are also associated with angular grains (factor of 3–5) and with imbricated grains (factor of 5–10). Both these experimental results help to explain the field data of Carling (1983) and Hammond et al. (1984).

Other sedimentary features, such as grain packing and cluster bedforms (Laronne and Carson, 1976; Brayshaw, 1983), must also contribute to the range of variation in the Shields' parameter and to the overall mobility of the bed. The effect of packing is analysed by Church (1978) in terms of the percentage of grains which are capable of being picked out of the bed without disturbing other grains. The effect of cluster bedforms has been examined both in terms of the mobility of individual grains (Naden and Brayshaw, 1987) and the changes seen in the drag and lift coefficients around obstacles (Brayshaw et al., 1983).

All these sedimentary characteristics of river beds seem to be important in their influence on sediment transport thresholds (Table 8.1). It is vitally important, therefore, that they are incorporated in some form into models for the initiation of sediment motion and that some standard measurements of these properties are assembled.

8.2.3 Flow

Flow-related factors revolve around two main points—the problem of measuring the appropriate bed shear stress and the issue of turbulence.

Detailed knowledge of the bed shear stress involves the accurate measurement of point velocity profiles in the bottom 10–20 per cent of the flow. The difference between such measurements and an average shear stress, calculated using the relationship $\tau = \rho gRs$ (where R is the hydraulic radius and s is the water surface slope), can be as much as two orders of magnitude (Bridge and Jarvis, 1976). This is largely due to the fact that the bed shear stress used in moving the sediment is that set up by friction between the flow and the grains alone. The average bed shear stress also includes the stresses set up by friction with the bedforms (both small and medium scale) and the channel.

Cheetham's (1979) experiments also suggest that using a time–mean velocity instead of an 'instantaneous' value may underestimate the transport threshold by a factor of ten. This may be particularly important in the case of large sediment unlikely to be moved by the mean bed shear stress. The generation of turbulence, as well as the local velocity profile, is also influenced by the bed roughness or relative depth, as incorporated in the Neill (1968) threshold, and by the presence of bedforms. It is also likely to be a factor in the difference observed by Carling (1983) between thresholds in narrow versus wide streams.

Two important components in the modelling of sediment transport thresholds are, therefore, the modelling of the velocity profile and the implementation of a stochastic element in the model to represent turbulence (Gessler, 1971).

8.2.4 Modelling Strategies

While it is useful to identify the factors which may influence the Shields' criterion, it is quite another proposition to build the sort of model which is simple enough to be useful while at the same time including at least some of these factors. The physical basis of any threshold model is a balance of the forces acting. In this case, the drag and lift forces provided by the flow are balanced against the gravity force tending to hold the grain in place. The threshold flow can be determined under two different sets of assumptions. The grain may be assumed to *roll* out of its place on the bed and the threshold derived by considering the turning forces (White, 1940; Helley, 1969) about some pivot point (Figure 8.2(a)). Alternatively, grain motion may be represented by a *slide* (Naden, 1987a) and a simpler balance of the downstream versus friction forces calculated (Figure 8.2(b)). Both processes have been observed (Hammond *et al.*, 1984) and would appear to be partly dependent on grain shape (Komar and Li, 1986). A brief description of the models of Helley (1969) and Naden (1987a) is perhaps a useful starting point for a fuller discussion of the field requirements for future modelling work in this field.

The Model of Helley (1969)

Helley (1969), in his analysis of turning moments, assumes that the centroid of the drag force acts at a height of 0.6 diameters up from the base of the particle. It is the velocity at this height above the general level of the bed which is quoted as the threshold. In calculating the length of the turning arms, Helley also assumes that the particle shape may be approximated by an ellipse in cross-section and that the orientation angle is less than 25°. The coefficient of drag is related to shape (U.S. Inter-Agency Committee of Water Resources, 1957) and the coefficient of lift is considered to be the constant provided by Einstein and El Samni (1949). Measurements required from the field are the three axes of the particle, the specific gravity of the sediment and the orientation angle of the particle (Figure 8.2(a)). The threshold criterion derived in this way (and written in terms of SI units) is

$$v_{0.6\alpha} = 1.81 \sqrt{\left(\frac{(\gamma s - 1) \, \gamma \, (\alpha + \beta)^2 \, MR_L}{C_D' \, \alpha\gamma \, MR_D + 0.178 \, \beta\gamma \, MR_L} \right)} \tag{1}$$

where $v_{0.6\alpha}$ is the velocity at 0.6α above the bed
 α, β, γ are the small, intermediate, and large axes of the particle
 C_D' is the corrected coefficient of drag
 MR_L is the turning arm of the lift force $(0.25\beta\cos\theta + 0.43\alpha\sin\theta)$
 MR_D is the turning arm of the drag force $(0.1\alpha\cos\theta + 0.43\alpha\cos\theta - 0.25\beta\sin\theta)$
 θ is the orientation angle of the particle.

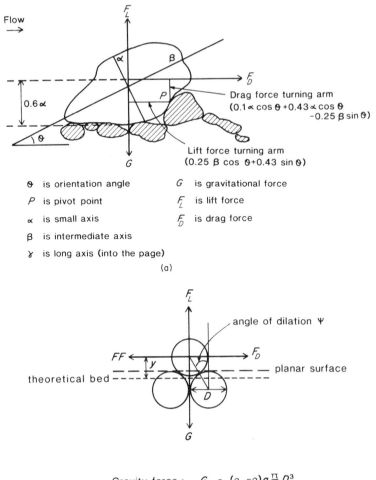

Flow →

0.6 α

θ is orientation angle

P is pivot point

α is small axis

β is intermediate axis

γ is long axis (into the page)

G is gravitational force

F_L is lift force

F_D is drag force

Drag force turning arm
$(0.1\,α\cos θ + 0.43\,α\cos θ - 0.25\,β\sin θ)$

Lift force turning arm
$(0.25\,β\cos θ + 0.43\,\sin θ)$

(a)

angle of dilation Ψ

planar surface

theoretical bed

Gravity force : $G = (ρ_s - ρ)g\frac{\Pi}{6}D^3$

Drag force : $F_D = C_D\,ρ\,D^2\,u_y^2$

Lift force : $F_L = C_L\,ρ\,D^2\,u_y^2$

Friction force : $FF = (G - F_L)\tan(ø_c + Ψ)$

($ø_0$ is angle of friction at zero dilation = $5°$)

(b)

Figure 8.2. (a) Definition diagram for turning moments (based on Helley, 1969); (b) Balance of forces acting on a single grain

The model, therefore, contains the effect of grain shape both in the calculation of turning moments and in the coefficient of drag used. There is also a limited degree of variation in the packing or positioning of the particle, as represented by the value of the orientation angle. The effect of changing the angle of the particle from 0° to 30° yields an almost linear increase of up to 50 per cent in the threshold velocity. However, the tilting of the particle seems to reflect increasing imbrication of the bed rather than the protrusion of

Table 8.2. Grain geometries and thresholds based on a sliding assumption (shaded grain is to be moved)

Threshold I

Grains in contact with the bed

$$0.157 \, (u_y^2 + 0.59u_t'^2 \tan\phi) = \left(\frac{\rho_s - \rho}{\rho}\right) \cdot g \cdot \frac{\pi}{6} \, D \, (\cos\beta\tan\phi - \sin\beta)$$

Threshold II

Grains within the bed

$$0.07 \, u_y^2 = \left(\frac{\rho_s - \rho}{\rho}\right) \cdot g \cdot \frac{\pi}{6} \, D \, \cos\beta$$

Threshold III

Grains within the bed but unsupported by a downstream grain

$$u_y^2 \, (0.12 + 0.07 \tan\theta) = \left(\frac{\rho_s - \rho}{\rho}\right) \cdot g \cdot \frac{\pi}{6} \, D \, (\cos\beta\tan\theta - \sin\beta)$$

where ϕ is angle of friction $(\phi_0 + \psi)$
β is average bed slope angle

individual particles and the effect on the threshold seems rather limited compared with the range of empirical values quoted in Section 8.2.2.

The Model of Naden (1987a)

The model of Naden (1987a), on the other hand, assumes spherical grains and attempts to build into the analysis both the effect of grain geometry, protrusion and pivoting angle, and the effect of turbulence.

Taking spherical grains of similar diameter, it is clear that essentially three types of grain geometry can be identified — grains in contact with the bed, grains within the bed, and grains within the bed but unsupported by a downstream grain (Table 8.2). Strictly speaking, a fourth geometry, of grains below the coplanar surface, also exists but the analysis used here will not extend to cover such cases. A balance of the 'instantaneous' forces acting on the grain provides the threshold equations for the three geometries as given in Table 8.2. Their derivation and the use of appropriate lift and drag coefficients are fully discussed in Naden (1987a).

The turbulence effects are built into the model by assuming that the instantaneous velocity is made up of a time–mean velocity and a velocity fluctuation:

$$u_y = \bar{u}_y + u'_t \tag{2}$$

where $\bar{u}_y$ is time-mean velocity at height y
u'_t is the velocity fluctuation:

The time-mean flow velocity at a height, y, above the bed is assumed to be adequately given by the logarithmic flow law, based on the Karman-Prandtl equation, with the roughness length taken to be equal to the median grain size. The velocity fluctuation is assumed to come from a normal distribution with a mean of zero and a standard deviation related to the height above the theoretical bed (analysis of McQuivey's (1973) lead shot data—see Naden, 1987a):

$$\sigma_u/\bar{u}_y = 0.16 \ (y/D)^{-0.65} \tag{3}$$

where σ_u is standard deviation of velocity

y is height above the theoretical bed (see Einstein and El Samni, 1949)

This simple view of turbulent velocity fluctuations ignores the work of Grass (1971), Falco (1977), Nakagawa and Nezu (1981), and Leeder (1983) on the burst-sweep cycle. However, in the absence of suitable data, it is the best description of turbulence available for use in the model.

The instantaneous velocity required for movement is then compared with the mean and standard deviation of the velocity at the required height for any given flow. The probability of erosion is obtained from tables of the probability density function for a standardized normal distribution. Alternatively, Shields' parameters for chosen probability levels can be calculated. For example, Christensen and Bush (1971) suggest as suitable a probability level of 0.01. Using this, calculations of the threshold velocities for the two extreme grain geometries of Table 8.2 yield Shields' parameters of 0.015 and 0.07 respectively—not dissimilar to the values of 0.01 and 0.1 quoted by Fenton and Abbott (1977) from their flume experiments.

Grain Geometry

Following up the question of grain geometries, it is clear from Table 8.2 that, in the model of Naden (1987a), the upstream geometry controls the height above the bed for which calculations are made, while the downstream geometry determines the angle of friction used. In other words, the model incorporates both the effect of relative protrusion and the question of pivoting angles. Extending the analysis to a range of grain sizes, this simply means that the appropriate heights and angles have to be calculated for the range of geometries concerned. Although in nature this is obviously a continuum, it is helpful to limit the analysis to some set geometries. Assuming a rectangular grid, 36 stable geometries can be drawn using just two grain sizes (Naden, 1985). These are shown in Table 8.3 along with their reference heights and angles of dilation. Given a bed shear velocity of 0.3 m s^{-1}, Figure 8.3 shows, in three dimensional form, their probabilities of movement based on the two parameters—relative protrusion and dilation angle. Three features are of interest. First, the probabilities cover the full range from 0 to 1. Second, in the case of larger grains the effect of relative protrusion is less pronounced than the effect of the dilation angle. Third, the effect of the mixed grains is to change the probabilities of movement such as to make small grains less easy to move (even when not hiding within crevices) and large grains easier to move. This is in agreement with the field data of Parker et al. (1982) and Andrews (1983), and with the discussion of Komar and Li (1986).

The model developed above, then, is seen to agree qualitatively with field observations and begins to make some sense of the plethora of factors contributing to variations in the

Table 8.3. Grain geometries in a simplified two-size mixture (shaded grain is to be moved)

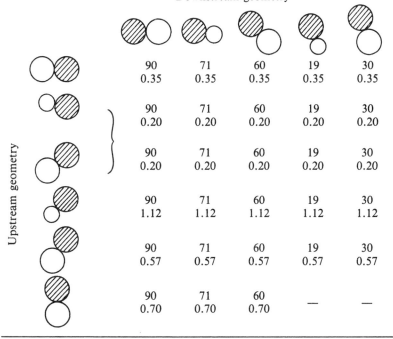

(a) **Smaller grains**

Downstream geometry

90	71	30	42	—dilation angle
0.35	0.35	0.35	0.35	—relative protrusion above
				the theoretical bed
90	71	30	42	
0.35	0.35	0.35	0.35	
90	71	30	42	
0.57	0.57	0.57	0.57	
90	71	30	42	
0.26	0.26	0.26	0.26	
90	—	—	—	
0.45				

(b) **Larger grains**

Downstream geometry

90	71	60	19	30
0.35	0.35	0.35	0.35	0.35
90	71	60	19	30
0.20	0.20	0.20	0.20	0.20
90	71	60	19	30
0.20	0.20	0.20	0.20	0.20
90	71	60	19	30
1.12	1.12	1.12	1.12	1.12
90	71	60	19	30
0.57	0.57	0.57	0.57	0.57
90	71	60	—	—
0.70	0.70	0.70		

(a) Smaller grains

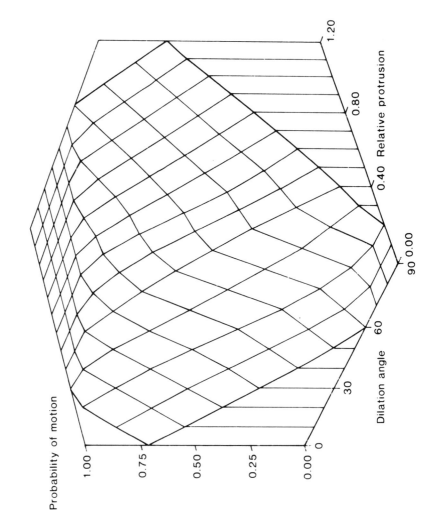

Probability of motion

1.00

0.75

0.50

0.25

0.00

Dilation angle

0

30

60

90 0.00

0.40 Relative protrusion

0.80

1.20

(b) Larger grains

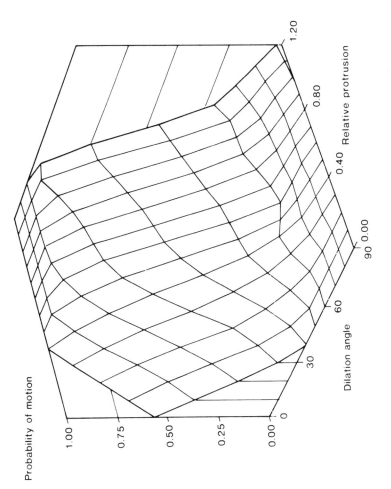

Figure 8.3. Probability of motion for small and large grains as related to angle of dilation and relative protrusion assuming a bed shear velocity of 0.3 m s⁻¹

Shields' parameter. However, having increased the complexity of the model, its field requirements have also increased. These field requirements are discussed below.

8.2.5 Field Requirements

First, there is obviously a need to characterize river beds not only in terms of their grain size distribution but also in terms of their grain geometry. The Helley (1969) model used one version of this in which orientation angle was considered. However, this accounted for relatively little variation in the predicted threshold. A more promising measure is that of relative protrusion as examined by Abbott and Francis (1977) or that of upstream and downstream protrusion as used in the Naden (1987a) model. A distribution of protrusion figures and friction angles for different grain sizes would certainly help to establish a link with the empirical results of Andrews (1983) based on the ratio of grain size to median grain size.

Secondly, there is the problem of the measurement of near-bed velocities and their turbulence intensity. As yet there is little data on this subject and, although not essential to implementing the Naden (1987a) model described above, rather tenuous assumptions regarding the flow parameters are implicit in this model. Equally important in the contemplation of stochastic models for the prediction of a probability of movement is the question of time and space scales. For example, given long enough, any grain with a probability of motion greater than zero will move. In terms of space, the scale of eddy capable of producing movement will depend upon grain size—the larger the grain size, the larger the eddy required to move the grain and the nearer the 'instantaneous' bed shear stress to the mean. Thus, a spectrum of turbulence energies would be most appropriate for the future development of this modelling approach.

Thirdly, more work needs to be done on bedforms both in sand and gravel bed rivers and their effect on thresholds. Two effects are immediately apparent: first, their influence on the flow regime and, second, any changes in individual grain geometry. The first point relates back to accurate descriptions of the near-bed velocities and to the work of Brayshaw et al. (1983) on the spatial variation of drag and lift coefficients around obstacles. The second relates to the potential for reducing the availability of grains through the development of bedforms (Naden and Brayshaw, 1987) and imbrication. The net effect of this is to delay the onset of incipient motion and to reduce the total grain movement. This also has a feedback on the release of fines from gravel bed rivers and the rates of bedload transport (see below).

Consequently, what might be envisaged as suitable data from which to develop these modelling ideas further is some sort of cross-tabulation between grain size, type of site (i.e. grain geometry and position on the bed), and distribution of bed velocity. In this way, it would be possible to derive probability limits for different timescales around some sort of modified Shields' criterion.

8.3 RATES OF SEDIMENT TRANSPORT

In addition to the problem of the initiation of movement, it is important to know how much sediment moves. This is a more difficult problem and it is useful to clarify a number of definitions before considering the different approaches to sediment transport modelling. Moving sediment may be classified either with respect to its source or to its mode of motion.

In the first case, bed material load is the sediment derived from the river bed regardless of grain size or means of travel, whereas washload is the inflowing sediment from the reach above and local tributary streams. Washload usually consists of fines derived from the catchment and, as such, its production is mainly covered in Chapters 9 and 10 on catchment yield. However, it may also include bed material load from just upstream of the reach in question and, therefore, can pose an important measurement problem. As a word of caution, readers should be aware that some authors use washload simply to denote silt–clay material which may be derived from the catchment or from temporary storage in the bed, and bedload to refer to sand and gravel (e.g. Pizzuto, 1985).

Alternatively, moving sediment can be classified in terms of its mode of motion. Thus, we can define bedload as that sediment load moving in contact with the bed and suspended load as that sediment which is wholly supported by the fluid. In this way there is a vertical differentiation between bedload and suspended load which might suggest which fluid parameters should become the focus of modelling. For example, in the case of bedload we are concerned with the shear stress at the bed, whereas in the case of suspended load the whole velocity profile is taken into account. This definition, however, is not without its problems as the actual mode of travel is not only related to grain size but also to transport stage (Abbott and Francis, 1977) and concentration of transported sediment (Leeder, 1979). In the case of bedload in particular, there is the transition from rolling and sliding motions to saltation where grains are not supported by the flow but movement may occur to a height of 2–3 grain diameters above the bed (Abbott and Francis, 1977).

The character of each of these types of sediment movement differs so much that a range of approaches to the problem of modelling sediment transport has evolved — some dealing with one particular sediment load type; others seeking, from alternative standpoints, the illusive model for total sediment load. Here, the models available are discussed under the two headings of suspended sediment and bedload.

8.4 SUSPENDED SEDIMENT

Irrespective of the source of suspended sediment, there have been three different approaches to modelling — empirical rating curve models, diffusion theory models, and stochastic models. While these have shown reasonable success in specific examples, they are not widely applicable due to their ignorance of the sediment supply side of the problem. Hence, more recent developments have revolved around the question of supply-based models (McCaig, 1981; VanSickle and Beschta, 1983). This section reviews the more common traditional models and then goes on to assess supply-based models and their demands on field data and experimentation.

8.4.1 Empirical Rating Curve Models

The simplest of the approaches is that of the rating curve, widely used by geomorphologists. Here overall suspended sediment concentration is empirically related to stream discharge by a simple power function:

$$C = aQ^b \qquad (4)$$

where C is suspended sediment concentration
       Q is stream discharge
       a,b are empirical constants

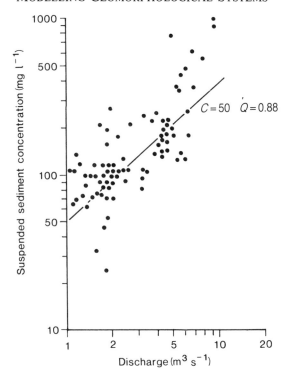

Figure 8.4. Typical scatter graph of suspended sediment against discharge (after Richards, 1982, reproduced by permission of John Wiley & Sons Ltd)

A typical example is given in Figure 8.4. Suspended sediment loads are then calculated by multiplying the water discharge by the suspended sediment concentration derived from the rating curve. Ferguson (1986) suggests that a correction factor should also be incorporated to counteract the statistical bias involved in these estimates. Other problems inherent in using a model of this type fall into two categories.

First, as with any empirical relationship, there is heavy reliance on the original data used to construct the model. In particular, the model should not be used outside the range of previously observed events and the accuracy of the model may well be limited by the accuracy of the original data. Probably the most common error in suspended sediment data is that caused by using an inadequate sampling programme. For example, Walling (1977) found that one-hourly sampling of continuous discharge and sediment concentration records was adequate to characterize the sediment loads for the River Creedy 1972–1974 whereas weekly data were not. Other inaccuracies in the data are often associated with limited spatial representation in the samples and, for a discussion of the magnitude of such errors, the reader is referred to Burkham (1985). As well as having regard for the accuracy and range of data used to construct the model, it is also clear that the model is site specific, i.e. the parameters *a* and *b* cannot be transferred to other rivers. Although a means of deriving these coefficients from the physical characteristics of the catchment are examined by Rannie (1978) for some 50 basins across the U.S., the data demands to calibrate this type of model are heavy.

The second problem with suspended sediment rating curve models relates to the scatter of the data. Fairly extreme examples of this are the envelope curves of Burt and Gardiner (1984). There are two methods of dealing with the scatter. One method is to simply preserve the scatter and assign probabilities to the values (Grenney and Heyse, 1985). However, looking in detail at suspended sediment dynamics, it is clear that one reason for the scatter is the presence of hysteresis (Bogen, 1980). In other words, the sediment wave is not synchronous with the water wave and either lags it (Heidel, 1956) or, more often, precedes it (Walling, 1977; Collins, 1979; Richards, 1984; Fenn et al., 1985). The second method of dealing with scatter about the regression line, then, is to explicitly model the hysteresis effects.

The mechanisms behind hysteresis are not, as yet, quantified but are known to relate, in general, either to water and sediment routing (Heidel, 1956), or to the exhaustion of sediment storage (Walling, 1978). The possible processes involved will be discussed later but even the recognition of hysteresis provides a simple means of reducing the scatter in single-valued rating curves. The modeller can either split the data into that associated with the rising limb of the hydrograph and that associated with the falling limb to give two distinct functions (Walling, 1977) or build a second independent variable, such as change in flow, into the regression equation (Richards, 1984). Other data splits aimed to reduce the scatter in the rating curve have been used. For example, seasonal variations might be identified (Walling, 1977) and, in a pro-glacial context, events governed by snowmelt versus rainfall (Richards, 1984) or other climatic controls (Fenn et al., 1985).

The existence of hysteresis also suggests that there is serial autocorrelation in the data. In other words, each data point is related to the others in the sequence. Yet another modelling approach, then, is to use time series models (Box and Jenkins, 1976) such as ARIMA (Gurnell and Fenn, 1984). Both time series models and the multiple regression models discussed above provide more satisfactory predictions of suspended sediment loads but tend to increase the problem of data collection.

8.4.2 Diffusion Theory Models

By contrast, the typical engineering approach, often incorporated into total sediment load equations, is a deterministic diffusion model to express the distribution of suspended sediment in the vertical. It is worth looking briefly at the derivation of the diffusion equation and the assumptions behind it in order to assess its relevance to the suspended sediment loads of natural river channels.

The approach assumes that the distribution of suspended sediment concentration in the vertical is due solely to the action of the flow on a mobile bed and that the two-dimensional turbulent flow is both steady and uniform. The equilibrium distribution of suspended sediment can then be derived using an analogy between mass transfer and momentum transfer. Consider two levels y_1 and y_2 in the flow a distance ℓ apart where ℓ is the average size of a macroturbulent eddy. As there is no net movement of sediment in the vertical over time, the weight of sediment moving from y_1 to y_2 is equal to the weight of sediment moving from y_2 to y_1 (see Yalin, 1977), i.e.

$$(C - dC/dy.\ell/2)(|v'| - V_g) = (C + dC/dy.\ell/2)(|v'| + V_g) \tag{5}$$

where C is concentration at $y=(y_1+y_2)/2$
$\quad$ $|v'|$ is average vertical velocity
$\quad$ V_g is settling velocity of sediment
$\quad$ ℓ is the average size of a macroturbulent eddy.

Thus

$$dC/dy + a_* C = 0 \qquad (6)$$

and

$$a_* = 2V_g/\ell|v'|$$

Integrating gives

$$C = C_0\, e^{-\int_{y_0}^{y} a_*\, dy} \qquad (7)$$

where C_0 is the concentration at some reference height y_0.

The quantity $\ell|v'|$ is proportional to the turbulent kinematic viscosity which can be expressed in terms of the turbulent shear stress and velocity gradient (Yalin, 1977) giving

$$a_* = \frac{V_g}{B}\frac{1}{\tau}\frac{du}{dy} \qquad (8)$$

where B is, in general, related to the Reynolds number.

Thus, the vertical distribution of suspended sediment reduces to a problem of the vertical distribution of shear stress and flow velocity.

The classic solution to this problem (Rouse, 1937; Einstein, 1950) is to assume a linear law for the shear stress distribution and a logarithmic law for the velocity distribution giving

$$\frac{C}{C_0} = \left[\frac{(1/\eta - 1)}{(1/\eta_0 - 1)}\right]^{V_g/\chi Bu_*} \qquad (9)$$

where η $= y/h$
$\quad$ h is total flow depth
$\quad$ η_0 $= 0.05$
$\quad$ χ is a constant
$\quad$ u_* is the shear velocity

This yields the series of curves shown in Figure 8.5 which show very good agreement with the data of Vanoni (1941) if $\chi B \approx 0.4$.

An alternative formulation by Hunt (1954) uses Von Karman's original distribution for velocity to yield

$$\frac{C(1-C_0)}{C_0(1-C)} = \left[\sqrt{\frac{(1-\eta_0)(B-/(1-\eta_0))}{(1/\eta)(B-/(1-\eta))}}\right]^{V_g/\chi Bu_*} \qquad (10)$$

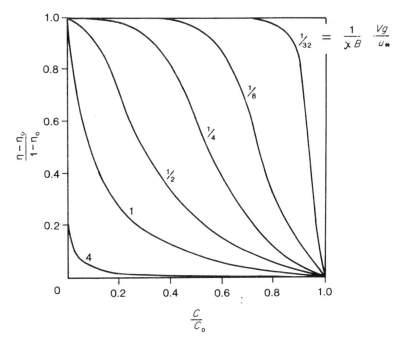

Figure 8.5. Solutions to the diffusion theory equations (based on Yalin, 1977)

which gives slightly better agreement with Vanoni's (1941) data especially at higher sediment concentrations.

However, the whole rationale of using a velocity profile based on clear water data is suspect (Coleman, 1981) and much recent work has concentrated on refining the model to reflect the effect of turbulence damping due to the presence of sediment and on the form of the sediment diffusion coefficient (van Rijn, 1984b). Other work has focused on the methods of calculating the reference concentration C_0—see, for example, Pizzuto (1984) for the case of dune beds. Due to inaccuracies in the main controlling parameters, however, the total sediment load *in conditions of unlimited sediment supply* still cannot be predicted with an inaccuracy less than a factor of 2 (van Rijn, 1984b).

Other contributions in a similar vein but using rather different assumptions, such as the gravitational theory (Velikanov, described in Yalin, 1977) and the more recent mixture theory (McTigue, 1981) yield similar functions for the suspended sediment concentration and show similar agreement with published data.

All these models, however, rely on an equilibrium between the buoyant weight of particles and a correlation associated with the fluid drag due to the vertical velocity fluctuations of the flow. Turbulence is assumed to be isotropic rather than corresponding to more recent findings (Grass, 1971; 1974) in which turbulence in the vertical is found to be asymmetric—the so-called burst–sweep cycle. Here, masses of fluid moving away from the general region of the bed into the turbulent boundary layer remove with them large amounts of bedload material. As the upward-directed fluid momentum decays, grains fall out towards the bed until they respond to another burst motion. Analysis of Brodkey *et al.*'s (1974) data by Leeder (1983) shows that the residual upward shear stress responsible

for the suspension of the grains is approximately equal to 0.3 times the shear stress on
the bed.

This is consistent with Bagnold's (1966) formulation of the suspended sediment equation.
Using data from upper stage plane beds of fine sand with a grain diameter less than 4 mm
(Guy *et al.*, 1966), Leeder (1983) finds that a transport equation of the form

$$q_s = 0.3\bar{u}/g\tau \tag{11}$$

where q_s is sediment transport rate in terms of immersed mass per unit width per unit
        time
   u is mean flow velocity
   g is acceleration due to gravity
   τ is bed shear stress

would seem to be a valid reply to the detailed analysis above and one worth comparing
with other models.

8.4.3 Stochastic Models

In so far as suspended sediment particles owe their position in the flow to its turbulent
velocity fluctuations, the distribution of these particles might be approached as a stochastic
problem. Consider the flow depth, h, divided up into m layers of unit length and thickness
Δy. Particles are distributed across the layers such that there is a transitional probability
of each particle moving from layer k to layer $k+1$ of

$$p(k/k+1) = 0.5\{1 + \psi[c[k-1,i) - c(k+1,i)]\} \tag{12}$$

and from layer k to $k-1$ of

$$p(k/k-1) = 0.5\{1 + \psi[c(k+1,i) - c(k-1,i)]\} \tag{13}$$

where ψ is a suitable weighting function

   $c(k,i)$ is concentration in layer k at time i

If concentrations in layers $(k+1)$ and $(k-1)$ are equal then there is a 50:50 chance of
the particle moving up or down. Otherwise, there is a greater chance that the particle will
move in the direction of the lower concentration. This type of model (Bechteler and Färber,
1985) can be used to derive the diffusion equation. However, instead of relying on shear
stress and velocity distributions for its results, the weighting function ψ must be determined.
The time step over which the random walk is calculated is also an important variable and
should be related to the flow conditions (Yalin, 1977).

In their numerical examples, Bechteler and Färber (1985) use the velocity components of
the flow to drive the model with turbulent velocity fluctuations drawn from a normal distribution. Their results show good agreement with the data of Vanoni (1941) but, as in the diffusion
case, the experimental evidence of the burst–sweep cycle is ignored. However, there is no

reason why this sort of model should not use a skewed distribution for the vertical velocities. Other advantages of this approach include the potential application of the model to both higher concentrations of sediment than those appropriate to the assumptions of diffusion theory and to non-equilibrium conditions caused by gradual changes in bed configuration.

The main limitations of the model amount to the lack of information about turbulence or velocity distributions in rivers and the fact that it is only applicable to rivers in which there is a uniform and unconstrained supply of sediment.

8.4.4 Supply-based Models

Throughout this discussion of approaches to modelling suspended sediment transport, it has been apparent that theoretical developments have related solely to sediment sources within the channel which are unlimited and provide uniform availability of sediment. This severely limits the application of these models to the field. The problem was neatly summarized by Einstein and Chien (1953) who found that the transport of washload particles, in this case simply denoting the silt–clay fraction, was solely determined by their availability. The need for supply-limited models, then, is clearly evident and, as will be apparent later, is also pertinent to the question of bedload especially in mixed and gravel bed rivers.

An early attempt to build in the question of sediment availability was that of Walling (1978) in which he proposed a multiple regression model to predict peak sediment concentrations from the hydrograph rise and a sediment availability index. Applied to the River Dart, the relationship suggested was

$$\log C = 2.4454 + 0.8825 \log Qr - 0.4905 \log Ai \tag{14}$$

where C is peak suspended sediment concentration for an event
Qr is hydrograph rise for the event
Ai is availability index given by

$$Ai = (Ai_{d-1}K) + Qr \tag{15}$$

where Ai is sediment availability index at end of day
Ai_{d-1} is sediment availability at end of previous day
K is a constant decay factor (0.9)

This type of availability index based on antecedent moisture conditions demands nothing of the measurement of sediment supply and is obviously open to calibration in other drainage basins. Its main limitation, however, is the fact that it does not include a mass balance model to account for the sediment stored nor consider the mechanisms of sediment release.

An alternative model which at least incorporates the concepts of sediment storage capacity, mass balance, and depletion of storage is that of VanSickle and Beschta (1983). The two model equations are

$$dS(t)/dt = -Q(t).C(t) \tag{16}$$

and

$$C(t) = aQ(t)^b \, p \, \exp[rS(t)/S_0] \tag{17}$$

where $S(t)$ is the sediment stored at time t
     $Q(t)$ is the discharge at time t
     $C(t)$ is sediment concentration at time t
     a, b are empirical rating curve parameters
     p, r are empirical washout function parameters
     S_0 is the sediment storage capacity

Thus, an exponential washout function related to the size of sediment store is built into the rating curve model. For an initial storage set to the capacity storage, successive events of the same magnitude yield less sediment depending on the ratio of $S(t)$ to S_0. A mechanism for replenishing the sediment store has yet to be added.

VanSickle and Beschta (1983) explore two versions of this model—a single store, i.e. a lump of sediment added to the channel between storms (mass soil erosion being the most significant input of sediment in the Pacific Northwest), and a distributed model in which the magnitude of the discharge determines how many of a series of sediment stores are tapped. For example, one can envisage progressive wetting up a vertical bank or progressive disruption of the bed releasing new pockets of sediment. The models were calibrated for Flynn Creek and showed good agreement with collected data. They also provide an excellent conceptual guide to the problem of sediment budgetting.

Along similar lines is the model proposed by McCaig (1981) in which he begins to derive functions for the different sediment release mechanisms and suggest the field data needed to put together a more process-based, supply-limited, suspended sediment model. The actual sources of sediment can be classified into two main types—catchment supplies and within-channel supplies. The relative importance of these sources varies considerably from catchment to catchment. Glymph (1957), in a survey of 113 watersheds in the U.S. proposed that sheet erosion accounted for between 11 and 100 per cent of the sediment, while Imeson (1974) estimated that 35 per cent of suspended sediment in Hodge Beck, North York Moors, was from bank erosion. Sources of sediment in the channel bed have received less quantitative attention although experiments on the intrusion of fine sand into gravel (Beschta and Jackson, 1979; Frostick et al., 1984) suggest that between 8 and 25 per cent of the bed may be composed of sediment capable of being suspended.

McCaig (1981), therefore, proposes three submodels for supply from the catchment, channel banks and channel bed. Supply from the catchment, based on dynamic contributing areas, really links into the material discussed in Chapters 9 and 10. Consequently, it will not be considered here, although it is recognized that this may be a dominant control of the sediment supply. With regard to the within channel sources, available sediment can be defined as that sediment stored on the banks or in backwater areas which simply requires an increase in discharge for its release. In the case of these supplies, a model of the form used by Walling (1978) may be appropriate but should include a mass balance equation to account for exhaustion effects. Semi-available sediment or, in other words, sediment which is only released when a given threshold is exceeded may also be defined. For example, this should include sediment stored behind cobbles or within the armour sublayer. To release this sediment, a model related to the erosion threshold equations presented above is needed.

Table 8.4. Data requirements for supply-based models of suspended sediment transport (sediment supply from channel boundary only) (based on McCaig, 1981)

Source	Data requirements
Bank sediments	area of bank bank material (particle size, erodibility) vegetation (effect on erosion, sediment supply) relation between discharge and wetted area of bank climatic factors generating sediment rates of generation of sediment by frost, wetting and drying cycles etc.
Backwater sediments	area of backwater reaches relationship between discharge and area of backwaters suspended sediment concentration settling rates of sediment relationship between velocity and erosion/deposition algal growth
Cobble lee sediments	number of cobble lee sites relationship between cobble size and sediment storage capacity relationship between discharge, cobble size, scale of lee side eddies and sediment stored deposition rates of sediment in cobble lee sites erosion thresholds for cobbles
Subarmour layer sediments	erosion threshold for disruption of armour layer rate of infiltration of fines fine-sediment storage capacity

Exhaustion effects are provided by a mass balance equation operating on the stores behind each cobble or within each unit area of the bed.

8.4.5 Field Requirements

The data required to implement this type of model is given in Table 8.4. The number of relations to be developed, at first, seems prohibitive. However, if these relations can be used as the basis for future theoretical developments of supply-based models, then the prediction of suspended sediment loads in natural channels should be greatly improved. The source of the improvement lies both in the incorporation of sediment supply limits and in the choice of relevant process assumptions behind the release and generation of stored sediment. (For the application of this type of model to bacterial concentrations in natural channels, see Jenkins *et al.*, 1984.)

8.5 BEDLOAD

The immediate source of material for bedload transport is the channel boundary. However, bedload transport is not exactly synonymous with bed material transport as some of the smaller sizes of bed material may go directly into suspended load. Other sources of bedload are the channel banks and mass movements directly into the channel. However, for the most part, in modelling bedload transport, it is assumed that the river channel is the sole

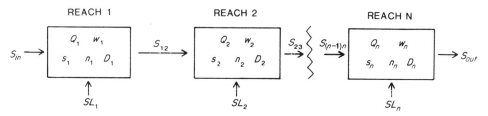

Figure 8.6. Sediment input and output for a series of uniform channel reaches as a basis for modelling bedload transport

source of bedload material. The question is how much sediment moves, where from, and where to.

There are two basic approaches to the problem of bedload transport. In the first case, use is made of empirical or semiempirical sediment transport formulae, whereas, in the second, the sediment transport process is built up from a consideration of individual or mean grain movements. Each of these approaches is discussed below.

8.5.1 Models Based on Sediment Transport Equations

Models based on sediment transport equations all have a similar overall outline. The channel acts as a conveyance for sediment transport and attention is focused on the total amounts of sediment moving, the associated scour or deposition on the bed, and the corresponding changes in the overall long profile. In order to do this, the river is divided into a number of reaches (Figure 8.6) with similar characteristics of discharge, channel width, slope, bed material, and roughness.

For each of these reaches, the initial step (Figure 8.7) is to calculate the flow parameters required. For this, use is made of one of the standard techniques for calculating gradually varied flow (Fread and Harbaugh, 1971; Humpidge and Moss, 1971). The assumptions behind these techniques are discussed in Chapter 6. They have been successfully applied to the scale of pools and riffles by Richards (1978) and provide the modeller with flow depths and velocities down the channel. However, sediment transport usually requires the prediction of near-bed velocities or bed shear stress which may not be so accurately predicted. For example, the mean shear stress calculated by $\tau = \rho gRs$ does not reflect the shear stress on the bed (Bridge and Jarvis, 1976) and the standard logarithmic flow law for the calculation of velocity profiles is not really applicable to flows governed by backwater curves (Bathurst, 1982a). However, the degree of error involved in such estimates has yet to be assessed.

Using the derived flow parameters, the potential sediment transport in each reach is calculated (Figure 8.7) using a standard sediment transport equation. There is a wide range of alternatives to choose from. The reader is referred to the review of White *et al.* (1973) for a detailed assessment of the fit of many of these equations to available data. There are, however, three main types of bedload equation each of which is based on a different set of assumptions which may be more or less applicable to different types of channel. The equations also make rather different demands on data collected or calibration, and, therefore, they merit a brief review (Table 8.5).

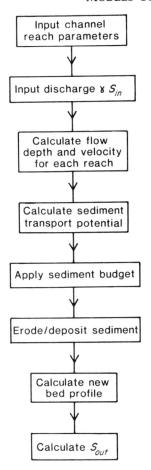

Figure 8.7. Generalized flow diagram depicting the steps in bedload modelling

Meyer-Peter and Müller Equation

The first type of equation is that in which bedload transport is related to an excess, above some threshold value, of either shear stress (du Boys, 1879; Meyer-Peter and Müller, 1948), near-bed velocity (Kalinske, 1947) or water discharge (Schoklitsch, 1950). Equations of this type are empirically based and suffer from the problem of having to define a threshold value akin to a Shields' parameter (see discussion above). Of the equations cited, the most popular among geomorphologists is the Meyer-Peter and Müller (1948) equation

$$q_s = 8 \, [g\{(\rho_s - \rho)/\rho\}D^3]^{0.5} \, [K\rho Rs/\{(\rho_s - \rho)D\} - 0.047]^{1.5} \tag{18}$$

where q_s is sediment transport rate per unit width

    R is hydraulic radius

    s is water surface slope

    K is a calibration constant based on the ratio of total to particle roughness

Table 8.5. The most commonly used sediment transport equations: their data requirements and range of application

Equation	Data required	'Correction factors'	Application
Meyer-Peter and Müller	ρ, ρ_s D_{50}, D_{90} R, s, u	K	large grains (>0.4 mm) low transport rates
Bagnold	ρ, ρ_s, v D_{50}, V_g R, s, u	$\dfrac{e_b}{\tan\phi}$	later stages of motion depth >150 mm $\left(\dfrac{\rho_s-\rho}{\rho}\right)=1.65$ $0.3 \leqslant u \leqslant 3.0$ m s^{-1} fully developed suspension
Einstein	ρ, ρ_s, v grain size distribution u (or velocity profile)	ξ, p, $\dfrac{B}{B_*}$	later stages of motion

The constant K varies between 0.5 and 1.0, and, for natural channels and fully developed turbulent flow, is given by

$$K = (\bar{u}\, D_{90}^{1/6}/26R^{2/3}s^{1/2})^{3/2} \tag{19}$$

where D_{90} is the grain size than which 90 per cent grains are finer
    $\bar{u}$ is mean flow velocity

The popularity of the Meyer-Peter and Müller equation is partly due to the fact that it is relatively easy to use and makes few demands on field data (Table 8.5). It is also useful in that it was originally derived using a wide range of bed material sizes including coarse materials (0.4 mm to 30 mm) whereas many of the other equations are derived from flume experiments using only the sand range of particle sizes. However, the equation suffers from the problem of defining a suitable constant threshold term. For instance, it makes use of what is essentially a Shields' parameter of 0.047. However, the Shields' parameter is known to vary with grain Reynolds numbers for small grain sizes (Figure 8.1). This means that the equation can only be applied to large grain sizes i.e. coarse sand and above. Furthermore, as discussed in section 8.2, the Shields' parameter is not a constant even for large grains but varies with grain exposure and packing. Due to the omission of a relative depth (depth/grain size) term, the equation is also limited to conditions of low sediment transport rates where motion is restricted to rolling and sliding modes (Yalin, 1977; Richards, 1982).

Bagnold Equation

Akin to the relation between sediment transport and discharge is the approach used by Bagnold (1956) in which he considers a stream power function. The theoretical background to Bagnold's equation is that sediment transport rate has the same units as rate of work or power expenditure. The river is, therefore, likened to an engine, and sediment transport is given by stream power multiplied by an efficiency factor

$$q_s = \omega e_b / \tan \phi \tag{20}$$

where ω is stream power per unit bed area
       e_b is bedload efficiency factor
       $\tan \phi$ is a dynamic friction coefficient

Stream power per unit bed area is variously quoted as $\rho g R s u$ or $\rho u u_*^2$. The equation has evolved through several refinements. In 1966, Bagnold extended the equation to include suspended sediment load derived from the bed material and carried by the power remaining after bedload movement. Total load was, therefore, given by

$$q_s = \omega \left[(e_b / \tan \phi) + (e_s (1 - e_b) \, u / V_g) \right] \tag{21}$$

where e_s is suspended sediment efficiency factor
       V_g is settling velocity of suspended sediment

Bagnold suggested values, based on his experimental evidence, of 0.15 for the bedload efficiency factor and 0.015 for suspended load. These constants assume the transport of sediment with a density of 2650 kg m^{-3} in water. Use of the equation provides good predictions of sediment transport for plane beds but systematically overestimates sediment transport at low stages. This is because the equation fails to take into account enhanced energy losses through friction with the bed. The equation should, therefore, only be applied to the later stages of sediment movement and for depths greater than 150 mm. Bagnold (1973, 1977, 1980) partly accommodates these criticisms by introducing a relative depth term and a threshold stream power in his later equations.

More recent experimental work has also suggested that bedload efficiency, far from being a constant, is related to grain size (Leopold and Emmett, 1976) and to grain size mix (White and Day, 1982; Reid et al., 1986). One of the main problems in testing these ideas, and applying this type of equation, is that the levels of efficiency quoted—5 per cent in East Fork River (Leopold and Emmett, 1976) to 0.05 per cent in Turkey Brook (Reid and Frostick, 1986)—are so low that they are well within the limits of measurement accuracy for stream power. Bagnold's equation, however, is, like the Meyer-Peter and Müller equation, simple to apply and, provided values of efficiency can be estimated or assumed, is again not unduly demanding of field data (Table 8.5).

Einstein Equation

The Einstein (1950) equation, most often used by engineers (for example, the HEC-6 model of the Hydrologic Engineering Center, California, 1974), is based on rather a different philosophy to the Bagnold equation but is again most relevant to established sediment transport. It was originally derived from flume experiments using sand-sized particles in which Einstein observed that there was a steady and continuous exchange of particles between the moving bedload and the bed. However, although the bedload moved steadily downstream, individual particles moved in a series of quick steps with relatively long intervening rest periods. Einstein (1950) observed that the step length was independent of the flow conditions, the transport rate or the bed composition and was constant at around 100 grain diameters. Different transport rates were achieved by changing the average length of time spent between jumps and by changing the thickness of the moving layer. Einstein's analysis of the problem was then to write down functions for the rate of erosion and for the rate of deposition of particles based on a statistical consideration of the lift forces acting. The final transport equation is derived by assuming an equilibrium between the moving bedload and the bed material i.e. rate of erosion = rate of deposition.

Calculations of sediment transport are made for each grain size fraction and are based on two dimensionless functions—the flow intensity index

$$\psi_* = \xi p (B/B_*)^2 [(\rho_s - \rho) D/(\rho u'^2_*)]$$ (22)

where ξ is a hiding factor
         p is a pressure correction term for use in the transition zone from smooth to rough boundaries
       B/B_* is a friction correction factor
       u'_* is the shear velocity with respect to the grain

and the transport intensity for the chosen grain size

$$\phi_* = (F/_B/F_b) (q_s \rho^{0.5}/[(\rho_s - \rho) g D]^{1.5})$$ (23)

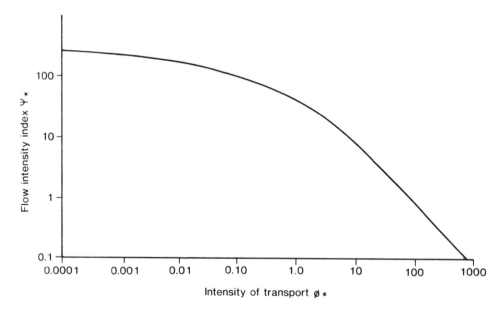

Figure 8.8. Einstein's (1950) bedload function

where F_B is fraction of bedload of a given size
F_b is fraction of bed material of a given size

The flow intensity index is calculated and the value of the intensity of transport read off a graph, such as that shown in Figure 8.8, in order to calculate the sediment transport rate, q_s, for each size fraction.

The problem in applying Einstein's equation is its complexity and the derivation of the various correction factors (Table 8.5) required for its implementation. Although ideally a velocity profile would be useful in determining the appropriate shear velocity for each grain size, a logarithmic flow law can be assumed under some flow conditions (Bathurst, 1982a), and the appropriate velocities derived. The equation then makes no more demands on field data than other equations (Table 8.5). The main question, as with Bagnold's equation, is the accuracy and reliability of the derived correction factors ξ, p and B/B_*.

The Einstein equation, however, does get over the problem of a threshold. Instead, this is represented by the very steep rise in transport rate with increasing shear velocity reflecting the increasing probability of motion. The equation has proved successful in sand-bed applications where the idea of a moving carpet is an appropriate one. For the case of gravel-bed rivers, it is worth noting that subsequent flume experiments, using a wider range of sand and gravel materials, have reported that particle step length is dependent on bed roughness (Ippen and Verma, 1953), sediment size (Grigg, 1970) and flow conditions (Grigg, 1970; Abbott and Francis, 1977). The use of such information might provide an interesting extension of Einstein's work.

Continuity Equation

Having chosen a suitable sediment transport equation, the final calculation is of a sediment budget for the reach. Written in its simplest form this reads

$$S_{out} = S_{in} + \text{bed erosion} - \text{bed deposition} + \text{washload} \tag{24}$$

where S_{out} is the sediment exported from the reach
S_{in} is the sediment input to the reach

or in its differential form

$$\frac{dQ_s}{dx} - \frac{w_s dy_s}{dt} = q_{sl} \tag{25}$$

where Q_s is total sediment transport rate
w_s is active bed width
y_s is bed surface elevation
q_{sl} is sediment input from lateral or tributary sources

In other words, downstream sediment transport is related to the change in bed elevation over time and any additional inputs of sediment from lateral or tributary sources. This formulation of the model already provides us with an additional measurement problem in terms of the amount of lateral or tributary input and, unless using the example of a spillway, the amount of sediment input from upstream.

Having outlined the general method of modelling sediment transport using the sediment transport equations, we can ask a number of questions, not least of which is how accurate are the model predictions of sediment discharge and river bed elevation? This question is difficult to answer, in that this type of model has largely been applied to engineering problems, and so the model has normally been fitted to a period of data for the river concerned and then applied to a proposed engineering scheme.

The overall accuracy of the model is limited by three factors. The first of these is the accuracy of the sediment transport and flow equations used to construct the model. These are dealt with respectively in the monograph of White *et al.* (1973) and in Chapter 6 above. The second factor which limits the accuracy of the model is the accuracy of the field data, either used as input to the model or used in calibrating the model. The third, and more serious, limitation on model accuracy is the question of whether there are any significant factors not accounted for by the model. Into this last category must come a great deal of the recent evidence from bedload transport experiments. The most serious of these observations is that of bed armouring and this is one factor which has been built into more recent models (Thomas, 1982; Holly and Karim, 1986).

8.5.2   Bed Armouring

One of the most important factors to be taken into account when applying sediment transport models to the field is the question of sediment supply. As described in the case of suspended sediment, supply of bedload is often limited either in absolute terms or because of bed armouring and paving (Gomez, 1984). This has two important effects. First, the grain size distribution of the surface layer differs markedly from that of the overall grain size distribution (Carling and Reader, 1982; Little and Mayer, 1976; Parker *et al.*, 1982). Secondly, the grain geometry and packing of the surface layers causes a change in the threshold of sediment transport. These two factors lead to a change in bedload transport, as shown by Gomez (1983) in terms of the bedload efficiency factor in Bagnold's equation. After a major disruption of the river bed, efficiency was shown to be around 10 per cent. However, as an armour layer developed with successive flows, this dropped to around 1 per cent, effectively causing a positive hysteresis in the sediment transport–stream power relationship. This is similar to the observations made by Newson (1980) on the effect of flood discharges on the rivers Tanllwyth and Cyff in Central Wales. In fact, Moore and Newson (1986) in a discussion of the sediment transport behaviour of these two rivers suggest that it is the channel storage of sediment that is the critical control on sediment yield.

The importance of bed armouring in controlling the sediment transport rate has long been recognized and has been built into the HEC–6 model by Thomas (1982). He defines the equilibrium bed elevation for zero sediment transport in terms of the discharge, Einstein's flow intensity parameter, and grain size. When this coincides with the actual bed elevation then the bed is considered to be fully armoured. Under these conditions, Gessler's (1971) probability function is used to give an effective grain size for the armour layer and grains smaller than this are no longer available for transport. Potential sediment transport is calculated, but only available sediment is removed from the reach. Applications of this model appear to give good predictions of bed elevation (Thomas and Prasuhn 1977;

Thomas, 1982) but how they compare with calculations derived from a model which does not simulate the armouring process is unknown.

8.5.3   Other Implications of Field Data

Other interesting field data has recently come from the techniques of Reid *et al.* (1980, 1984) for continuously monitoring bedload. Two features of their data pertain to the question of modelling sediment transport. First, is the reported wave-like nature (Figure 8.9) of sediment movement (Reid *et al.*, 1985) over which there is much speculation. However, in terms of the models presented here, the averaging of the sediment load over a reach and over a chosen time period means that, although interesting, this phenomena should not be a significant problem. Locally, though, and with respect to sediment transport data from a chosen cross-section, this may be significant and may suggest a kinematic wave approach to modelling (see below).

The second feature reported from the field is the existence of substantially different initial and final thresholds of motion (Reid and Frostick, 1984). This has the effect of prolonging sediment movement to flows well below the level at which movement should have ceased. Consequently, more sediment is being carried over the course of a hydrograph than otherwise predicted. This could well be a significant factor in sediment transport totals especially for events which are close to the threshold. However, it is something which could easily be built into future simulation models.

Another factor which relates to the field situation but which is not adequately covered by the simple models is the gross morphology of the channel. This may be reflected in the bed topography and relate to pools and riffles or channel bars, or it may be the plan

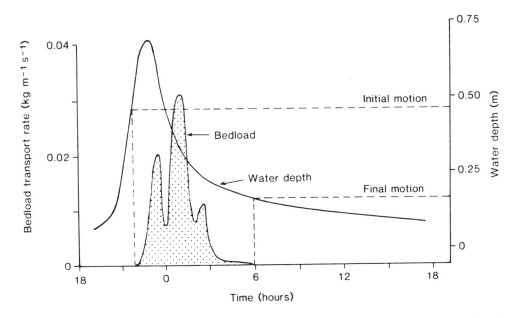

Figure 8.9. Typical pattern of bedload transport over a hydrograph — data from Turkey Brook (after Reid *et al.*, 1985, reproduced by permission of John Wiley & Sons Ltd)

geometry of meander bends and braided channels which poses the problem. As yet, little work has been done on the routing of sediment through pools and riffles apart from the study of Jackson and Beschta (1982). Similarly, sediment transport in meander bends has received some attention from Hooke (1975), Bridge (1977), and Dietrich and Smith (1984). However, these findings have yet to be applied more widely. They certainly make additional demands on field data and it is unknown whether this is justified by the improvement in predictions.

The approach from the sediment transport equations is, then, attempting to build into simple models something of the complications found in the field. Much of this new work, however, still requires testing and many of the governing equations are still being formulated. However, as an engineering tool, such models probably do not perform too badly in simple cases. The alternative route to sediment transport models is to take the complications first, model them, and then simplify to the level of usable functions. This approach is still very much in its infancy and, in most cases, far from the stage of river applications. However, as a conceptual–computational device, it is worth exploring in more detail.

8.5.4 Models Based on a Grain-by-Grain Process

Experimental work using the single-grain fixed-bed technique (Francis, 1973; Abbott and Francis, 1977; and others) has triggered off much of the modelling work from the view of single grain processes. They observed that there were three modes of motion for moving grains—rolling, saltation, and suspension. The rolling mode is limited either to an initial phase of subsequent saltation or to very low levels of sediment transport. The suspension mode that they define is not strictly that where the fluid motions support the grain (see Leeder, 1979) but simply saltation trajectories with slight upward perturbations. Thus, the focus for the work on single grain modelling has been on the saltation process.

Two examples of such models bear consideration—that by van Rijn (1984a) and that by Naden (1985, 1987b). It is hoped to point out the very different demands of these models for field work compared to the models based on sediment transport equations.

The Model of van Rijn (1984a)

Van Rijn (1984a) has modelled saltation as a result of two differential equations

$$m\ddot{x} - F_L\,(\dot{z}/v_r) - F_D\,[(u - \dot{x})/v_r] = 0 \tag{26}$$

and

$$m\ddot{z} - F_L\,[(u - \dot{x})/v_r] + F_D\,(\dot{z}/v_r) + G = 0 \tag{27}$$

where m is the particle and added fluid mass
    v_r is the particle velocity relative to the flow
    u is the local flow velocity
    $\dot{x}$ and $\dot{z}$ are the longitudinal and vertical particle velocities
    $\ddot{x}$ and $\ddot{z}$ are the longitudinal and vertical particle accelerations

The model is calibrated using the experimental results of Fernandez Luque (1974), as van Rijn (1984a) considers that the results of Abbott and Francis (1977) are not representative due to the shallow depths and large grain size used in their experiments. The dimensions of the saltation trajectories and the particle velocity are derived as a function of the transport stage and a dimensionless particle parameter. A bedload transport equation is then formulated by assuming that the transport rate can be represented by the product of the concentration, particle velocity and the thickness of the moving layer (i.e. the trajectory height). Using an empirical function for the bedload concentration, this gives

$$q_s = 0.053 \, [\, (\rho_s - \rho)g/\rho \,]^{0.5} D_{50}^{1.5} \; T^{2.1}/D_*^{0.3} \tag{28}$$

where T is the transport stage parameter $(u_*^2 - u_{*cr}^2)/u_{*cr}^2$
D_* is the particle parameter $D_{50}[\, (\rho_s - \rho)g/\rho v^2]^{1/3}$

The transport stage parameter uses the Shields' (1936) curve to derive the critical shear velocity. The effective shear velocity is given by

$$u_* = g^{0.5}\bar{u}/18 \, \log(4R/D_{90}) \tag{29}$$

The resulting bedload transport rates compare favourably with other sediment transport equations for selected data for a grain size range of 0.2 to 2 mm. Some 77 per cent of results lie within 0.5 to 2.0 of the measured values. The main limiting factor on the accuracy of the model is its calibration using trajectory data and much more information is required in this field in order to develop this sort of model further.

Van Rijn's (1984a) model is instructive from the point of view of the use that can be made of single grain experiments in deriving sediment transport models. However, the model does not make full use of the advantages of modelling the grain-by-grain process in that it ignores the details of individual particle trajectories and the interaction of the particle with the bed. Hence, the erosion and deposition processes are not modelled explicitly and there is no account taken of either bedforms or grain geometry. The simulation model described below, although not reaching the end product of a bedload transport rate prediction, shows something of the value of considering individual grains.

The Model of Naden (1985, 1987b)

Naden (1985, 1987b) was interested in modelling bedform geometry for gravel-bed rivers as a result of the sediment transport process. Thus, the relation between sediment transport and bed topography is explicitly built into the model. However, the demands on field data to test the assumptions of the model are as yet unsatisfied and this example of the single grain approach to sediment transport is only at the conceptual–computational stage of development. Sediment transport calculations are simply a spin-off of the model but they do have some interesting qualitative similarities with recent field data and represent a kinematic wave approach to the problem of sediment transport.

The model is only applicable to relatively low sediment transport rates near the threshold for movement. For simplicity, it assumes a two-dimensional reach one grain diameter wide and 1000 grain diameters long made up of two sizes of grain, one twice the diameter of

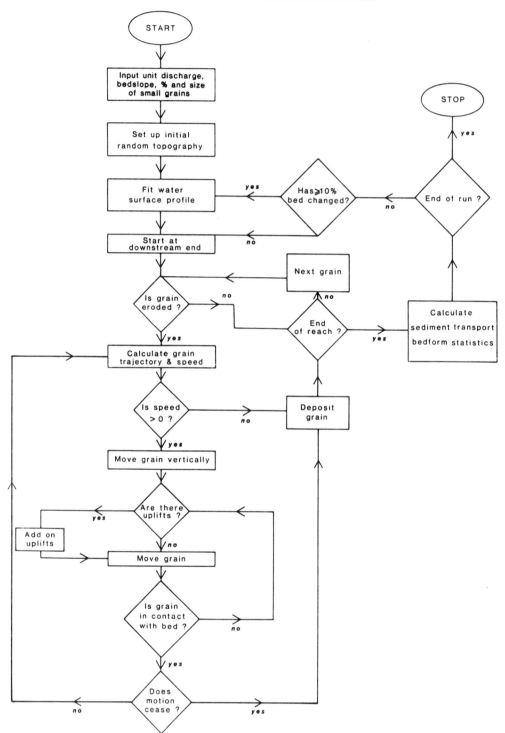

Table 8.6. Summary of the equations describing the movement of grains used in the model of Naden (1985)

Description	Equation	Source
Particle trajectories	$\dfrac{H_{max}}{D} = 0.43\ u_*^2\ D\left(\dfrac{\rho_s - \rho}{\rho}\right) + 0.663$	Abbott and Francis (1977)
	$\dfrac{L}{D} = 10\ \dfrac{H}{D} - 3.5 + 0.01\ m\ \dfrac{\bar{U}}{D}$	
Grain speed	$\dfrac{\bar{U}}{V_g} = 11.8\ \dfrac{u_*}{V_g} - 0.44$	Francis (1973)
Deposition criterion	$\bar{U} < \dfrac{\sqrt{\left(\dfrac{\rho_s - \rho}{\rho_s}\right) g\ D \sin \theta}}{\sin(\theta - \epsilon)}$	Gordon *et al.* (1972)
Erosion thresholds	$F_D + G \sin \beta = (G \cos \beta - F_L) \tan \phi$	Naden (1987a)

where H_{max} is maximum height of grain trajectory (m)
L is length of grain trajectory (m)
m is number of uplifts experienced by grain in flight
θ is angle of impact of grain (degrees), ϵ is angle of descent of grain (degrees)

the other. The outline of the model is presented in Figure 8.10 and the equations on which it is based are summarized in Table 8.6. The main thrust of the model is towards the individual grain and its motion. This is determined by an erosion threshold model (see section 8.2.4) and an empirical description taken from the experimental results of Abbott and Francis (1977) of grain trajectories and speeds. A deposition criterion based on the experiment of Gordon *et al.* (1972) accommodates the problem of different initial and final thresholds of motion (Reid and Frostick, 1984). Flow parameters are calculated using the standard methods for fitting a water surface profile applied over the bed topography grain by grain and a simple logarithmic velocity profile. This whole procedure is rather suspect in that the bed roughness does not change in response to the development of bedforms and the logarithmic flow profile is not necessarily applicable to the high bed roughness and high Froude numbers produced in the simulations.

Sample model results are given in Figure 8.11 for the bed topography developed by the model. This shows two scales of bedform—the small scale, particle cluster and the larger scale, step-pool or antidune bedform (see Naden and Brayshaw, 1987). More relevant to the discussion here is the pattern of sediment movement shown in Figure 8.12. Being a two-dimensional model, one grain wide, the amount of sediment transport at a chosen point is minimal. Consequently, the sediment transport is shown in terms of the total number of grains moved multiplied by their distance of travel. It is, therefore, a measure of the total transport activity of the reach. Over the course of a hydrograph, it shows the same sort of wave pattern as reported in the field and an initial motion threshold higher than that for final motion.

Figure 8.10. *(opposite)* Flow diagram of the grain-by-grain sediment transport process used in the model of Naden (1985) (Reproduced by permission of John Wiley & Sons Ltd)

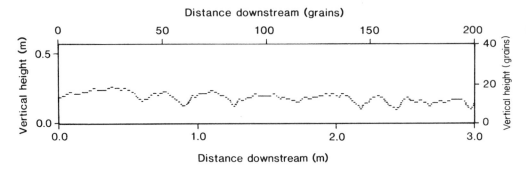

Figure 8.11. An example of the bed topography generated by the model of Naden (1985)— $q = 0.011$ cumecs m^{-1}, $s = 0.025$, $D = 0.015$ m, 80 per cent grains are small. (Reproduced by permission of John Wiley & Sons Ltd)

As yet, this sort of model is rather restricted in its application being limited both by the complexity of the model and the need for more process information from which to formulate and test the functions used. It does, however, offer an alternative approach to the question of sediment transport. One of the keys to its success will lie in the way in which the model can be subsequently simplified for river applications without losing the advantages of the more detailed process description.

8.5.5   New Research Needed

With respect to the field, several demands can be made relating to the calibration of this type of model, testing of its assumptions and application. Leaving aside the particular assumptions made, the general form of the process relationships represented in the model of Naden (1985) are given in Figure 8.13 (based on Leeder (1983) but annotated with reference to the model described here). The definition of poorly defined linkages in this diagram helps to clarify the need for more flume and field experimentation.

The water surface profile, resistance equation, turbulence link is perhaps the weakest of the three. In fact, this has been a recurring theme throughout this chapter. The modelling of sediment transport ultimately demands a reliable model of channel flow (discussed in part in Chapter 6).

The link between sediment transport and turbulent flow is the one that is central to the discussions here. Notably, any feedback from sediment transport to the flow has been omitted from the model. This is largely due to lack of information and, at low concentrations of sediment transport, is not thought to be important. In terms of the movement of grains by the flow, the erosion routines are fairly well founded on theory (see section 8.2 above) and supported by both field and flume experiments. However, as noted by van Rijn (1984a), information on grain pathways is relatively scarce and more work needs to be done on the relation between grain speed, trajectory dimensions, and grain size, especially for large grain sizes and at a wider range of depth to grain size ratios. Initially, evidence for this is expected to come from further flume experimentation, but it is hoped that the techniques developed by Reid et al. (1984) may also be used to calibrate such relations in the field. Intergranular collisions, added for completion, were found to be rare in the model runs and could be omitted. However, at higher rates of sediment transport, for which the model

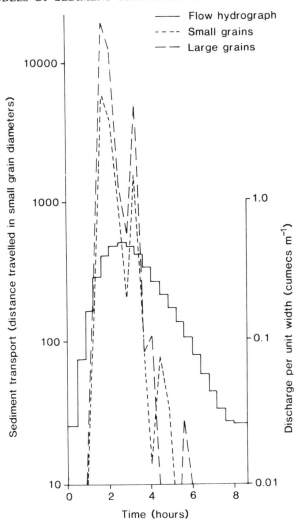

Figure 8.12. Sediment transport activity over a hydrograph as generated by the model of Naden (1985) — $s = 0.064$, $D = 0.03$ m, 50 per cent grains are small. (Reproduced by permission of John Wiley & Sons Ltd)

is not applicable, they should become more important and begin to affect the particle trajectory dimensions.

The bedform–sediment transport link is the most important contribution of the grain-by-grain model of sediment transport. Here, the grain geometry is used to control the availability of sediment for transport via the erosion thresholds. However, this is somewhat limited by the choice of two grain sizes so that important processes such as bed armouring are not adequately represented. Increasing stability of the bed is, instead, reflected purely in the development of cluster bedforms and the reduction in the number of more mobile grains (Naden and Brayshaw, 1987). As regards deposition, the threshold used here does differ from the erosion threshold and so coincides with the field observations of

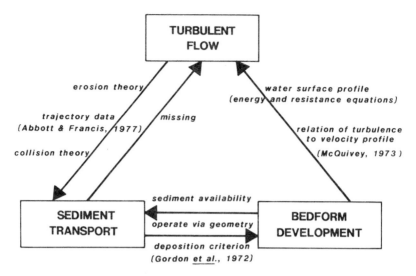

Figure 8.13. Feedback loop between turbulent flow, sediment transport and bedform development — based on Leeder (1983) but annotated with respect to the processes modelled in Naden (1985). (Reproduced by permission of John Wiley & Sons Ltd)

Reid and Frostick (1984). However, the characteristics of the grain–bed collision need to be better understood and the criterion used tested against experimental data.

8.6   FUTURE DEVELOPMENTS

Sediment transport modelling, then, revolves around three main issues — the initiation of sediment movement, rates of suspended sediment transport and the question of bedload movement. As has been amply demonstrated, current developments in these fields pinpoint a number of major gaps in our knowledge and understanding of the sediment transport process in rivers.

The first of these is the question of fluid parameters — in terms of both the velocity profile and turbulence. As was seen in section 8.2 in the case of the initiation of sediment movement and in section 8.5.4 describing the grain-by-grain process of sediment movement, neither the mean bed shear stress acting on the grains, its spatial variation over bed topography (whether individual grains or bedforms), nor its temporal variation is adequately known or modelled. This is particularly the case in geomorphological modelling where it is important ultimately to model the processes *in the channel*. Channels are three-dimensional and may be quite complex in form with braided or meandering patterns, and yet all the sediment transport models discussed apply to two-dimensional channels or those of relatively simple design. In particular, problems of flow convergence and divergence, the presence of backwater curves, and the question of flow in boulder bed channels (Bathurst, 1982b) have not been tackled. Instead, they are issues that are masked by recourse to stream power or discharge as driving parameters for sediment transport models. Indeed, we really do not even know whether bed shear stress increases with increasing discharge, especially for over-bankfull flows.

The second major issue is the adequate description of *bed topography*. Grain geometry is clearly an important variable in the initiation of movement and in the supply of fines from the bed. However, as yet no satisfactory means of describing grain geometry — relative protrusion and pivoting angle versus grain size — has been formulated. In particular, we still have no idea as to the range and distribution of these variables in channels with different bed materials. This problem also feeds back into the question of bed roughness and the related problem of bed-form development. Sediment transport models which include the feedback between sediment transport and bedforms are relatively rare. However, the actual moulding of the bed is crucial to the friction on the flow and, therefore, the bed shear stress calculation and the generation of turbulence. It is also vitally important in the availability of sediment for transport.

Partly linked to this question of bed topography, then, is that of *sediment supply*. This is another crucial question for geomorphologists and one largely ignored until recently. It is important in suspended sediment models as evidenced in the list of modelling requirements outlined in Table 8.4. It is also important in gravel-bed rivers in terms of the armouring process.

While it is important to understand the processes of bed armouring and the infiltration of fines into the bed, another aspect of the sediment supply problem is simply to quantify the amounts of sediment available in channel storage, bank storage, and catchment storage. Add to this, the event thresholds needed to release this sediment and the functions required to replenish the sediment stores, and we have a far more relevant model for sediment yield than a simple application of the sediment transport equations.

In addition to the three areas for research outlined above, there is still a great need for reliable field data — of sediment transport rates and relevant fluid parameters for a wide range of channel and sediment types. Other areas, not discussed, about which relatively little is known are those of cohesive sediment (Parthenaides, 1965; Nicholson and O'Connor, 1986) and of superconcentrated flows (Costa, 1984; Carling, 1986).

Finally, as regards modelling, I think that the key developments will come from three directions. First, the more complex process models probably hold some promise for the growth of our understanding of sediment transport mechanisms. However, in order to make them accessible and usable in a river context, they need to be simplified and the key interactions extracted. Secondly, I think there will be a greater demand for stochastic models. Whereas the erosion and sediment transport processes are essentially deterministic, the driving parameter of turbulent flow is more appropriately described by a stochastic function. Equally, the representation of the spatial distribution of both sediment and fluid parameters may well be accommodated more neatly within a stochastic variable rather than a bald average. The greatest contribution from geomorphology in the field of sediment transport modelling, however, should be in the linkage between sediment transport and channel behaviour. This ought to be a major growth point for modelling in future decades.

REFERENCES

Abbott, J. E. and Francis, J. R. D. (1977). Saltation and suspension trajectories of solid grains in a water stream. *Philosophical Transactions of the Royal Society of London*, **284A**, 225–254.

Andrews, E. D. (1983). Entrainment of gravel from naturally sorted river-bed material. *Bulletin of the Geological Society of America*, **94**, 1225–1231.

Bagnold, R. A. (1956). The flow of cohesionless grains in fluids. *Philosophical Transactions of the Royal Society*, **29A**, 235–297.

Bagnold, R. A. (1966). An approach to the sediment transport problem from general physics. *United States Geological Survey Professional Paper*, **422 I**.

Bagnold, R. A. (1973). The nature of saltation and of 'bed-load' transport in water. *Proceedings of the Royal Society of London*, **332A**, 473–504.

Bagnold, R. A. (1977). Bedload transport by natural rivers. *Water Resources Research*, **13**, 303–312.

Bagnold, R. A. (1980). An empirical correlation of bedload transport rates in flumes and natural rivers. *Proceedings of the Royal Society*, **372A**, 453–473.

Baker, V. R. and Ritter, D. F. (1975). Competence of natural rivers to transport coarse bedload material. *Bulletin of the Geological Society of America*, **86**, 975–978.

Bathurst, J. C. (1982a). Theoretical aspects of flow resistance. In Hey, R. D., Bathurst, J. C., and Thorne, C. E. (Eds.), *Gravel-bed Rivers*, Wiley, Chichester, 83–108

Bathurst, J. C. (1982b). Flow resistance in boulder-bed streams. In Hey, R. D., Bathurst, J. C., and Thorne, C. E. (Eds.), *Gravel-bed Rivers*, Wiley, Chichester, 443–465.

Bechteler, W. and Färber, K. (1985). Stochastic model of suspended solid dispersion. *Journal of Hydraulic Engineering, American Society of Civil Engineers*, **111**, 64–78.

Beschta, R. L. and Jackson, W. L. (1979). The intrusion of fine sediments into a stable gravel bed. *Journal of the Fisheries Research Board of Canada*, **36**, 204–210.

Bogen, J. (1980). The hysteresis effect of sediment transport systems. *North Geografisk Tidsskrift*, **34**, 45–54.

Box, G. E. P. and Jenkins, G. M. (1976). *Time Series Analysis, Forecasting and Control*, Holden-Day, San Francisco.

Brayshaw, A. C. (1983). *Bed microtopography and bedload transport in coarse-grained alluvial channels*, Ph.D. thesis, University of London.

Brayshaw, A. C., Frostick, L. E., and Reid, I. (1983). The hydrodynamics of particle clusters and sediment entrainment in coarse alluvial channels. *Sedimentology*, **30**, 137–143.

Bridge, J. S. (1977). Flow, bed topography, grain size and sedimentary structure in open channel bends: a three dimensional model. *Earth Surface Processes*, **2**, 401–416.

Bridge, J. S. and Jarvis, J. (1976). Flow and sedimentary processes in the meandering river South Esk, Glen Clova, Scotland. *Earth Surface Processes*, **1**, 303–337.

Brodkey, R. S., Wallace, J. M., and Ecklemann, H. (1974). Some properties of truncated turbulence signals in bounded shear flows. *Journal of Fluid Mechanics*, **63**, 209–224.

Burkham, D. E. (1985). An approach for appraising the accuracy of suspended-sediment data. *United States Geological Survey Professional Paper*, **1333**.

Burt, T. P. and Gardiner, A. T. (1984). Runoff and sediment production in a small peat-covered catchment: some preliminary results. In Burt, T. P. and Walling, D. E. (Eds.), *Catchment Experiments in Fluvial Geomorphology*, Geobooks, Norwich.

Carling, P. A. (1983). Threshold of coarse sediment transport in broad and narrow natural streams. *Earth Surface Processes and Landforms*, **8**, 1–18.

Carling, P. A. (1986). A terminal debris-flow lobe in the Northern Pennines, United Kingdom. In Macklin, M. G. and Rose, J. (Eds.), *Quaternary River Landforms and Sediments in the Northern Pennines, England. Field Guide*, British Geomorphological Research Group/Quaternary Research Association.

Carling, P. A. and Reader, N. A. (1982). Structure, composition and bulk properties of upland stream gravels. *Earth Surface Processes and Landforms*, **7**, 349–365.

Cheetham, G. H. (1979). Flow competence in relation to stream channel form and braiding. *Bulletin of the Geological Society of America*, **90**, 877–886.

Christensen, B. A. and Bush, P. W. (1971). Statistically-based determination of depth and width ratios in alluvial watercourses. In Chiu, C. L. (Ed.), *Stochastic Hydraulics. Proceedings of 1st International Symposium on Stochastic Hydraulics*, University of Pittsburgh, 402–425.

Church, M. A. (1978). Palaeohydrological reconstructions from a Holocene valley fill. In Miall, A. D. (Ed.), *Fluvial Sedimentology. Canadian Society of Petroleum Geologists Memoir*, **5**, 743–772.

Coleman, N. L. (1981). Velocity profiles with suspended sediment. *Journal of Hydraulic Research*, **19**, 211–229.

Collins, D. N. (1979). Sediment concentration in melt waters as an indication of erosion processes beneath an Alpine glacier. *Journal of Glaciology*, **23**, 247–257.

Costa, J. E. (1983). Palaeohydraulic reconstructions of flash-flood peaks from boulder deposits in the Colorado Front Range. *Bulletin of the Geological Society of America*, **94**, 986–1004.

Costa, J. E. (1984). Physical Geography of debris flows. In Costa, J. E. and Fleisher, P. J. (Eds.), *Developments and Applications of Geomorphology*, Springer-Verlag, Berlin, 268–317.

Dietrich, W. E. and Smith, J. D. (1984). Bed load transport in a River Meander. *Water Resources Research*, **20**, 1355–1380.

Du Boys, M. P. (1879). Études du regime et l'action exercée par les eaux sur un lit à fond de graviers indefinement affouillable. *Annals des Ponts et Chaussées, Series 5*, **18**, 141–195.

Einstein, H. A. (1950). The bedload function for sediment transportation in open channel flows. *Technical Bulletin, United States Department of Agriculture*, **1026**.

Einstein, H. A. and Chien, N. (1953). Can the rate of washload be predicted from the bed-load function? *Transactions of the American Geophysical Union*, **34**, 876–882.

Einstein, H. A. and El-Samni, E. A. (1949). Hydrodynamic forces on a rough wall. *Reviews of Modern Physics*, **21**, 520–524.

Falco, R. E. (1977). Coherent motions in the outer region of turbulent boundary layers. *The Physics of Fluids*, **20**, S124–S132.

Fenn, C. R., Gurnell, A. M., and Beecroft, J. R. (1985). An evaluation of the use of suspended sediment rating curves for the prediction of suspended sediment concentration in a proglacial stream. *Geografiska Annaler*, **67A**, 71–82.

Fenton, J. D. and Abbott, J. E. (1977). Initial movement of grains on a stream bed: the effect of relative protrusion. *Proceedings of the Royal Society of London*, **352A**, 523–537.

Ferguson, R. I. (1986). River loads underestimated by rating curves. *Water Resources Research*, **22**, 74–76.

Fernandez Luque, R. (1974). *Erosion and transport of bed-load sediment*, Dissertation, Krips Repro B.V., Meppel, The Netherlands.

Forchheimer, P. (1914). *Hydraulics*, Teuber, Leipzig.

Francis, J. R. D. (1973). Experiments on the motion of solitary grains along the bed of a water stream. *Proceedings of the Royal Society of London*, **332A**, 443–471.

Fread, D. L. and Harbaugh, T. E. (1971). Open channel profiles by Newton's iteration technique. *Journal of Hydrology*, **13**, 70–80.

Frostick, L. E., Lucas, P. M., and Reid, I. (1984). The infiltration of fine matrices into coarse-grained alluvial sediments and its implications for stratigraphical interpretations. *Journal of the Geological Society London*, **141**, 955–965.

Gessler, J. (1971). Beginning and ceasing of sediment motion. In Shen, H. W. (Ed.), *River Mechanics*, vol. 1, H. W. Shen, Fort Collins, Colorado.

Glymph, L. M. (1957). Importance of sheet erosion as a source of sediment. *Transactions of the American Geophysical Union*, **38**, 903–907.

Gomez, B. (1983). Temporal variations in bedload transport rates: the effect of progressive bed armouring. *Earth Surface Processes and Landforms*, **8**, 41–54.

Gomez, B. (1984). Typology of segregated (armoured/paved) surfaces: some comments. *Earth Surface Processes and Landforms*, **9**, 19–24.

Gordon, R., Carmichael, J. B., and Isackson, F. J. (1972). Saltation of plastic balls in a 'one-dimensional' flume. *Water Resources Research*, **8**, 444–459.

Grass, A. J. (1971). Structural features of turbulent flow over smooth and rough boundaries. *Journal of Fluid Mechanics*, **50**, 223–256.

Grass, A. J. (1974). Transport of fine sand on a flat bed: turbulence and suspension mechanics. *Euromech.*, **48**, 33–34.

Grenney, W. J. and Heyse, E. (1985). Suspended sediment—river flow analysis. *Journal of Environmental Engineering, American Society of Civil Engineers*, **111**, 790–803.

Grigg, N. S. (1970). Motion of single particles in alluvial channels. *Journal of the Hydraulics Division, American Society of Civil Engineers*, **96**, 2501–2518.

Gurnell, A. M. and Fenn, C. R. (1984). Box–Jenkins transfer function models applied to suspended sediment concentration–discharge relationships in a proglacial stream. *Arctic and Alpine Research*, **16**, 93–106.

Guy, H. P., Simons, D. B., and Richardson, E. V. (1966). Summary of alluvial channel data from flume experiments 1956–1961. *United States Geological Survey Professional Paper*, **462 I**.

Hammond, F. D. C., Heathershaw, A. D., and Langhorne, D. N. (1984). A comparison between Shields' threshold criterion and the movement of loosely packed gravel in a tidal channel. *Sedimentology*, **31**, 51–62.

Heidel, S. G. (1956). The progressive lag of sediment concentration with flood waves. *Transactions of the American Geophysical Union*, **37**, 56–66.

Helley, J. R. (1969). Field measurement of the initiation of large bed particle motion in Blue Creek near Klamath, California. *United States Geological Survey Professional Paper*, **562 G**.

Hjulström, F. (1936). Studies of the morphological activity of rivers as illustrated by the River Fyris. *Bulletin of the Geological Institute, University of Uppsala*, **25**, 221–527.

Holly, F. M. and Karim, M. F. (1986). Simulation of Missouri River bed degradation. *Journal of Hydraulic Engineering, American Society of Civil Engineers*, **112**, 497–517.

Hooke, R. LeB. (1975). Distribution of sediment transport and shear stress in a meander bend. *Journal of Geology*, **83**, 543–566.

Humpidge, H. B. and Moss, W. D. (1971). The development of a comprehensive computer program for the calculation of flow profiles in open channels. *Proceedings of the Institute of Civil Engineers*, **50**, 49–64.

Hunt, J. N. (1954). The turbulent transport of suspended sediment in open channels. *Proceedings of the Royal Society*, **224A**, 322–335.

Hydrologic Engineering Center (1974). *Scour and Deposition in rivers and reservoirs, HEC–6 Users Manual*, HEC, Davis, California.

Imeson, A. C. (1974). The origin of sediment in a moorland catchment with particular reference to the role of vegetation. In Gregory, K. J. and Walling, D. E. (Eds.), *Fluvial Processes in Instrumented Watersheds. Institute of British Geographers, Special Publication*, **6**, 59–72.

Ippen, A. T. and Verma, R. P. (1953). The motion of discrete particles along the bed of a turbulent stream. *International Association of Hydraulics Research, 5th Congress*, Minneapolis, 7–20.

Jackson, W. L. and Beschta, R. L. (1982). A model of two-phase bedload transport in an Oregon Coast Range Stream. *Earth Surface Processes and Landforms*, **7**, 517–527.

Jenkins, A., Kirkby, M. J., McDonald, A., Naden, P., and Kay, D. (1984). A process based model of faecal bacterial levels in upland catchments. *Water Science Technology*, **16**, 453–462.

Kalinske, A. A. (1947). Movement of sediment as bedload in rivers. *Transactions of the American Geophysical Union*, **28**, 615–620.

Komar, P. D. and Li, Z. (1986). Pivoting analyses of the selective entrainment of sediments by shape and size with application to gravel threshold. *Sedimentology*, **33**, 425–436.

Laronne, J. B. and Carson, M. A. (1976). Inter-relationships between bed morphology and bed material transport for a small gravel-bed channel. *Sedimentology*, **23**, 67–86.

Leeder, M. R. (1979). 'Bedload' dynamics: grain–grain interactions in water flows. *Earth Surface Processes*, **4**, 229–240.

Leeder, M. R. (1983). On the interactions between turbulent flow, sediment transport and bedform mechanics in channelized flows. In Collinson, J. D. and Lewin, J. (Eds.), *Modern and Ancient Fluvial Systems. International Association of Sedimentologists Special Publication*, **6**, 5–18.

Leliavsky, S. (1955). *An Introduction to Fluvial Hydraulics*, Constable, London.

Leopold, L. B. and Emmett, W. W. (1976). Bedload measurements, East Fork River, Wyoming. *Proceedings of the National Academy of Sciences*, **73**, 1000–1004.

Li, Z. and Komar, P. D. (1986). Laboratory measurement of pivoting angles for applications to selective entrainment of gravel in a current. *Sedimentology*, **33**, 413–423.

Little, W. C. and Mayer, P. G. (1976). Stability of channel beds by armouring. *Journal of the Hydraulics Division, American Society of Civil Engineers*, **102**, 1647–1661.

Mavis, F. T. and Laushey, L. M. (1948). A reappraisal of the beginning of bed movement–competent velocity. *International Association of Hydraulics Research*, 2nd Meeting, Stockholm.

McCaig, M. (1981). Modelling storm period fluctuations in suspended sediment—an appraisal. *Working Paper 311*, School of Geography, University of Leeds.

McQuivey, R. S. (1973). Summary of turbulence data from rivers, conveyance channels and laboratory flumes. *United States Geological Survey Professional Paper*, **802–B**.

McTigue, D. F. (1981). Mixture theory for suspended sediment transport. *Journal of Hydraulics Division, American Society of Civil Engineers*, **107**, 659–673.

Meyer-Peter, E. and Müller, R. (1948). Formulas for bed-load transport. *Proceedings of the International Association of Hydraulic Research, 3rd Annual Conference, Stockholm*, 39–64.

Miller, M. C., McCave, I. N., and Komar, P. D. (1977). Threshold of sediment motion under unidirectional currents. *Sedimentology*, 24, 507–527.

Moore, R. J. and Newson, M. D. (1986). Production, storage and output of coarse upland sediments: natural and artificial influences as revealed by research catchment studies. *Journal of the Geological Society of London*, 143, 921–926.

Naden, P. S. (1985). *Gravel bedforms — the development of a sediment transport model*, Ph.D. thesis, University of Leeds.

Naden, P. S. (1987a). An erosion criterion for gravel-bed rivers. *Earth Surface Processes and Landforms*, 12, 83–93.

Naden, P. S. (1987b). Modelling bed topography from sediment transport. *Earth Surface Processes and Landforms*, 12, 353–367.

Naden, P. S. and Brayshaw, A. C. (1987). Small and medium scale bedforms in gravel-bed rivers. In Richards, K. S. (Ed.), *River Channels: Environment and Process*, Institute of British Geographers Special Publication, Basil Blackwell, Oxford.

Nakagawa, H. and Nezu, I. (1981). Structure of space–time correlations of bursting phenomena in an open-channel flow. *Journal of Fluid Mechanics*, 104, 1–43.

Neill, C. R. (1967). Mean velocity criterion for scour of coarse, uniform bed material. *International Association for Hydraulics Research, 12th Congress, Fort Collins, Colorado*, 3, 46–54.

Neill, C. R. (1968). Note on initial movement of coarse uniform bed material. *Journal of Hydraulics Research*, 6, 173–176.

Newson, M. D. (1980). The geomorphological effectiveness of floods — a contribution stimulated by two recent events in Mid-Wales. *Earth Surface Processes*, 5, 1–16.

Nicholson, J. and O'Connor, B. A. (1986). Cohesive sediment transport model. *Journal of Hydraulic Engineering, American Society of Civil Engineers*, 112, 621–640.

Parker, G., Klingeman, P. C., and McClean, D. G. (1982). Bedload and size distribution in paved gravel-bed streams. *Journal of the Hydraulics Division, American Society of Civil Engineers*, 108, 544–571.

Parthenaides, E. (1965). Erosion and deposition of cohesive soils. *Journal of Hydraulics Division, American Society of Civil Engineers*, 91, 105–139.

Pizzuto, J. E. (1984). An evaluation of methods for calculating the concentration of suspended bed material in rivers. *Water Resources Research*, 20, 1381–1389.

Pizzuto, J. E. (1985). Bank sediment type and suspended sediment transport in sand-bed streams. *Journal of Sedimentary Petrology*, 55, 222–225.

Rannie, W. F. (1978). An approach to the prediction of suspended sediment rating curves. In Davidson-Arnott, R. and Nickling, W. (Eds.), *Research in Fluvial Systems*, Geo-Abstracts, Norwich, 149–167.

Reid, I., Brayshaw, A. C., and Frostick, L. E. (1984). An electromagnetic device for automatic detection of bedload motion and its field applications. *Sedimentology*, 31, 269–276.

Reid, I. and Frostick, L. E. (1984). Particle interaction and its effect on the thresholds of initial and final bedload motion in coarse alluvial channels. In Koster, E. H. and Steel, R. J. (Eds.), *Sedimentology of Gravels and Conglomerates. Canadian Society of Petroleum Geologists Memoir*, 10, 61–68.

Reid, I. and Frostick, L. E. (1986). Dynamics of bedload transport in Turkey Brook, a coarse-grained alluvial channel. *Earth Surface Processes and Landforms*, 11, 143–155.

Reid, I., Frostick, L. E., and Layman, J. T. (1985). The incidence and nature of bedload transport during flood flows in coarse-grained alluvial channels. *Earth Surface Processes and Landforms*, 10, 33–44.

Reid, I., Layman, J. T., and Frostick, L. E. (1980). The continuous measurement of bedload discharge. *Journal of Hydraulics Research*, 18, 243–249.

Richards, K. S. (1978). Simulation of flow geometry in a riffle–pool stream. *Earth Surface Processes*, 3, 345–354.

Richards, K. S. (1982). *Rivers: Form and Process in Alluvial Channels*, Methuen, London.

Richards, K. S. (1984). Some observations on suspended sediment dynamics in Storbregrova, Jotunheimen. *Earth Surface Processes and Landforms*, 9, 101–112.

Rijn, L. C. van (1984a). Sediment transport, part I: bed load transport. *Journal of Hydraulic Engineering, American Society of Civil Engineers*, 110, 1431–1456.

Rijn, L. C. van (1984b). Sediment transport, part II: suspended load transport. *Journal of Hydraulic Engineering, American Society of Civil Engineers*, **110**, 1613–1641.

Rouse, H. (1937). Modern concepts of the mechanics of fluid turbulence. *Transactions of the American Society of Civil Engineers*, **102**, 436–505.

Schoklitsch, A. (1934). Geschiebetrieb und die Geschiebefracht. *Wasserkraft und Wassersirtsch*, **39**.

Schoklitsch, A. (1950). *Handbuch des Wasserbauers*, volume 1, 2nd ed., Springer-Verlag, Vienna.

Shields, A. (1936). Anwendung der Aehnlichkeitsmechanik und der turbulenzforschung auf die geschiebebewegung. *Mitteilung der Preussischen versuchsanstalt fuer Wasserbau und Schiffbau*, **26**, Berlin.

Simons, D. B. and Sentürk, F. (1977). *Sediment Transport Technology*, Water Resources Publications, Fort Collins, Colorado.

Thomas, W. A. (1982). Mathematical modelling of sediment movement. In Hey, R. D., Bathurst, J. C., and Thorne, C. R. (Eds.), *Gravel-Bed Rivers*, Wiley, Chichester, 487–511.

Thomas, W. A. and Prasuhn, A. L. (1977). Mathematical modelling of scour and deposition. *Journal of the Hydraulics Division, American Society of Civil Engineers*, **103**, 851–863.

US Inter-Agency Committee on Water Resources (1957). Measurement and analysis of sediment loads in streams: some fundamentals of particle size analysis. *Project Report No. 12*, Washington, D.C.

Vanoni, V. A. (1941). Some experiments on the transportation of suspended load. *Transactions of the American Geophysical Union*, Section on Hydrology.

VanSickle, J. and Beschta, R. L. (1983). Supply-based models of suspended sediment transport in streams. *Water Resources Research*, **19**, 768–778.

Walling, D. E. (1977). Assessing the accuracy of suspended sediment rating curves for a small basin. *Water Resources Research*, **13**, 531–538.

Walling, D. E. (1978). Suspended sediment and solute response characteristics of the River Exe, Devon, England. In Davidson-Arnott, R. and Nickling, W. (Eds.), *Research in Fluvial Geomorphology*, Geo-Abstracts, Norwich, 169–197.

White, C. M. (1940). The equilibrium of grains on the bed of a stream. *Proceedings of the Royal Society of London*, **174A**, 322–338.

White, S. J. (1970). Plane bed thresholds of fine grained sediments. *Nature*, **228**, 152–153.

White, W. R. and Day, T. J. (1982). Transport of graded gravel bed material. In Hey, R. D., Bathurst, J. C., and Thorne, C. R. (Eds.), *Gravel-Bed Rivers*, Wiley, Chichester, 181–223.

White, W. R., Milli, H. and Crabbe, A. D. (1973). Sediment transport: an appraisal of available methods. *Report No. INT 119*, vols 1 and 2, Hydraulics Research Station, Wallingford.

Williams, G. P. (1983). Paleohydrological methods and some examples from Swedish fluvial environments. I. Cobble and boulder deposits. *Geografiska Annaler*, **65A**, 227–243.

Yalin, M. S. (1977). *Mechanics of Sediment Transport*, Pergamon Press, Oxford.

Modelling Geomorphological Systems
Edited by M. G. Anderson
©1988 John Wiley & Sons Ltd.

Chapter 9

Stochastic models of sediment yield

V. P. SINGH

Department of Civil Engineering and Louisiana Water Resources Research Institute

P. F. KRSTANOVIC

Department of Civil Engineering, Louisiana State University

and

L. J. LANE

Southwest Rangeland Watershed Research Center, Agricultural Research Service, Tucson

9.1 INTRODUCTION

Sediment yield is defined as the total sediment outflow from a watershed measurable at a point of reference during a specified period of time (American Society of Civil Engineers, 1970, 1982). The sediment outflow from the watershed is induced by processes of detachment, transportation, and deposition of soil materials by rainfall and runoff. Estimates of sediment yield are required in a wide spectrum of problems such as design of reservoirs and dams, transport of sediment and pollutants in rivers, lakes and estuaries, design of stable channels, dams and debris basins, undertaking cleanup following floods, protection of fish and wildlife habitats, determination of the effects of watershed management, and environmental impact assessment.

Two aspects of watershed sediment yield are of interest: chemical and hydrologic. The former examines sediment as the particles associated with various constituents such as pesticides, radioactive materials or nutrients, and may be needed to determine whether the influence of sediment yield upon a certain area is beneficial or detrimental. For example, sediment can sometimes serve as an effective scavenger in removing chemicals from surrounding waters, e.g. it absorbs phosphorus in sewage effluent. On the other hand, in the U.S.A. it has been named the greatest single pollutant of the Nation's streams and its strict control is therefore required to keep the water quality in the range of tolerable limits. Specifically, the Federal Water Pollution Control Act (PL 92–500, 1972) has directed

259

states to develop 'Water Quality Management Plans' that will achieve specified water quality goals. As a consequence, there has been an upsurge in attempts to model sediment yield.

9.2   SEDIMENT YIELD SOURCES

Sediment yield sources can be broadly classified as upland watershed sources (both hillslopes and bottomlands) and lowland channel sources (rivers and streams). The former are characterized by dominant overland flow and concentrated flow erosion in ephemeral gullies (Foster *et al.*, 1975), and the sources can be either natural or artificial (Simons *et al.*, 1979). Major factors effecting sediment yield are soil characteristics, climate, vegetation, topography, and human activities. Natural sources include various phases of erosion, sheet erosion or removal of a sediment layer of uniform thickness from the entire area, rill erosion or development of small channels of runoff concentrations and gully erosion or massive removal of soil by large concentration of runoff. It has been shown (Meyer *et al.*, 1975) that as downslope distance increases, the relative importance of a sediment yield source moves from sheet erosion to gully erosion.

Artificial sources are primarily initiated by human activities, some dominant ones are (Guy, 1974): agricultural tillage, domestic animal overgrazing, highway construction, and maintenance, timber cutting, mining, urbanization and recreational land development. These activities may alter ground slope and flow paths, remove vegetation and disturb the soil surface, and thus increase sediment yield.

The channel sources receive sediment yield from upland sources, but as they grow in size they develop sediment sources of their own and the influence of upland sources slowly diminishes as in the case of large rivers. Contributing most to sediment yield of these sources are stream bed and banks which are affected by bank erosion, channel changes (e.g. accretion or avulsions), construction of hydraulic structures, channelization, dredging, dam breaches and mining of gravels from channels.

There may also exist another source of sediment yield. It may appear to be random but can create conditions significantly different from the normal environment and may even surpass other sources in production of sediment yield. This third source can be the result of several natural disasters such as extra-large storms creating excessive floods with extraordinary sediment amounts, earthquakes causing slides and mudflows and loss of structures, fires changing the watershed hydrology, and droughts affecting vegetation cover and sediment accumulation.

9.3   MATHEMATICAL MODELS OF SEDIMENT YIELD

Perhaps the first quantitative estimate of sediment yield was made by Dole and Stabler (1909). Numerous studies have since been conducted. Comprehensive summaries of all major world rivers' sediment have been reported by Holeman (1968), and of U.S. rivers' sediment by Curtis *et al.* (1973). Under the auspices of the International Hydrologic Decade (IHD), a program was initiated to prepare a global map of the sediment yield to the Ocean (Fournier, 1969). However, the global picture of continental denudation is of little interest in studying the local processes that occur at the watershed scale. Many processes that occur over large timescales are slow, extremely difficult to measure, and many occur so slowly that they require years of observations to acquire data of any significance. In this chapter, we confine ourselves only to sediment yield at the watershed scale.

Sediment yield $Y(t)$ at a given point in space (say, watershed outlet) can be represented as

$$Y(t) = \bar{Y}(t) + \epsilon(t) \tag{1}$$

in which $\bar{Y}(t)$ is the mean value or deterministic component of $Y(t)$, and $\epsilon(t)$ is the error from or fluctuation around the mean value or stochastic component of $Y(t)$. The relative contribution of $\bar{Y}(t)$ or $\epsilon(t)$ to $Y(t)$ depends on the watershed and space–time scales. Clearly, $Y(t)$ encompasses the full range of variability from being entirely deterministic to being entirely stochastic. All sediment yield models are special cases of equation 1. In other words, they specialize in modelling $\bar{Y}(t)$, $\epsilon(t)$ or both. This chapter is principally concerned with treatment of $Y(t)$ as $\epsilon(t)$. Equation 1 can also serve as a basis for classification of sediment yield models, of which previously reported various classifications are special cases. For example, Williams (1982) classified sediment yield models according to significant water resources problems: erosion control planning (for agricultural fields, construction sites, reclaimed mines, etc.), water resources planning and design (hydraulic structures), and water quality modelling. However, all these models were for computing $\bar{Y}(t)$ and were thus deterministic models. Alonso (1980) evaluated a number of sediment transport formulas of bed load as well as total load. Again, he dealt with deterministic models only. Li *et al.* (1973) considered both deterministic and stochastic models for small watersheds.

The deterministic models can be distinguished as empirical or conceptual. Most of the empirical models are related to the Universal Soil Loss Equation (USLE) (Wischmeier and Smith, 1960) and its latter modifications (Williams and Berndt, 1972; Williams, 1975). These models usually require long data records, so that average annual sediment yield can be determined. The conceptual models combine mechanics of sediment transport with empirical relationships (Foster *et al.*, 1975; Shirley and Lane, 1978; Singh and Regl, 1983; Prasad and Singh, 1982; Singh, 1983). Both empirical and conceptual models approximate the physical processes controlling sediment yield.

Another way to represent the complex sediment behaviour is to interpret a sequence of sediment yield measurements as random. If the processes governing sediment yield such as soil particle detachment, entrainment, transport, and deposition are assumed to be stochastic and thus governed by the laws of probability, the sediment yield can be described by a stochastic process and associated probability distribution functions (*pdf*).

Deterministic models such as described earlier can be applied across a broad range of temporal and spatial scales. For example, overland flow models often assume broad sheet flow and then rill and interrill erosion processes are assumed to occur on spatial scales from small plots (on the order of a square metre) to hillslopes (on the order of 10 to 100 m in length) to small watersheds (on the order of a few hectares). At each scale, processes occur in seemingly random sequences. For example, raindrop sequences striking a given area of soil surface may appear to be entirely random in time. Similar observations can be made with respect to the movement of sediment particles on the beds of rills and small channels. The stochastic assumption for movement of sediment particles in streamflow was found to be quite useful in describing sediment transport processes (Einstein, 1950). At a given time, spatial distributions of soil properties affecting runoff and erosion can be visualized as randomly distributed (beyond the significant spatial correlation distance) in space.

These considerations lead to stochastic models describing sediment yield on spatial and temporal scales of interest herein. Main properties of stochastic models which may make them preferable to wholly deterministic models of sediment yield include: (1) Stochastic models can explicitly include time sequences of sediment yield and thus provide insight into expected variability of sediment yield over various time intervals; (2) Stochastic models are amenable to computer simulation studies and to comparison of various sequences used to study changes in hydrologic systems; and (3) Stochastic models force the consideration of uncertainty in governing processes, model input, and model output. This forced consideration of uncertainty in sediment yield is of significant value in designing measurement apparatus and sampling schemes, in determining required length of record in monitoring, in experimental design, and in the interpretation of historical data.

Some sediment yield models contain both deterministic and stochastic elements. A classical example is the relationship between sediment yield and runoff, represented by a line in a logarithmic plot. This is deterministic part $\bar{Y}$ of the model. When the measurements are plotted, they encircle this line and most often will not lie directly on it. Thus, the line represents only the mean trend of the sediment yield–runoff relationship, and the fluctuations $\epsilon(t)$ above and below may be considered stochastic. A successful model will have to include a deterministic component or the mean behaviour as well as a stochastic component or fluctuations around it.

9.4   STOCHASTIC MODELS OF SEDIMENT YIELD

Let us consider a watershed subjected to stochastic inputs and outputs distributed in time and space. The input to the watershed is represented by a precipitation flux $R(x,y,z,t)$. Output from the watershed consists of direct runoff $Q(t)$, sediment outflow $Q_s(t)$, infiltration $I(x,y,z,t)$, and evapotranspiration $E(x,y,z,t)$. If we denote instantaneous measurement of sediment outflow in time at the watershed outlet by $\{Q_s(t), t \epsilon T\}$, then sediment yield $Y(t)$ is obtained by integration in time over a specified time interval,

$$Y(t) = \int_0^t Q_s(w)\,\mathrm{d}w \tag{2}$$

Depending upon the relationship of $Y(t)$ and/or $Q_s(t)$ with $R(x,y,z,t)$, $I(x,y,z,t)$ and $E(x,y,z,t)$, four groups of stochastic sediment yield models can be distinguished:

1. Regression models which relate $Y(t)$ empirically to rainfall $R(t)$ and runoff $Q(t)$. Spatial variability of these variables is not considered. Stochasticity is represented by variations around the mean trend.
2. Time series models where a watershed is considered as spatially lumped system. Deterministic relationships between $R(t)$, $Q(t)$ and $Y(t)$ are represented by a transfer function and stochasticity is modelled as an autoregression process.
3. Entropy models where the *pdf* of $Y(t)$ is obtained using constraints based on observed values of $Y(t)$ and/or $Q(t)$. Spatial variability of the variables is not accounted for.
4. Probability models where sediment yield $Y(t)$ is treated as a stochastic process, and so may be rainfall $R(x,y,z,t)$ and runoff $Q(t)$. The behaviour of $Y(t)$ is described by its *pdf* or its joint probability density function with other stochastic sequences.

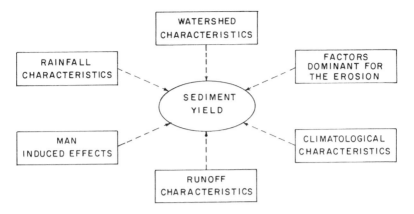

Figure 9.1. Composition of regression models

9.4.1   Regression Models

With this method sediment yield, Y, is related to one or more input factors and often one or more watershed characteristics (Figure 9.1). The coefficients of the resulting regression equation represent the deterministic portion of the model and the stochastic component explains the remainder of the variation in sediment yield. An excellent source for assumptions and derivations of regression techniques is given in Draper and Smith (1966).

If Y is the observed or actual sediment yield with mean $\bar{Y}$ for n sample data points, and Y_c is the sediment yield computed by a regression model then

$$SSR = \sum (Y_c - \bar{Y})^2 \tag{3}$$

$$SSE = \sum_{i=1}^{n} e_e^2 = \sum_{i=1}^{n} (Y - Y_c)^2 \tag{4}$$

and

$$SST = \sum (Y - \bar{Y})^2 = \sum (Y - Y_c)^2 + \sum (Y_c - \bar{Y})^2$$

$$= SSE + SSR \tag{5}$$

SSE is the sum of squared errors or stochastic (unexplained) part of the regression model, and measures the dispersion of the actual values about their corresponding computed values. SSR is the dispersion of the computed values about the sample mean, and can be considered as deterministic component. SST is the dispersion of observed values about their mean, and is thus seen to be the sum of both the stochastic and deterministic components. The relative contribution of stochastic or deterministic component to SST is measured by the coefficient of determination given as

$$r^2 = \frac{SSR}{SST} \tag{6a}$$

Table 9.1. Regression models

Author	Year	Equation	R^2	Variables
Dragoun	1962	$Y = EI_{30}(1 + P_A - Q_A)$	0.713–0.925	Y = sediment yield (tons) E = kinetic energy of the rainfall (ft-tons/acre-inch) I_{30} = max. storm rainfall intensity for 30 min duration (inches/hr) P_A = weighted antecedent rainfall volume for a 5-day period (in.) Q_A = antecedent runoff volume for a 5-day period (in.)
Tatum	1963	$X_0 = X_{10}A_n + X_v A_b$	not given	X_0 = observed debris production (yards³) X_{10} = rate of debris production 10 or more years after the burn (yards³ miles⁻²) X_v = rate of debris production for v year (1 to 10) after 100% burn (yards³ miles⁻²) A_n = portion of drainage area not burned (miles²) A_b = portion of drainage area burned (miles²)
Branson and Owen	1970	$Y = 40.87\, X_1 + 0.03\, X_2 - 1.27$	0.74	Y = sediment yield (acre-feet miles⁻²) X_1 = relief ratio X_2 = per cent bare soil
Flaxman	1972	$\log(Y + 100) = 6.21301 - 2.19113 \log(X_1 + 100)$ $+ 0.06034 \log(X_2 + 100)$ $- 0.01644 \log(X_3 + 100)$ $+ 0.04250 \log(X_4 + 100)$	0.920	Y = sediment yield (acre-ft miles⁻²) X_1 = ratio of average precipitation (inches) to average annual temperature (°F) X_2 = watershed average slope (%) X_3 = soil particles coarser than 1.0 mm (%) X_4 = soil aggregation index (%)
Jansen and Painter	1974	$\log Y = a + b \log D + c \log A + d \log L_r$ $+ e \log T + f \log H + g \log P$ $+ h \log V_I + j \log G$	0.579	$a, \ldots, j$ = regression coefficients A = basin area (km²) D = denudation rate (yield unit area⁻¹) L_r = relief/length ratio (m km⁻¹) T = average annual temperature (°C) H = altitude above sea level (m) P = average annual rainfall (mm) V_I = vegetation protection G = rock softness to erosion
Rendon-Herrero	1974	$S_{Di} = [(S_{Ti})(Q_{Ti}) - (S_{Bi})(Q_{Bi})]0.0027$ $ES = \dfrac{1}{12} \sum_{i=1}^{2n-1} \dfrac{S_{Di}}{A(26.9)}$ = excess sediment yield (tons miles⁻²)	0.75–0.86	S_{Di} = direct sediment discharge (tons per day) S_{Ti} = total sediment discharge (parts per million) S_{Bi} = base sediment discharge (parts per million) Q_{Ti}, Q_{Bi} = total and base water discharge (ft³ s⁻¹)

$$USO = \frac{DI}{ES} = \text{unit sediment graph (tons miles}^{-2})$$

A = watershed area (miles^2)
n = number of 2-hour increments of the hydrograph time base

McPherson 1975

$\log TSE = 0.8797 \log MLS - 0.690 \log BD + 0.6462 \log MCL + 1.1035$ — 0.85

$\log DSE = 8.25 MLS - 1.50 BD + 15.54$ — 0.65

$\log SSE = 0.5532 \log MLS - 0.5859 \log SCA + 0.8332 \log MCL + 1.37878$ — 0.25

TSE = total sediment yield (tons miles^{-2}) = $DSE + SSE$
DSE = dissolved sediment yield (tons miles^{-2})
MLS = mean land slope
MCL = main channel length (miles)
BD = basin diameter (miles)
SCA = sediment contribution area (miles^2)
SSE = suspended sediment yield (tons miles^{-2})

Dendy and Bolton 1976 — 0.75

$Y^* = A^{*-0.16}$
$Y^* = 1.07 V^*$, for $V < 2''$
$Y^* = 1.19 e^{-0.11} (V/V_R)$, for $V > 2''$
$A^* = A/A_R$, $Y^* = Y/Y_R$, $V^* = V/V_R$

A = drainage area (miles^2)
Y_R = reference sediment yield = 1,645 t miles^{-2} year^{-1}
A_R = references area = 1 mile^2
V_R = reference runoff depth (2'')
V = mean annual runoff depth (inches)

Yorke and Herb 1976 — 0.40–0.92

$\log Y = b_0 + b_1 X_1 + \ldots + b_{10} X_{10}$

$X_1, \ldots, X_{10}$ = total runoff (ft^3 s-days); storm antecedent discharge (ft^3 s^{-1}), antecedent discharge (ft^3 s^{-1}), antecedent days (number of days between storms), total precipitation (inches), rainfall intensity (in/hour), number of peaks during the storm, time (months) between the beginning of the record and the storm, construction (% of the area), peak ratio (1)
Y = total sediment yield (tons)

Hindall 1976 — —

$Q_s = a A^{b_1} Q_a^{b_2} Q_{25}^{b_3} S^{b_4}$

A = drainage area (miles^2)
Q_a = average discharge (ft^3 s^{-1})
Q_{25} = 25-year flood (ft^3 s^{-1})
S = main channel slope (feet mile^{-1})
$b_1, \ldots, b_4$ = regression coefficients
a = regression const.

Singh and Chen 1982 — 0.60–0.95, 0.82–0.85

$\log Y = \log a + b \log V$
$\log a = -2.5814 - 0.0038 L - 0.20 S_0 - 0.0003 A_f + 0.0042 H_L + 8.6451 K$

a, b = regression coefficients
V = runoff volume
L = main channel length (km)
S_0 = main channel slope (m km^{-1})
H_L = mean basin elevation (m)
A_f = forest area (km^2)
K = erodibility factor

or

$$1 - r^2 = \frac{SSE}{SST} \tag{6b}$$

Equation 6a gives the amount of variance explained by the regression model, whereas equation 6b gives the amount of unexplained variance.

Many regression models have been developed for both sedimentation in reservoirs and sedimentation in channels and streams as listed in Table 9.1; only four of them are discussed here. The main disadvantages of these models are that they require large amounts of data for determining model parameters and the resulting equations often cannot be transferred from one watershed to another because of differences in the input, watershed and output not considered in developing the regression equation.

Flaxman Model

Flaxman (1972) studied 27 rangeland watersheds ranging in size from 12 to 54 mi^2 in the western United States for reservoir design purposes. He related sediment yield Y (tons/mi^2) to five dominant erosion factors as

$$\log(Y + 100) = 6.21301 - 2.19113 \log(X_1 + 100) + 0.06034 \log(X_2 + 100)$$
$$- 0.01644 \log(X_3 + 100) + 0.04250 \log(X_4 + 100) \tag{7}$$

where X_1 is ratio of average annual precipitation (in.) to average annual temperature (°F), X_2 is watershed average slope (%), X_3 is soil particles greater than 1.0 mm (%), and X_4 is soil aggregation index (%). The coefficient of determination was 0.92. On analysing Flaxman's data, Singh (1973) found that the sediment yield Y could be obtained equally accurately by a linear regression analysis without logarithmic transformation.

Dendy–Bolton (DB) Model

Dendy and Bolton (1976) analysed over 800 reservoirs from all over the United States. However, they introduced an error by equating sediment deposition rate to watershed sediment yield, thus assuming 100 per cent sediment trap efficiency. The sediment yield in their model is related to the size of the drainage area and mean annual runoff V. The watershed area ranges from 1 mi^2 to 30,000 mi^2 and the runoff from nearly 2 to 13 in./year. The relationship between the annual sediment yield Y and watershed area A is:

$$Y_* = A_*^{-0.16}, \ Y_* = Y/Y_R, \ A_* = A/A_R \tag{8}$$

where the subscript R stands for a reference value: $Y_R = 1{,}645$ tons/mi^2 and $A_R = 1$ mi^2. Similarly, the relationship between the annual sediment yield Y and the runoff V is:

$$V < 2'': \ Y_* = 1.07 \, V_*^{0.46}, \ V_* = V/V_R$$

$$V \geq 2'': \ Y_* = 1.19 \exp[-0.11 \, V_*] \tag{9}$$

where V is in inches, and $V_R = 2$ in.

The influence of both drainage area and runoff on sediment yield is expressed as:

$$V < 2'': \ Y_* = 1.07 \ V_*^{0.46} \ [1.43 - 0.26 \ \log A_*]$$

$$V \geq 2'': \ Y_* = 1.19 \ \exp[-0.11 \ V_*] \ [1.43 - 0.26 \ \log A_*] \tag{10}$$

The coefficient of determination for equation 10 is 0.75. These equations express the general relationships between sediment, runoff and drainage area. On the average, sediment yield per unit area is inversely proportional to the 0.16 power of net drainage area for the watersheds considered. Sediment yield per unit area increases rapidly to about 1,860 tons/mi² per year as runoff increases from 0 to about 2 inches, and then decreases as runoff increases from 2 to about 50 inches.

Onstad et al. (1977) cautioned that the DB model might not be appropriate for a specific location because of widely varying local factors. Later it was found that it was probably applicable throughout the southwestern United States.

Jansen-Painter (JP) Model

Jansen and Painter (1974) developed linear regression models for annual average sediment yield from large river basins with area greater than 5000 km² which may be applicable throughout the world. Data from 79 basins grouped into four major climatic zones were used to express average annual suspended sediment yield (ton km⁻²) as a function of up to eight parameters relating to climate and topography. Within each group, the factors controlling erosion would vary in local climate, erodibility of geology and soils, topography, vegetation, and man-made changes. Average values of these factors can then be used or the factors can be expressed as a parameter for multiple-linear regression analysis. The regression equations developed are:

1. Climate A with 14 basins (tropical, rainy, where the coldest month has an average temperature greater than 18°C):

$$\log Y = 4.354 + 1.527 \ \log D - 0.302 \ \log A + 0.296 \ \log L_r - 3.417 \ \log T \tag{11}$$

2. Climate B with 14 basins (dry, for which a parameter $w = (P/T) + c < 2$, where P is average annual rainfall (cm), T is average annual temperature (°C) and $c = 0$ (rainfall mainly in winter), $c = 7$ (rainfall uniform throughout the year), $c = 14$ (rainfall mainly in summer)):

$$\log Y = 12.133 - 0.340 \ \log D + 1.590 \ \log H + 3.704 \ \log P +$$
$$0.936 \ \log T - 3.495 \ \log V_I \tag{12}$$

3. Climate C with 25 basins (humid, mesothermal, where the coldest month has an average temperature above 0°C but below 18°C, and the warmest month above 10°C):

$$\log Y = -3.078 + 1.003 \ \log D + 0.686 \ \log L_r$$
$$+ 4.287 \ \log T - 5.031 \ \log V_I \tag{13}$$

4. Climate D with 26 basins (humid, microthermal, where the coldest month has an average temperature below 0°C, and the warmest month above 10°C):

$$\log Y = -5.073 + 0.514 \log H + 2.195 \log P - 3.706 \log V_I + 1.449 \log G \qquad (14)$$

In these equations P is average annual rainfall (mm), T is average annual temperature (°C), D is discharge (m^3 km^{-2}), G is rock softness (proneness) to erosion parameter estimated as: Palaeozoic $G = 3$, Mesozoic $G = 5$, Cainozoic $G = 6$, Quarternary $G = 2$, H is altitude above sea level (m), L_r is relief-length ratio (m km^{-1}) equivalent to main channel slope, A is basin area (km^2) and V_I is vegetation protection index (forest $V = 4$, grass $V = 3$, steppe $V = 2$, desert $V = 1$). Table 9.2 summarizes the variance accounted for the significance level of equations 11–15 and each parameter. Parameters not significant at 90 per cent level are eliminated. On a global basis, sediment increases with runoff, altitude, relief, precipitation, temperature and rock softness, but decreases with increasing area and protective vegetation. Three anomalies, however, remain: (1) Climate A: inverse relationship of sediment yield with temperature; (2) Climate B: inverse relationship of sediment yield with discharge; and (3) Climate C: inverse relationship of sediment yield with relief. These models can be used to predict the scale of the sediment problem in river basins where no sediment records exist.

When all climates are combined, the resulting regression equation is

$$\begin{aligned} \log Y = &-2.032 + 0.1 \log D - 0.314 \log A + 0.75 \log H + 1.104 \log P + 0.368 \log T \\ &- 2.324 \log V_I + 0.786 \log G \end{aligned} \qquad (15)$$

which accounted for 57.9 per cent of the variance.

Singh–Chen (SC) Model

Based on the work of Rendon-Herrero (1974), and Rendon-Herrero, *et al.* (1980), Singh and Chen (1982) established on log–log paper a linear relation between sediment yield (metric tons) and volume of surface runoff V per unit area (cm) as

$$\log Y = \log a + b \log V \text{ or } Y = a\, V^b \qquad (16a)$$

in which $\log a$ is the intercept and b is the slope of the line. The amount of variance explained by this relationship varied from 61 per cent to 95 per cent for 21 watersheds 45 to 2,056 km^2. The slope b varied from 1.03 to 1.86, and the coefficient a varied from 0.22 to 37.3. A regression analysis of a and b with geomorphic characteristics produced:

$$\log a = -2.5814 - 0.0038 L - 0.215 S_0 - 0.0003 A_f + 0.0042 H_L + 8.6451 K \qquad (16b)$$

where L is main channel length (km), S_0 is main channel slope (m km^{-1}), H_L is mean basin elevation (m), K is erodibility factor, and A_f forest area (km^2). The coefficient of determination was 82 per cent and increased to 85 per cent with inclusion of shape factor, storage area and watershed area.

Table 9.2. Variance and significance of equations and parameters for the JP model

Climate	Variance Accounted for (%)	Multiple Correlation Coefficient	Significance Levels								
			eq.	D	A	H	L_r	P	T	V_1	G
A	93.5	0.967	99.5	98.5	99.5		90		90		
B	86.0	0.927	99.5	97.5		95		99.5	90	99.5	
C	62.8	0.792	99.5	99.5			90		99.5	90.5	
D	64.5	0.803	99.5			90		99.5		90.0	97.5

Notations: D = discharge (m^3 km^{-2}), A = watershed area (km^2), H = altitude above sea level (m), L_r = relief-length ratio (m km^{-1}), P = average annual rainfall (mm), T = average annual temperature (°C), V_1 = vegetation protection index, and G = rock softness (proneness) index.

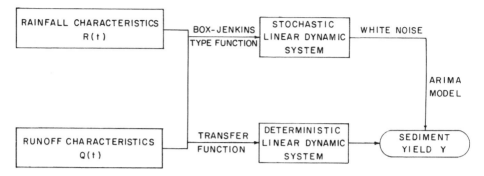

Figure 9.2. Composition of time series models

An interesting application of the SC model is in distinguishing similar watersheds in terms of sediment production. Singh and Chen (1982) identified four groups based on the value of b: (1) 1.03 to 1.25; (2) 1.26 to 1.40; (3) 1.41 to 1.70; and (4) 1.71 to 2.0, and found that the correlation between the intercept a and geomorphic parameters increased to as much as 0.999 for each group. Because of simplicity of equation 16a, the SC model can be easily coupled with another stochastic model.

9.4.2   Time Series Models

These models consider watershed as a spatially lumped linear dynamic system. Rainfall $R(t)$ and runoff $Q(t)$ constitute input, sediment yield $Y(t)$ is output, and the watershed system is represented by transfer function (deterministic part) and an appropriate Box–Jenkins type noise function (stochastic part) as shown in Figure 9.2. The Box–Jenkins type functions usually account for the white noise component. A time series model was probably first employed by Rodriguez-Iturbe and Nordin (1968) to analyse a series of sediment measurements on the Rio Grande river in New Mexico. They found that cyclic components of time series were useful in prediction of fluctuations in both water and sediment discharges and that the annual cycles of water discharges were highly correlated with those of the sediment. Several models using time series analysis have since been developed but only two representative models are discussed here.

Sharma–Dickinson (SD) Model

Discrete–time series models of daily and monthly watershed sediment yield have been developed by Sharma and associates (Sharma, 1977; Sharma and Dickinson, 1979; Sharma and Dickinson, 1980; Sharma et al., 1979). Referring to Figure 9.2, sediment yield Y_t at time t is expressed as a function of discharge Q_t and rainfall intensity R_t of previous times:

$$Y_t = f(Q_t, Q_{t-1}, \ldots, Q_{t-m}, R_t, R_{t-1}, \ldots, R_{t-n}), \\ m \geq 1, \ n \geq 1 \tag{17}$$

Equation 17 can be expanded in Taylor's series in such a way that its nonlinearity is preserved:

$$Y_t = A \prod_{j=0}^{m} Q_{t-j}^{c_j} \prod_{j=0}^{n} R_{t-j}^{d_j} \tag{18}$$

where A, c_j, d_j, $j = 0, \ldots, m$ or n, are constants. Taking the logarithmic transformation and considering errors in data:

$$y_t = A_0 + \sum_{j=0}^{m} c_j \, q_{t-j} + \sum_{j=0}^{n} d_k \, P_{t-j} + e_t \tag{19}$$

where y_t, A_0, q_{t-j}, P_{t-j} are the logarithms of Y_t, A, Q_{t-j} and R_{t-j} and ϵ_t are the white noise or error terms normally distributed with zero mean and homogeneous variance and uncorrelated with discharge and rainfall intensity. However, error terms could be autocorrelated and may be modelled by the Box–Jenkins type models (Box and Jenkins, 1976). Sharma *et al.* (1979) found, using multiple correlation analysis, that monthly rainfall contributed little, while the monthly runoff was the dominant factor in explaining the variance in Y_t. If one introduces back-shift operator B as

$$q_{t-1} = B \, q_t, q_{t-m} = B^m \, q_t$$

and the impulse response function of the system as

$$G(B) = \sum_{j=0}^{m} c_j \, B^j, \tag{20}$$

then equation 19 with introduction of b as delay parameter becomes:

$$y_t = A_0 + G(B) \, q_{t-b} + e_t \tag{21}$$

If y_t and q_t are expressed in terms of deviations from their respective means $\bar{y}$ and $\bar{q}$ then equation 21 becomes

$$(y_t - \bar{y}) = G(B) \, B^b \, (q_t - \bar{q}_t) + e_t \tag{22}$$

Denoting $(y_t - \bar{y})$ as w_t and $(q_t - \bar{q})$ as x_t, equation 22 takes the form

$$w_t = G(B) \, B^b \, x_t + e_t = G(B) \, x_{t-b} + e_t \tag{23}$$

Equation 23 represents two systems: one representing the deterministic linear system with x_t as input and w_t as output; and the other representing the stochastic linear dynamic system with white noise a_t as input and noise e_t as output. Errors due to measurements, modelling hypothesis and discretization can be regarded as equivalent to the noise in the watershed fluvial system.

After analysis of daily and monthly data of runoff and sediment, Sharma *et al.* derived two models: one for predicting monthly watershed sediment yield as:

$$y_t - d_1 \, y_{t-1} = A_1 + g_0 \, q_t - g_1 \, q_{t-1} + e_t \tag{24}$$

and the other for daily watershed sediment yield as:

$$y_t - d_1 \, y_{t-1} - d_2 \, y_{t-2} = A_1 + g_0 \, q_t - g_1 \, q_{t-1} - g_2 \, q_{t-2} + e_t \tag{25}$$

where g_0, g_1 and g_2 are autoregressive coefficients, $A_1 = f(\bar{q}, \bar{y}, d(B), g(B))$, and e_t is represented by a white noise sequence with zero mean and homogeneous variance.

Equation 24 represents a first order time-series model with five parameters. Its deterministic part explained 81 to 91 per cent of the variance in the tested data and the white noise component the remaining 9 to 19 per cent. Equation 25 is second order time series model with seven parameters. Its deterministic part explained 95 to 97 per cent of the variance in tested data and the white noise sequence the remaining 3 to 5 per cent. The parameters were determined using least squares method.

These models have the advantage that they can be easily calibrated and even used to extend short-term records of a watershed. The main disadvantage of this model is that it does not include the sediment yield sources (erosion sources) in any way and therefore cannot be used for agricultural planning or evaluation of land treatments.

Caroni–Singh–Ubertini (CSU) Model

Another time series model has recently been developed by Caroni *et al.* (1984), with reference to flashy streams where a large amount of total sediment transport is concentrated in short-duration intervals. Analogous to the SD model, the CSU model has two parts: (a) a transfer function, and (b) a noise term. Transfer between input series x_t and sediment yield output y_t both standardized using their means, is represented by

$$y_t = v(B)\, x_t = [(v_0 + v_1 B + v_2 B^2 + \ldots)\, B^b]\, x_t \tag{26}$$

where B is the backward shift operator, b is magnitude of the shift and v_i, $i = 1, 2, \ldots$, are weights which constitute the impulse response function. The noise term e_t, representing the stochastic part, is treated as the output of a linear filter in the form of an autoregressive-integrated moving average (ARIMA) model with a random (white noise) input sequence a_t as

$$(1 - B)^d\, e_t = \frac{1 - \Theta_1 B - \ldots - \Theta_n B^n}{1 - \Phi_1 B - \ldots - \Phi_m B^m}\, a_t \tag{27}$$

where Φ_i, Θ_i are autocorrelation parameters and m, d, n are the orders of ARIMA-noise model. The input x_t was represented separately by rainfall and runoff, thus giving rise to two models. The models were calibrated and tested on an experimental watershed in the Pigeon Roost Creek basin, near Oxford, Mississippi, U.S.A. The time interval for calibration and prediction was taken as 30 minutes. The model parameters were fairly stable and their average values were therefore suggested for the development of an 'average' model. However, the 'average model' might lead to incorrect results as it differs from the real model.

9.4.3 Entropy Models

Singh and Krstanovic (1985) developed stochastic models for sediment yield from watersheds, reservoirs or channels using the principle of maximum entropy (POME). These models can be entirely stochastic or be comprised of both deterministic and stochastic components. The stochastic component is derived using the entropy $H(x)$, joint entropy $H(x,y)$ and conditional entropy $H(y|x)$ defined respectively as

$$H(x) = - \int_{-\infty}^{+\infty} f(x) \log f(x) \, dx \tag{28}$$

$$H(x,y) = - \int_{-\infty}^{\infty} \int_{-\infty}^{+\infty} f(x,y) \log f(x,y) \, dx \, dy \tag{29}$$

and

$$H(y|x) = - \int_{-\infty}^{+\infty} \int_{-\infty}^{+\infty} f(x,y) \log \left[\frac{f(x,y)}{f(x)} \right] \, dx \, dy \tag{30}$$

where x and y are random variables representing the runoff volume and the sediment yield, respectively. Equations 28–30 are related as

$$H(y|x) = H(x,y) - H(x) \tag{31}$$

For known values of runoff (available information) we seek values of the sediment yield or probability density function (pdf) and cumulative density function (cdf) of y given x. In order to derive these density functions, the following constraints are assumed:

$$\sigma_x^2 = \int_{-\infty}^{+\infty} \int_{-\infty}^{+\infty} f(x,y) \, x^2 \, dx \, dy = \text{var}(x)$$

$$\sigma_y^2 = \int_{-\infty}^{+\infty} \int_{-\infty}^{+\infty} f(x,y) \, y^2 \, dx \, dy = \text{var}(y)$$

$$\sigma_{xy} = \int_{-\infty}^{+\infty} \int_{-\infty}^{+\infty} f(x,y) \, xy \, dx \, dy = \text{cov}(x,y) \tag{32}$$

$$\int_{-\infty}^{+\infty} \int_{-\infty}^{+\infty} f(x,y) \, dx \, dy = 1$$

in which $\sigma_w^2 = \text{var}(w)$ is variance of w and $\sigma_{xy} = \text{cov}(x,y)$ is covariance of x and y. The variables x and y are defined about their mean values, i.e. $x = x - E[x]$ and $y = y - E[y]$. Depending upon the available information, the constraints can be defined differently, and different forms of pdf and cumulative density function (cdf) of y given x (written $y|x$) consequently obtained. The constraints are usually computed from available data as:

$$S_x^2 = \frac{1}{n-1} \sum (x_i - \bar{x})^2$$

$$S_y^2 = \frac{1}{n-1} \sum (y_i - \bar{y})^2$$

$$S_{xy} = \frac{1}{n-1} \sum (x_i - \bar{x})(y_i - \bar{y})$$

where n is the sample size, S_w^2 is sample variance of w, and S_{xy} is sample covariance of x and y.

Using the method of Lagrange multipliers, $H(x,y)$ can be maximized subject to equation 32 resulting finally in

$$f(y|x) = \frac{\sigma_x}{\sqrt{2\pi(\sigma_x^2 \sigma_y^2 - \sigma_{xy}^2)}} \exp\left\{-\left[\frac{\sigma_x^2 y^2 - 2\sigma_{xy} xy + x^2 \frac{\sigma_{xy}^2}{\sigma_x^2}}{2(\sigma_x^2 \sigma_y^2 - \sigma_{xy}^2)}\right]\right\} \tag{33}$$

and

$$F(y|x) = \frac{1}{\sqrt{2\pi}} \int_{-\infty}^{b} \exp(\frac{-z^2}{2}) \, dz \tag{34}$$

where

$$b = (y - \frac{\sigma_{xy}}{\sigma_x^2} x) \frac{\sigma_x}{\sqrt{\sigma_x^2 \sigma_y^2 - \sigma_{xy}^2}}$$

Equation 34 is the integral form of the standard normal distribution.

For easy application, the stochastic part of the model was derived in dimensionless form above (Singh and Krstanovic, 1987) by using the beta coefficients: $\beta_1 = x/\sigma_x$, $\beta_2 = y/\sigma_y$ and $\beta_3 = \beta_1\beta_2/r$, where r is correlation coefficient derived from the records. Thus, equations 33 and 34 become

$$f(y|x) = \frac{\exp\{\beta_1^2(0.5 - [2(1-r^2)]^{-1})\}}{[2\pi \sigma_y^2(1-r^2)]^{0.5}} \exp\{-\beta_2^2[2(1-r^2)]^{-1}[1-2\beta_1^2\beta_3^{-1}]\} \tag{35}$$

$$F(y|x) = g(\beta_1,\beta_3) \, B(\beta_1,\beta_3) \tag{36}$$

where

$$g(\beta_1,\beta_3) = \frac{\beta_1}{r(\beta_1^2 - \beta_3)} [\frac{1-r^2}{2\pi}]^{0.5} \tag{37}$$

and

$$B(\beta_1,\beta_3) = \exp\left[-\frac{(\beta_1 - \frac{\beta_3}{\beta_1})^2}{2(\frac{1}{r^2} - 1)}\right] - \exp\left[-\frac{\beta_1^2}{2(\frac{1}{r^2} - 1)}\right] \tag{38}$$

The functions $g(\beta_1,\beta_3)$ and $B(\beta_1,\beta_3)$ can be plotted against β_1 for various values of β_3 and r.

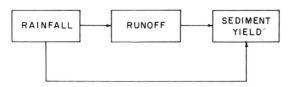

Figure 9.3. Rainfall–runoff–sediment relation

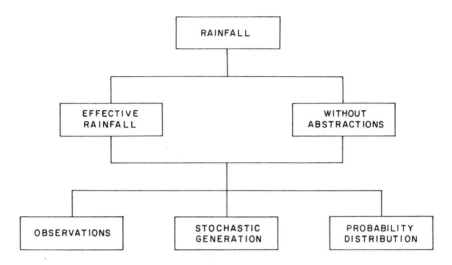

Figure 9.4. Treatment of rainfall

The deterministic part can be represented by an appropriate model. Singh and Krstanovic (1985) used the logarithmic relationship between sediment yield and runoff volume of equation 15 discussed earlier where the parameters a and b were obtained by the least squares method but can be determined from watershed characteristics. The β coefficients can be expressed in terms of these parameters as

$$\beta_3 = \frac{\beta_1 \beta_3}{r} = \frac{1}{r} \beta_1 \frac{a x^b}{\sigma_y} \tag{39}$$

On simplifying,

$$\beta_3 = \frac{a \, \beta_1^{b+1} \, \sigma_x^b}{r \, \sigma_y} \tag{40}$$

From known $B(\beta_1, \beta_3)$ and $g(\beta_1, \beta_3)$, one can evaluate $F(y|x)$ and then from tables of standard normal distribution $N(0,1)$ one can obtain the argument of the function in equation 34, and finally evaluate β_2. Thus, the sediment yield is

$$y = \beta_2 \, \sigma_y \tag{41}$$

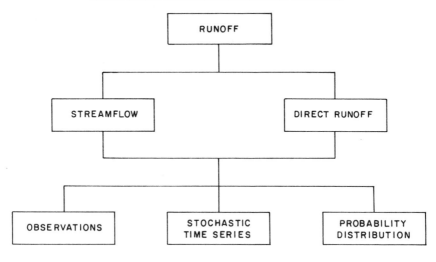

Figure 9.5. Treatment of runoff

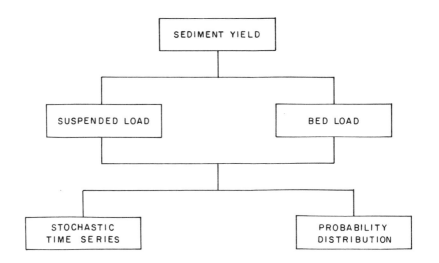

Figure 9.6. Treatment of sediment yield

The coupling of deterministic and stochastic components provides an objective way of evaluating their relative contributions. If $r=1$, $F(y|x)=0$, and the model becomes completely deterministic. For $r<1$, both components are present and should be accounted for. A comparison of the coupled model showed that it was superior to some other models (e.g. bivariate normal distribution).

9.4.4 Probability Models

Probability models, in general, stochastically treat rainfall as input, and runoff and sediment yield as output, or directly consider runoff as input and sediment yield as output as shown

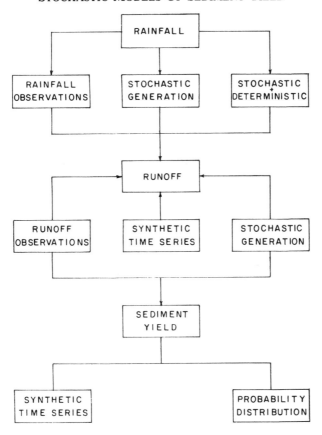

Figure 9.7. Composition of stochastic sediment yield models

in Figure 9.3. Rainfall has been treated in various ways as shown in Figure 9.4. Abstractions have been considered to obtain effective rainfall in some models. Its observations have been used in some other models, and it has been stochastically generated or its probability distribution has been derived in some others. In a similar vein, either total runoff or direct runoff has been used. Some models use runoff observations, some generate it stochastically, and some derive its probability distribution as shown in Figure 9.5. These descriptions of rainfall and runoff are variously combined to develop different types of stochastic sediment yield models as shown in Figure 9.6.

The manner in which sediment yield is related to runoff and/or rainfall varies with the type of the sediment yield model. Figure 9.7 shows how rainfall, runoff and sediment yield have been variously combined to produce the various sediment yield models. The connection between any two of these three variables may be deterministic or stochastic. Thus, a stochastic sediment yield model may be the result of stochastic relationship or both the stochastic and deterministic relationships. This point is further illustrated in Table 9.3, where references are given for several stochastic sediment yield models.

Table 9.3. Probabilistic sediment yield models

Author	Year	Rainfall					Runoff			Sediment yield		
		No. of Events	Intensity	Depth	Infiltration	Evaporation	No. of Events	Discharge	Volume	No. of Events	Yield/ Event	Yield (0,t)
Murota & Hashino	1969, 1971	X	X	—	—	—	—	—	—	X	X	—
Kisiel et al.	1971	X	—	X	X	X	X	X	X	—	X	—
Jacobi	1971, 1975	—	—	—	—	—	—	—	—	—	X	—
Woolhiser & Todorovic	1974	X	X	—	X	X	X	—	X	X	X	X
Woolhiser & Blinco	1975	X	X	—	X	X	X	X	X	X	X	X
Renard & Lane	1975	—	—	—	—	—	X	X	X	X	X	—
Renard & Laursen	1975	—	—	X	X	—	X	X	X	—	X	—
Duckstein et al.	1977	X	—	X	X	X	—	—	X	—	X	—
Smith et al.	1977	X	X	—	X	X	—	X	X	X	X	X
Mills	1980, 1981	—	—	—	X	X	—	X	X	—	X	X
Soares et al.	1982	—	—	—	X	—	—	X	—	—	—	X
Van Sickle	1982	—	—	—	X	X	X	X	X	X	X	X
Moore & Clark	1983	—	—	X	X	—	—	—	X	—	—	X
Moore	1984	—	—	X	X	—	—	X	X	—	—	X
Thomas	1985	—	—	—	—	—	—	X	—	—	—	X

X denotes inclusion of the components

Woolhiser–Todorovic (WT) Model

Woolhiser and Todorovic (1974) developed a stochastic model treating sediment yield in some time interval $(0,t)$ from a watershed characterized by ephemeral streamflow as the sum of a random number of random variables. Its conceptual as well as mathematical formulation is more enlightening than most stochastic models reported to date. The distribution of the sediment yield per event can be obtained empirically. The number of sediment events in $(0,t)$ is equal to the number of runoff events and is related to the number of rainfall events. Thus, the sediment yield counting process is related to the rainfall counting process by a rainfall–runoff model. Three progressively more complicated rainfall–runoff models were postulated: (1) the pure threshold model, (2) the general threshold model, and (3) the infiltration model. The ensuing discussion considers the pure threshold model, for it is the simplest of the three.

Spatially averaged rainfall intensity $R(t)$, and runoff rate $Q(t)$ and total sediment outflow $Q_s(t)$ at the outlet of the watershed are considered random variables. A rainfall event is defined to be any continuous period of rainfall for which $R(t) > 0$. Likewise, a runoff event is any continuous period of surface runoff where $Q(t) > 0$. A sediment yield event is defined similarly. Each runoff event is assumed to cause a sediment yield event. A k-th runoff event can be characterized by its duration D_k, time of ending T_k, and volume V_k. For example,

$$V_k^R = \int_{T_{k-1}^R}^{T_k^R} R(s) \, ds$$

$$V_k^Q = \int_{T_{k-1}^Q}^{T_k^Q} Q(s) \, ds$$

$$V_k^S = \int_{T_{k-1}^S}^{T_k^S} Q_S(s) \, ds$$

Similarly, $D_k^w = T_k^w - T_{k-1}^w$ with w denoting R for rainfall, Q for water discharge, and S for sediment discharge. The duration and time of ending are the same for both the runoff and sediment yield events. Thus, computation of sediment yield requires computation of the number $N_S(t)$ of sediment yield events in $(0,t)$ and $Q_s(t)$ for each event; both these variables are random variables.

Number of Sediment Yield Events

The stochastic counting process can be defined as

$$N_w(t) = \sup\{k; \ T_k^w \le t\}, \quad w = R, Q, S \tag{42}$$

Let $N_R(t)$, $N_Q(t)$ and $N_S(t)$ define the number of complete rainfall events, runoff events and sediment yield events respectively. By assumption, $N_S(t) = N_Q(t)$. By virtue of the physical relationship between rainfall and runoff, $N_Q(t)$ depends on $N_R(t)$ and for all $t \ge 0$,

$$N_Q(t) \le N_R(t) \tag{43}$$

The number of rainfall events is described by the time-dependent (nonhomogeneous) Poisson process as

$$P[N_R(t) = n] = [\Lambda(t)]^n \exp(-\Lambda(t))/n! \tag{44}$$

where

$$\Lambda(t) = \int_0^t \lambda(s) \mathrm{d}s$$

in which $\lambda(s)$ is the intensity function and $\Lambda(t)$ is the expected value function, i.e. $E[N_R(t)] = \Lambda(t)$. If $\Lambda(t)$ is not a time-dependent function, then $\Lambda(t) = \Lambda \cdot t$, with Λ as parameter.

From the definition of the counting process, it is obtained that

$$P[T_k^w \le t] = \sum_{i=k}^{\infty} P[N_w(t) = i]; \ k = 0, 1, 2, \dots \tag{45}$$

It then follows that

$$P[N_S(t) = k] = P[N_Q(t) = k] = \sum_{n=k}^{\infty} P[N_R(t) = n, N_Q(t) = k]$$

$$= \sum_{n=k}^{\infty} F_n(t,k) P[N_R(t) = n] \tag{46}$$

where

$$F_n(t,k) = P[N_Q(t) = k | N_R(t) = n], \ n = k, \ k+1, \ k+2, \dots k = 0, 1, 2, \dots \tag{47}$$

Clearly, the conditional probability $F_n(t,k)$ depends upon the probabilistic mechanism of selecting a runoff event from a rainfall event. To this end, the pure threshold model is considered. This model assumes that a certain threshold amount of rainfall V_0 must be exceeded before runoff occurs. In other words, a k-th rainfall event will be a runoff event only if $V_k^R > V_0$. Therefore, the probability of runoff occurring for the k-th rainfall event is

$$p = P[V_k^R > V_0] \tag{48}$$

where V_0 is constant. If V_0 is considered to be a random variable independent of rainfall then the probability of a runoff event for a given rainfall event is

$$p = 1 - \int_0^{\infty} \int_0^{V_k^R = V_0} f(V_k^R, V_0) \, \mathrm{d}V_k^R \, \mathrm{d}V_0 \tag{49}$$

in which $f(V_k^R, V_0)$ is the joint pdf of V_k^R and V_0.

If V_k^R, $k = 1, 2, \dots$, are independent, identically distributed random variables then the conditional probability $F_n(t,k)$ can be obtained from the binomial distribution,

$$F_n(t,k) = \binom{n}{k} p^n (1-p)^{n-k} \tag{50}$$

Under these assumptions, the runoff or sediment yield counting process can be determined from equations 44, 47 and 50,

$$P[N_Q(t)=n] = \exp(-\Lambda(t)) \sum_{n=k}^{\infty} \binom{n}{k} p^n (1-p)^{n-k} \Lambda(t)/n! \tag{51}$$

Equation 51 represents a Poisson process with parameter $p\Lambda(t)$ (Woolhiser and Todorovic, 1974). The pure threshold model renders the runoff process Poissonian if the rainfall process is Poissonian.

The Sediment Yield Process

The sediment yield process is comprised of a series of events and can be approximated by

$$V_S(t) = \sum_{k=0}^{N_S(t)} V_S^k, \quad V_S^0 = 0 \tag{52}$$

An estimate of sediment yield V_S for each k-th runoff event now is needed. A simple way is to assume that the sediment yield per event is independent, identically distributed random variable with pdf $f(v)$, independent of $N_S(t)$ for all t. Then

$$F_S(v|t) = P[N_S(t)=0] + \sum_{k=1}^{\infty} P[N_S(t)=k] \int_0^v [f(u)]^{k*} \, du \tag{53}$$

where $F_S(v|t) = P[V_S(t) \le v]$, $f(v)$ is the pdf of sediment yield per event for $v > 0$, and $k*$ represents $k*$-th convolution of $f(v)$ with itself. The mean and variance of $V_S(t)$ can be shown to be

$$E[V_S(t)] = E[N_S(t)] E[V_S] \tag{54}$$

$$\text{var}[V_S(t)] = E[N_S(t)] \, \text{var}[V_S] + \text{var}[N_S(t)] E^2[V_S] \tag{55}$$

Equations 54–55 show that the mean and variance of sediment yield in $(0,t]$ can be obtained from the mean and variance of the runoff counting process and the mean and variance of sediment yield per event. Woolhiser and Todorovic have suggested, based on data from two agricultural watersheds, that short sediment yield records can be combined with longer runoff records to improve estimates of mean sediment yield.

The WT model has a general framework of which some of the reported models can be considered as special cases. The model presented by Renard and Lane (1975) follows a similar approach, with a different stochastic process of runoff events and a deterministic runoff-sediment yield relationship. Van Sickle (1982) presented a simplification of the WT model, with a compound Poisson process representing the occurrence times of events and their discharge levels exceeding a threshold value. These discharge levels are modelled by a two-parameter gamma pdf.

9.5 DISCUSSIONS AND CONCLUSIONS

Four different types of stochastic sediment yield models have been discussed. Of these, probability models seem to provide the greatest insight into the sediment yield and its relationship to rainfall and runoff processes. Sediment yield is treated as a stochastic process. For plots and small watersheds, most often interpretation of sediment yield is on an event basis, where the sediment yield of a small watershed is proportional to that of a plot. Sediment yield events are either independent identically distributed random variables or are obtained from correlation with hydrologic characteristics of rainfall and runoff. For larger watersheds, consideration of sediment yield for each storm is not appropriate because the watershed response time is no longer short and dependency among the events is introduced. In such cases, sediment yield is treated on seasonal or annual basis and may not be kept as a stochastic part of the model. Either rainfall or runoff sequences are considered stochastic and are generated by simulation techniques. Their relationship with sediment yield is usually kept deterministic. This scheme is also used in cases of intermittent events. Many of the time series characteristics such as inter-arrival times among events, duration of the event, frequency of occurrence, and the volume of the accumulated yield are interpreted successfully by probability distribution functions which would not be possible by any other method.

Regression models are the easiest way to relate sediment yield to various physical characteristics of the watershed. Stochasticity in these models is seen only through variances and coefficient of determination that express fluctuations around the mean trend.

Treatment of the sediment yield as a time series model is mathematically correct but, because of complexity, may loose the advantage to other models which are more efficient.

Another way of stochastically interpreting sediment yield is by POME models. Although only in early developmental stage, they appear to have advantages over other models in that they require limited data.

REFERENCES

Alonso, C. V. (1980). Selecting a formula to estimate sediment transport capacity in nonvegetated channels. In *CREAMS: A field scale model for chemicals, runoff and erosion from agricultural management systems*, Vol. 3, Supporting Documentation, Chapter 5, US Department of Agriculture Conservation Research Report.

American Society of Civil Engineers (1970). Chapter V. Sediment sources and sediment yields. *Journal of Hydraulics Division, ASCE*, **96**, HY6, 1283–1330.

American Society of Civil Engineers (1982). Relationships between morphology of small streams and sediment yield. *Report of an ASCE Task Committee, Journal of the Hydraulics Division, Proceedings of the ASCE*, **108**, HY11, 1328–1365.

Box, G. E. P. and Jenkins, G. M. (1976). *Time Series Analysis, Forecasting and Control*, Holden Day, San Francisco, California.

Branson, F. A. and Owen, J. B. (1970). Plant cover, runoff and sediment yield relationships on Mancos shale in western Colorado. *Water Resources Research*, **6**, 3, 783–790.

Caroni, E., Singh, V. P., and Ubertini, L. (1984). Rainfall–runoff sediment yield relation by stochastic modeling. *Journal of Hydrological Sciences (des Sciences Hydrologiques)*, **29**, 2, 203–218.

Curtis, W. F., Culbertson, J. K., and Chase, E. B. (1973). Fluvial–sediment discharge to the oceans from the contiguous United States. *US Geological Survey Circular*, **610**, 17pp.

Dendy, F. E. and Bolton, G. C. (1976). Sediment yield–runoff–drainage area relationships in the USA. *Journal of Soil and Water Conservation*, **31**, 6, 264–266.

Dole, R. B. and Stabler, H. (1909). Denudation. In *Papers and Conservation of Water Resources. US Geological Survey-Water Supply Paper*, **234**, 79-93pp.

Dragoun, F. Y. (1962). Rainfall energy as related to sediment yield. *Journal of Geophysical Research*, **67**, 4, 1495-1502.

Draper, N. R. and Smith, H. (1966). *Applied Regression Analysis*, John Wiley and Sons, Inc., New York, 407pp.

Duckstein, L., Szidarovszky, F. F., and Yakowitz, S. (1977). Bayes design of a reservoir under random sediment yield. *Water Resources Research*, **13**, 4, 713-719.

Einstein, H. A. (1950). Bed load function for sediment transportation in open channel flow. *USDA Technical Bull. No. 1026*, US Department of Agriculture, Washington, D.C.

Flaxman, E. M. (1972). Predicting sediment yield in western US *Journal of Hydraulic & Division, ASCE*, **98**, 12, 2073-2085.

Foster, G. R., Meyer, L. D., and Onstad, C. A. (1975). Mathematical simulation of upland erosion mechanics. In *Present and Prospective Technology for Predicting Sediment Yields and Sources*, Proceedings of the Sediment Yield Workshop, USDA Sedimentation Laboratory, Oxford, Mississippi.

Fournier, F. (1969). Transports solides effectues par les cours d'eau. *International Association of Scientific Hydrology Bulletin*, **3**, 7-49.

Guy, H. P. (1974). An overview of urban sedimentology. *Proceedings*, National Symposium on Urban Rainfall and Runoff and Sediment Control, University of Kentucky, Lexington, Kentucky.

Herb, W. J. and Yorke, T. H. (1976). Storm-period variables affecting sediment transport from urban construction areas. *Proceedings*, 3rd Federal Inter-Agency Sedimentation Conference, 1.181-1.192.

Hindall, S. M. (1976). Prediction of sediment yields in Wisconsin streams. *Proceedings*, 3rd Federal Inter-Agency Sedimentation Conference, 1.205-1.218.

Holeman, J. N. (1968). The sediment yield of major rivers of the world. *Water Resources Research*, **4**, 4, 737-767.

Jacobi, S. (1971). *Economic worth of sediment load data in a statistical decision framework*, PhD Dissertation, Colorado State University, Fort Collins, Colorado.

Jacobi, S. (1975). Economic optimum record length defined in the context of a statistical decision approach. *Nordic Hydrology*, **6**, 28-42.

Jansen, J. M. L. and Painter, R. B. (1974). Predicting sediment yield from climate and topography. *Journal of Hydrology*, **21**, 4, 371-380.

Kisiel, C. C., Duckstein, L., and Fogel, M. M. (1971). Analysis of ephemeral flow in arid lands. *Journal of Hydraulics Division, ASCE*, **97**, HY10, 1699-1717.

Li, R. M., Simons, D. B., and Stevens, M. A. (1973). Review of literature: cooperative study of development of models for predicting sediment yield from small watersheds. *Report prepared for US Rocky Mountain Forest and Range Experimental Station*, Flagstaff, Arizona, CER72-73 RML-DBS-MAS, Civil Engineering Department, Colorado State University, Fort Collins, Colorado, 25pp.

McPherson, H. J. (1975). Sediment yields from intermediate-sized stream basins in southern Alberta. *Journal of Hydrology*, **25**, 314, 243-257.

Meyer, L. D., Foster, G. R., and Romkens, M. J. M. (1975). Source of soil eroded by water from upland slopes. In *Present and Prospective Technology for Predicting Sediment Yields and Sources*, Proceedings of the Sediment Yield Workshop, USDA Sediment Sedimentation Laboratory, Oxford, Mississippi.

Mills, W. C. (1980). Coupling stochastic and deterministic hydrologic models for decision making. Dept. of Hydrology and Water Resources, University of Arizona, *Natural Resources System Report 36*, Tucson, Arizona.

Mills, W. C. (1981). Deriving sediment yield probabilities for evaluating conservation practices. *Transactions of ASAE*, **24**, 5, 1199-1203.

Moore, R. J. (1984). A dynamic model of basin sediment yield. *Water Resources Research*, **20**, 1, 89-103.

Moore, R. J. and Clarke, R. T. (1983). A distribution function approach to modeling basin sediment yield. *Journal of Hydrology*, **65**, 239-257.

Murota, A. (1980). An proposal of stochastic evaluation for watershed sediment yields. In Shen, H. W. and Hikkara, W. (Eds.), *Application of Stochastic Processes in Sediment Transport*, Water Resources Publications, Littleton, Colorado.

Murota, A. and Hashino, M. (1969). Simulation of river bed variation in mountainous basin. *Proceedings, 13th Congress IAHR, Kyoto, Japan*, **5**, 1, 245–248.

Murota, A. and Hashino, M. (1971). An application of the simulated rainfall models to forecasting of the long term variation of river bed. In *Systems Approach to Hydrology*, Proceedings of the First Bilateral US–Japan Seminar in Hydrology, Water Resources Publications, Fort Collins, Colorado, 326–345.

Onstad, C. A., Mutchler, C. K., and Bowie, A. J. (1977). Predicting sediment yields. *Proceedings, National Symposium on Soil Erosion and Sedimentation by Water*, ASAE Publication 4–77, 43–58.

Prasad, S. N. and Singh, V. P. (1982). A hydrodynamic model of sediment transport in rill flows. In *Recent Developments in the Explanation and Prediction of Erosion and Sediment Yield*, Proceedings of the Exeter Symposium, July 1982, IAHS Publication No. 137, 293–301.

Renard, K. G. and Lane, L. J. (1975). Sediment yield as related to a stochastic model of ephemeral runoff. In *Present and Prospective Technology for Predicting Sediment Yields and Sources, ARS-S-40*, Agricultural Research Service, USDA, Washington, D.C., 253–264.

Renard, K. G. and Laursen, E. M. (1975). Dynamic behavior model of ephemeral stream. *Journal of the Hydraulics Division, Proceedings of the ASCE*, **101**, HY5, 511–526.

Rendon-Herrero, O. (1974). Estimation of washload produced on certain small watersheds. *Journal of the Hydraulics Division, Proceedings of the ASCE*, **100**, HY7, 835–848.

Rendon-Herrero, O., Singh, V. P., and Chen, V. J. (1980). ER–ES watershed relationships. *Proceedings*, International Symposium on Water Resources Systems held Dec. 20–22, 1980, at the University of Roorke, India, **1**, 11-8-41–11-8-47.

Rodriguez-Iturbe, I. and Nordin, D. F. (1968). Time series analysis of water and sediment discharges. *Hydrological Science Bulletin*, **13**, 2, 69–78.

Sharma, T. C. (1977). *A discrete dynamic model of watershed sediment yield*, unpublished PhD Dissertation, University of Guelph, Guelph, Ontario.

Sharma, T. C. and Dickinson, W. T. (1979). Discrete dynamic model of watershed sediment yield. *Journal of Hydraulics Division, ASCE*, **105**, HY5, 555–571.

Sharma, T. C. and Dickinson, W. T. (1980). System model of daily sediment yield. *Water Resources Research*, **16**, 3, 501–506.

Sharma, T. C., Hines, W. G. S., and Dickinson, W. T. (1979). Input–output model for runoff-sediment yield processes. *Journal of Hydrology*, **40**, 292–322.

Shirley, E. D. and Lane, L. J. 1978). A sediment yield equation from an erosion simulation mdoel. *Hydrology and Water Resources in Arizona and the Southwest*, Office of Arid Land Studies, University of Arizona, Tucson, **8**, 90–96.

Simons, D. B., Ward, T. J., and Li, Ruh-Ming (1979). Sediment sources and impacts on the fluvial system. In Shen, H. W. (Ed.), *Modeling of Rivers*, Wiley International, 7.1–7.27.

Singh, V. P. (1973). Predicting sediment yield in western United States—a discussion. *Journal of the Hydraulics Division, Proceedings of the ASCE*, **99**, HY10, 1891–1894.

Singh, V. P. (1983). Analytical solutions of kinematic equations for erosion on a plane, II: Rainfall of finite duration. *Advances in Water Resources*, **6**, June, 88–95.

Singh, V. P. and Chen, V. J. (1982). On the relationship between sediment yield and runoff volume. In Singh, V. P. (Ed.) *Modeling Components of Hydrological Cycle*, Water Resources Publications, Littleton, Colorado, 555–570.

Singh, V. P. and Krstanovic, P. F. (1985). A stochastic model for sediment yield. In Shen, H. W. (Ed.), *Multivariate Analysis of Hydrologic Processes*, Proceedings of the Fourth International Hydrology Symposium, Fort Collins, Colorado.

Singh, V. P. and Krstanovic, P. F. (1987). A stochastic model for sediment yield using the principle of maximum entropy (POME). *Water Resources Research*, **23**, 5, 781–793.

Singh, V. P. and Regl, R. R. (1983). Analytical solutions of kinematic equations for erosion on a plane, I. Rainfall of indefinite duration. *Advances in Water Resources*, **6**, March, 2–10.

Smith, J., Davis, D., and Fogel, M. (1977). Determination of sediment yield by transferring rainfall data. *Water Resources Bulletin*, **13**, 3, 529–541.

Soares, E. F., Unny, T. E., and Lenox, W. C. (1982). Conjunctive use of deterministic and stochastic models for predicting sediment storage in large reservoirs—Part I: a stochastic sediment storage model. *Journal of Hydrology*, **59**, 1/2, 49–82.

Tatum, F. E. (1963). A new method of estimating debris–storage requirement for debris basins. *US Department of Agriculture Misc. Publ.*, **970**, 886–898.

Thomas, R. B. (1985). Estimating total suspended sediment yield with probability sampling. *Water Resources Research*, **21**, 9, 1381–1388.

Van Sickle, J. (1982). Stochastic predictions of sediment yields from small coastal watersheds in Oregon, USA. *Journal of Hydrology*, **56**, 3/4, 309–323.

Williams, J. R. (1975). Sediment yield predictions with universal equations using runoff–energy factor. In *Present and Prospective Technology for Predicting Sediment Yields and Sources*, Proceedings of the Sediment Yield Workshop, USDA Sedimentation Laboratory, Oxford, Mississippi, 244–252.

Williams, J. R. (1982). Mathematical modeling of watershed sediment yield. In Singh, V. P. (Ed.) *Modeling Components of Hydrologic Cycle*, Water Resources Publications, Littleton, Colorado, 499–514.

Williams J. R. and Berndt, H. D. (1972). Sediment yield computed with universal equation. *Journal of Hydraulics Division, ASCE*, **HY12**, 2.

Wischmeier, W. H. and Smith, D. D. (1960). A universal soil-loss equation to guide conservation farm planning. *7th International Congress on Soil Science Transactions*, **1**, 418–425.

Woolhiser, D. A. and Blinco, P. H. (1975). Watershed sediment yield—a stochastic approach. In *Present and Prospective Technology for Predicting Sediment Yields and Sources*, Proceedings of the Sediment Yield Workshop, USDA Sedimentation Laboratory, Oxford, Mississippi.

Woolhiser, D. A. and Todorovic, P. (1974). A stochastic model of sediment yield for ephemeral streams. In *Proceedings USDA-IASPS Symposium on Statistical Hydrology. US Dept. of Agriculture Misc. Publication*, **1275**, 232–246.

Modelling Geomorphological Systems
Edited by M. G. Anderson
©1988 John Wiley & Sons Ltd.

Chapter 10

Modelling erosion on hillslopes

L. J. LANE and E. D. SHIRLEY

Southwest Rangeland Watershed Research Center, Agricultural Research Service, Tucson

and

V. P. SINGH

Department of Civil Engineering, Louisiana State University

10.1 INTRODUCTION

Surface runoff on upland areas such as hillslopes is often accompanied by soil erosion. Soil particles may be detached when the impact of raindrops exceeds the soil's ability to withstand the impulse at the soil surface. Detachment may also occur when shear stresses caused by flowing water exceed the soil's ability to resist these erosive forces. Vegetation as canopy and ground cover, and other surface cover such as gravel and rock fragments, protect the soil surface from direct raindrop impact, and also provide hydraulic resistance, reducing the shear stresses acting on the soil. Plant roots, incorporated plant residue, and minerals increasing cohesion tend to protect the soil by reducing the rate of soil particle detachment by flowing water and raindrop impact.

Once detachment has occurred, sediment particles are transported by raindrop splash and by overland flow. Conditions which limit raindrop detachment limit the sediment supply available for transport by splash and flow mechanisms. Vegetative canopies intercept splashed sediment particles and limit sediment transport by splash. The rate of sediment transport by overland flow is influenced by the factors controlling the amount of sediment available for transport, the sediment supply, and by hydraulic processes occurring in overland flow such as raindrop impacts, depth of flow, velocity, and accelerations due to microtopographic flow patterns. Obviously, the steepness, shape, and length of slopes affect both flow patterns and the resulting sediment transport capacity of the flowing water.

After sediment particles are detached from soil areas above, between, and near locations of small flow concentrations, they may enter the flow concentration areas for subsequent transport downslope by hydraulic processes. Throughout the remainder of the chapter, the flow concentration areas are called rills, and the areas between the rills are called interrill areas. Together, these interrill and rill areas make up the overland flow surface. Sediment particles detached in the interrill areas move to the rills by the processes of splash as the result of raindrop impact, and by suspension and saltation in overland flow. The rate of delivery of water and sediment to the rills affects the rates of sediment detachment, transport, and deposition in the rills.

Deposition occurs in overland flow when sediment particles come to rest on the soil surface, which occurs when sediment load in the flow exceeds the flow's capacity to transport the sediment. The rate of deposition is determined by both flow characteristics affecting energy, momentum, and turbulence and sediment particle characteristics, including particle interactions, affecting fall or settling velocity.

Thus, the processes controlling sediment detachment, transport, and deposition on the hillslope scale, lumped under the term erosion processes, are complex and interactive. This complexity leads to the need for upland erosion models as tools in resource management. Erosion models and observations are superior to observations alone, because simultaneous observation and measurement of all the processes controlling surface runoff and erosion are beyond the current and foreseeably available technology. Moreover, observations and measurements are particularly difficult, due to the small temporal and spatial scales necessary during a runoff and erosion event. Quite often, after the fact observations are the best that can be obtained. Almost as often, these post event observations reveal little of the actual mechanisms causing the erosion.

Ideally, an erosion model should represent the essential mechanisms controlling erosion, and the model parameters should be directly related to measurable physical properties. However, under real conditions, all models are more or less incorrect, because all models are abstractions and simplifications of the actual physical processes. Moreover, model parameters are often impossible or difficult to directly measure, and thus are always, to some extent, data based rather than predetermined. These real-world problems with mathematical models of overland flow and erosion have resulted in three main types of models: those that are primarily empirically based; those that are partially conceptually based and partially empirically based; and those that are partially process based or physically based and partially empirically based. As will be illustrated in subsequent discussions, these three main types of models are typified by the Universal Soil Loss Equation (USLE) as described by Wischmeier and Smith (1978), by the unit sediment graph (i.e. Rendon-Herrero, 1978; Williams, 1978), and by coupled overland flow–erosion equations based on the concepts of kinematic flow and separable rill and interrill erosion processes (i.e. Foster, 1982 and his example models listed on pp. 370–372). Some examples of the three types of models are shown in Table 10.1.

10.1.1 Developments Resulting in the USLE

Recently, Meyer (1984) and Nyhan and Lane (1986) summarized the evolution of the USLE, and the latter divided its development into four historical periods. The first period (1890s–1940) was described as a period wherein a basic understanding of most of the factors affecting erosion was obtained in a qualitative sense (Cook, 1936). This period included the rainfall studies of Laws (1940) and the analyses of the action of raindrops in erosion reported by Ellison (1947).

Table 10.1. Some examples of empirical and conceptual models

Model Type	Model	Author
Empirical	Musgrave Equation	Musgrave (1947)
	Universal Soil Loss Equation (USLE)	Wischmeier and Smith (1978)
	Modified Universal Soil Loss Equation (MUSLE)	Williams (1975)
	Sediment Delivery Ratio Method	Renfro (1975)
	Dendy–Boltan Method	Dendy and Boltan (1976)
	Flaxman Method	Flaxman (1972)
	Pacific Southwest Interagency Committee (PSIAC) Method	Pacific Southwest Interagency Interagency Committee (1968)
	Sediment Rating Curve	Campbell and Bauder (1940)
	Runoff–Sediment Yield Relation	Rendon-Herrero (1974), Singh, Baniukiwicz and Chen (1982)
Conceptual	Sediment Concentration Graph	Johnson (1943)
	Unit Sediment Graph	Rendon-Herrero (1978)
	Instantaneous Unit Sediment Graph	Williams (1978)
	Discrete Dynamic Models	Sharma and Dickinson (1979)
	Renard–Laursen Model	Renard and Laursen (1975)
	Sediment Routing Model	Williams and Hann (1978)
	Muskingum Sediment Routing Model	Singh and Quiroga (1986)
Physically based	Quasi-Steady State Erosion	Foster, Meyer and Onstad (1977)
	Kinematic Wave Models	Hjelmfelt, Piest and Saxton (1975), Shirley and Lane (1978), Singh and Regl (1983)
	Continuum Mechanics Model	Prasad and Singh (1982)

During the period 1940–1954, work in the Corn Belt of the United States resulted in a soil loss estimation procedure incorporating the influence of slope length and steepness (Zingg, 1940), conservation practices (Smith, 1941; Smith and Whitt, 1947), and soil and management factors (Browning *et al.*, 1947). In 1946, a national committee reappraised the Corn Belt factor values, included a rainfall factor, and produced the resulting Musgrave equation (Musgrave, 1947).

During the period 1954–1965, the USLE was developed by the United States Department of Agriculture (USDA), Agricultural Research Service in cooperation with the USDA–Soil Conservation Service and state agricultural experiment stations. Plot data from natural storms and from rainfall simulator studies formed the USLE data base. During the 1965–1978 period, additional data and experimental results were incorporated, resulting in the current USLE (Wischmeier and Smith, 1978).

The USLE in equation form is:

$$A = RKLSCP \qquad (1)$$

where:

A = the computed soil loss per unit area (tons per acre–yr),

R = the rainfall and runoff factor (hundreds of ft–tons–in per acre–hr–yr),

K = the soil erodibility factor (tons–acre–hr per hundreds of acre–ft–tons–in),

LS = the slope length–steepness factor (1.0 on uniform 72.6 ft slope at 9 per cent steepness),

C = the cover-management factor (1.0 for tilled, continuous fallow), and

P = the supporting practices factor (1.0 for up and down hill tillage, etc.).

The original USLE was presented in English units, hence their usage here. The unit plot (where LS, C and P are all equal to 1.0) is defined as a clean tilled, up and down slope, 72.6 ft long plot with a uniform 9 per cent slope. For slope lengths of 30 to 300 ft and steepness from 3 to 18 per cent, the LS factor ranges from a low of about 0.2 to a high of about 6. Values of the C factor range from a low of about 0.003, for near complete grass cover, to 1.0 for the unit plot. Values of the P factor range from 0.5 for contouring to 1.0 for the unit plot. Values of the R factor range from under 20 to over 550 in the continental United States, with some values outside these limits in other parts of the world representing greater climatic extremes. Wischmeier and Smith (1978, pp. 8–11) list values of the soil erodibility factor, K, ranging from 0.03 to 0.69, with most values in the range 0.2 to 0.4. With appropriate values of the above factors, the USLE is intended to predict the long-term average annual soil loss from uniform slopes, or from nonuniform slopes without deposition (Foster and Wischmeier, 1974).

That the USLE remains the most widely used tool in predicting upland erosion supports the description of upland erosion processes as complex and interactive. The state of the art is such that more conceptual and processes based erosion prediction equations for practical applications are just emerging, and do not yet have wide usage.

10.1.2 Development of Conceptual Models

The conceptual models lie somewhere between empirically and physically based models, and are based on spatially lumped forms of continuity equations for water and sediment and some other empirical relationships. Although highly simplified, they do attempt to model the sediment yield, or the components thereof, in a logical manner. To summarize, conceptual models of sediment are analogous in approach to those of surface runoff, and hence, embody the concepts of the unit hydrograph (UH) theory. Rendon-Herrero (1974, 1978) was probably the first to have extended this theory to derive a unit sediment graph (USG) for a small watershed. The sediment load considered in the USG is the wash load only. Rendon-Herrero (1974) expressed the following to define the USG:

A form of a unit sediment graph was indeed developed whose standard unit was 1.0 ton (910 kg) for a given duration, distributed over the watershed, analogous in unit–hydrograph analysis to 1.00 in. (25 mm) of excess (effective) rainfall over the same area.

In light of this definition, the USG and UH are similar in their derivations. To discuss the derivation of the USG by Rendon-Herrero, the following steps are outlined.

1. Select an isolated rainfall–runoff event of a desired duration in accordance with the requirement of the UH for which the sediment concentration graph C is known.

2. Separate the baseflow Q_b from the runoff hydrograph Q_T using a standard hydrograph separation technique to obtain the direct runoff hydrograph Q,

$$Q(t) = Q_T(t) - Q_b(t) \tag{2}$$

3. Using the same baseflow separation technique, separate out the sediment concentration due to baseflow. It should be noted that Rendon-Herrero assumed that the maxima of runoff and sediment concentration occurred at the same time.
4. Compute sediment discharge Q_s due to direct runoff by noting that sediment discharge is the product of water discharge and sediment concentration,

$$Q_s = Q_T C_T - Q_b C_b \tag{3}$$

5. Compute the volume of direct runoff, which is the area under the direct runoff hydrograph.

$$V_Q = \int_0^\infty Q(t) dt \tag{4}$$

6. Compute the sediment yield, which is the area under the sediment graph due to direct runoff.

$$V_s = \int_0^\infty Q_s \, dt \tag{5}$$

7. Divide the ordinates of the sediment graph by the sediment yield to obtain ordinates of the USG, H_s,

$$H_s = \frac{Q_s}{V_s} \tag{6}$$

The USG varies somewhat with the intensity of the effective rainfall. It can be used to generate a sediment graph for a given storm if the wash load produced by that storm is known. A relationship between V_s and V_Q was proposed. Using this relation, V_s can be determined. Therefore, Q_s can be determined by multiplying H_s with V_s. It must be noted that the duration of the USG chosen to determine Q_s must be the same as that of the effective rainfall generating V_Q. This USG method was tested on a small wash load-producing watershed, Bixler Run Watershed, near Loysville, Pennsylvania.

Rendon-Herrero (1974) proposed the use of the so-called 'series' graph to determine the sediment hydrograph. This method has the advantage that the duration of the effective rainfall is neglected altogether, but requires construction of the series graphs beforehand. Thus, this method cannot be extended to ungauged basins. Williams (1978) and Singh et al. (1982), among others, have used the USG to model watershed sediment yield.

10.1.3 Development of Physically Based Erosion Models

Fundamental erosion mechanics were of interest to scientists and engineers as early as 1936 (Cook, 1936), and were described in terms of subprocesses by Ellison (1947). Negev (1967) included an erosion component in the Stanford Watershed Model (Crawford and Lindsley, 1962). Meyer and Wischmeier (1969) presented relationships for the major erosion subprocesses, and incorporated them in a model of overland flow erosion, which formed the conceptual basis of most subsequent erosion modelling efforts.

Foster and Meyer (1972) published a paper on a closed-form soil erosion equation for overland flow, which demonstrated the ability of models in this class to provide insight into the spatial variability of erosion on hillslopes and into the separable interrill and rill erosion processes. This analysis assumed steady state conditions, and emphasized spatially variable processes. However, it set the stage for subsequent analyses of spatially varying and unsteady overland flow and erosion.

Hjelmfelt, Piest, and Saxton (1975) solved the coupled partial differential equations for overland flow with interrill and rill erosion and constant and uniform rainfall excess. However, they solved them only for the rising and steady state portions of the overland flow hydrograph. Shirley and Lane (1978) solved the equations for constant and uniform rainfall excess of finite duration over the entire overland flow hydrograph using the method of characteristics, and then integrated the equations to produce a sediment yield equation for the entire runoff hydrograph. Singh and Prasad (1982) advanced the models by formulating the partial differential equations for overland flow and erosion on an infiltrating plane, and then presented analytic solutions, by the method of characteristics, for the special case of constant and uniform rainfall and infiltration, or constant and uniform rainfall excess, on a sloping plane. Also, see Singh (1983) for a more complete description of the methods of solution. Solution domains and analytic solutions of the overland flow and interrill and rill erosion equations for the special case of constant and uniform rainfall excess on a plane are given in the Appendix. These solutions can be examined in analytic form to illustrate changes in sediment concentration in time and space for the case of unsteady and spatially variable overland flow.

Subsequent investigators examined various approximations to the analytic solutions described above (e.g. Rose *et al.*, 1983a), and their fit to measured data (Rose *et al.*, 1983b). Lane and Shirley (1982) also discussed the fit of the coupled overland flow and interrill and rill erosion equations to time varying runoff and sediment concentration data from plots and a small watershed on the Walnut Gulch Experimental Watershed in southeastern Arizona, USA. Blau (1986) examined the parameter identifiability of the overland flow–erosion model for the special case of constant and uniform rainfall excess on a plane. He concluded that, because of parameter interactions in the model, parameter values were difficult to obtain by least squares optimization using measured data.

As indicated above, the research reported on the more physically based overland flow–erosion equations are representative of mathematical derivations and manipulations, or of efforts to determine parameter values by fitting the models to measured data.

10.1.4   Scope and Limitations

Subsequent discussions will be primarily limited to erosion processes occurring in overland flow on plots and hillslopes. Although some of the more major assumptions and approximations used in deriving solutions to the governing equations are described, the main emphasis is on their solutions after the simplifying assumptions and the mathematical and practical significance of the approximating equations and their solutions.

10.1.5 Purpose

The first purpose of this chapter is to describe the evolution and status of erosion models for hillslopes based upon the kinematic wave equations for overland flow, and on the interrill and rill terms for erosion. The second purpose is to examine a particular erosion model for which analytic solutions can be obtained, and then to discuss the mathematical properties and implications of the solutions as they relate to experimental design and interpretation of experimental data.

10.2 OVERLAND FLOW AND EROSION EQUATIONS

The development of improved erosion equations for overland flow is based upon prior development of improved flow equations. That is, the development of methodology for simulation of unsteady and spatially varying overland flow made the subsequent simulation of interrill and rill erosion possible.

10.2.1 The Shallow Water Equations

Unsteady and spatially varying and one-dimensional flow per unit width on a plane was described by Kibler and Woolhiser (1970) using the following equations:

$$\frac{\partial h}{\partial t} + \frac{u\partial h}{\partial x} + \frac{h\partial u}{\partial x} = R \tag{7}$$

and

$$\frac{\partial u}{\partial t} + \frac{u\partial u}{\partial x} + \frac{g\partial h}{\partial x} = g(S_o - S_f) - (R/h)(u - v) \tag{8}$$

where

h = local depth of flow (dimension of length, L),
u = local mean velocity (L/T),
t = time (T),
x = distance in the direction of flow (L),
R = lateral inflow rate per unit area (L/T),
g = acceleration of gravity (L/T^2),
S_o = slope of the plane,
S_f = friction slope, and
v = velocity component of lateral inflow in the direction of flow (L/T).

Equation 7 is the continuity of mass equation, and equation 8 is the one-dimensional momentum equation. In general, equations 7 and 8 must be solved numerically. Modelling real overland flow with one-dimensional equations represents significant abstractions and simplifications. Real overland flow occurs in complex mixes of sheet flow and small concentrated flow areas. The routes of concentrated flow are often determined by irregular microtopographic features which vary in the downstream direction (x) and in the lateral direction (y). Definitive analyses of the influences of such simplifications upon hydraulic and erosion paramaters are nonexistent.

The lateral inflow, R, in equations 7 and 8, is, in reality, a complex function of time and space representing all the variations in rainfall input and in infiltration. It is often represented as the positive difference between instantaneous rates of rainfall and infiltration, or as zero if infiltration rate exceeds rainfall rate. This positive difference is called rainfall excess. In solving equations 7 and 8, a typical assumption is that a block of rainfall can be divided into infiltration and rainfall excess. Rainfall excess is then routed as if the surface were impervious, which is a significant simplification (Smith and Woolhiser, 1971). Moreover, infiltration is usually assumed to be uniform over the overland flow surface, while in reality, infiltration rates vary significantly in space. The assumption of spatially uniform infiltration, and thus rainfall excess, is a serious limitation in most current modelling approaches, and may preclude accurate prediction of overland flow under many natural conditions (i.e. Lane and Woolhiser, 1977).

The velocity component, v, in equation 8, is almost always assumed to be zero. This assumption may be reasonable on natural overland flow surfaces for distances on the order of a meter or larger. The validity of this assumption has not been tested on a smaller scale, on the order of a centimetre or so, and v may be quite significant in raindrop impact and sediment detachment and transport processes at this scale.

10.2.2 The Kinematic Wave Equations

If all terms in the momentum equation, equation 8, are assumed to be small in comparison with the $g(S_0 - S_f)$ term and can be neglected, then the shallow water equations become the kinematic wave equations. The kinematic wave equations for overland flow per unit width on a plane are:

$$\frac{\partial h}{\partial t} + \frac{\partial q}{\partial x} = R \tag{9}$$

and

$$q = Kh_m \tag{10}$$

where:

$\quad q$ = the local runoff rate per unit width ($L^2 T^{-1}$),
$\quad K$ = the stage-discharge coefficient ($L^{m-1} T^{-1}$), and
$\quad m$ = the exponent dependent upon the friction law assumed.

The exponent m is 3/2 for the Chezy equation and 5/3 for the Manning equation. Throughout the remainder of this chapter, the Chezy form will be used so that $m = 3/2$ and

$$K = C\sqrt{S} \tag{11}$$

where C is the Chezy resistance coefficient ($L^{1/2} T^{-1}$), and S is the slope of the plane surface.

Lighthill and Whitham (1955) introduced the kinematic wave theory for flood routing in rivers and for overland flow. Iwagaki (1955) used the kinematic assumptions and a method of characteristics for unsteady flow in rivers. Henderson and Wooding (1964) used the kinematic wave equations for steady rain of finite duration and for flow over a sloping plane. Woolhiser and Liggett (1967) showed that solutions to the kinematic wave equations are a good approximation to the solutions to the shallow water equations, provided the kinematic flow number is larger than about 20. It is important to note that this refers to the accuracy with which the kinematic wave solutions approximate solutions to the shallow water equations for sheet flow on a plane. The kinematic flow number says nothing about how well the shallow water equations, with one-dimensional flow and spatially uniform parameters, approximate overland flow on natural surfaces.

10.2.3 Equations for Erosion by Overland Flow

The sediment continuity equation, with the kinematic assumptions, is quite similar to the water continuity equation on the left hand side. The right hand side of the sediment continuity equation is commonly separated into an interrill erosion term, E_I, and the rill erosion term, E_R. With these assumptions, the continuity equation for sediment is:

$$\frac{\partial(ch)}{\partial t} + \frac{\partial(cq)}{\partial x} = E_I + E_R \tag{12}$$

where:

c = sediment concentration (M L^{-3}),
E_I = interrill erosion rate per unit area per unit time (M L^{-2} T^{-1}), and
E_R = net rill erosion (or deposition) rate (M L^{-2} T^{-1}),

and the other variables are as described earlier. The procedure is to solve the flow equations first, and then solve equation 12 for sediment concentration. Total sediment yield for a storm, V_s, is then found by integrating the product cq over the period of runoff.

The interrill term, E_I

The rate of interrill erosion is a function of the rate of detachment by raindrop impact and the rate of transport from the point of detachment to a rill.

As discussed in the introduction, interrill erosion is, by definition, caused by raindrop detachment and the rate of transport in the shallow interrill flow. On steep slopes, the rate of detachment by raindrop impact limits interrill erosion, whereas transport capacity in interrill flow limits the rate of delivery on flat slopes (Foster, Meyer, and Onstad, 1977). These authors, and others, document the dependence of interrill erosion on soil characteristics, slope steepness, and canopy and ground cover. In equation form, this can be expressed as

$$E_I = f(I, S, C, Soil) \tag{13}$$

where I, S, and C are rainfall intensity, slope of the land surface, and cover effects, respectively. *Soil* refers to the soil characteristics, primary particle-size distribution, type and amount of clay and crusting, and land use influencing soil properties, such as density and aggregation, which affect raindrop detachment and shallow flow. Following are some selected interrill erosion terms.

A simple functional form incorporating rainfall intensity, I, as a measure of the erosivity of raindrop impact is

$$E_I = aI^2 \tag{14}$$

where a is a coefficient to be determined experimentally. If the production of rainfall excess is related to I, and the transport capacity of shallow flow is, in turn, related to the rainfall excess, then a simple interrill erosion equation is

$$E_I = bR \tag{15}$$

where b is a coefficient to be determined. If the rate of detachment is related to the rainfall intensity squared, and the flow transport capacity is related to the ratio of rainfall excess to rainfall intensity, then a simple form of the interrill erosion equation is

$$E_I = cI^2(R/I) = cIR \tag{16}$$

where c is a coefficient to be determined. Additional information on a number of expressions for interrill erosion rates is given by Foster *et al.* (1982).

The rill term, E_R

There are two common ways of expressing soil detachment in rills, and one common way of expressing the rate of sediment deposition in rills. While more expressions or functional forms for detachment and deposition are available, the following material is indicative of modern erosion science.

If the rate of soil detachment in a rill is assumed to be a function of the shear stress in excess of a critical shear stress, then the following equation describes the rate of rill erosion:

$$Er = d(\tau - \tau_c)^e \tag{17}$$

where d is a coefficient to be determined, τ is the average shear stress in the cross-section, τ_c is a critical shear stress that must be exceeded to initiate soil detachment, and e is an exponent to be determined.

A second major class of rill erosion equations results when one assumes the rate of rill erosion is proportional to the amount the flow transport capacity, T_c, is in excess of the existing sediment load, cq. These equations are of the form

$$Er = f(T_c - cq) \tag{18}$$

where f is a coefficient to be determined, and the other variables are as described above.

Two issues are involved in selecting a rill erosion equation of the type discussed here. The first issue is whether or not one assumes an interaction among rill erosion, sediment load, and transport capacity. Meyer and Wischmeier (1969) neglected the interaction, and their model represents the first major class of rill erosion models. Foster and Meyer (1972) assumed an interaction, and their model represents the second major class of rill erosion models. The second issue is whether or not one assumes a critical shear stress in determining the rate of detachment, as in equation 17, or the transport capacity used in equation 18.

In the event that more sediment is delivered to the channel segment from upstream and from lateral inflow than its transport capacity, then sediment deposition will occur in the rill segment at a rate proportional to the deficit in transport capacity. This means that equation 18 can describe the rate of deposition if the coefficient f is a deposition coefficient. The deposition coefficient is primarily a function of particle characteristics, and is often calculated as a function of the particle fall velocity and the steady-state discharge rate (Foster, 1982).

10.2.4 Numerical Solutions

As stated earlier, equations 7 and 8 are solved numerically. Finite difference techniques are usually used (i.e. see Kibler and Woolhizer, 1970). If R, in equation 9, varies in space and time, then equations 9 and 10 must be solved numerically. If R in equation 10 varies, or if E_I and E_R in equation 12 are complex functions, then equation 12 must be solved numerically. The advantage of numerical techniques in solving the above equations is that one need not make as many assumptions as is required for analytic solutions, and the rainfall excess term can vary in time and space.

The disadvantages of numerical techniques, compared with analytic solutions, is that the former usually require much more computer time, the solutions are approximations of the real solutions, and the mathematics required for sensitivity analysis, limits, and other manipulations may be unavailable or very complex and difficult.

10.2.5 Analytic Solutions

Equations 9 and 10 can be solved analytically (by the method of characteristics) if R is uniform over the plane, and the temporal variation in R is described by a series of step functions. However, to obtain an analytic solution for equation 12, R in equation 9 must be uniform and constant for a finite or infinite duration. Equations 9 and 10 must be solved first to substitute into equation 12. Also, the form of T_c, in equation 18, should be simple, for example, a linear function of q, to obtain an analytic solution.

As stated earlier, the disadvantages of analytic solutions, in comparison with numerical solutions, is that they usually require much more restrictive and simplifying assumptions. The main advantages of analytic solutions include the ease with which they can be implemented on a computer, the speed with which they can be evaluated, the simplicity of sensitivity analysis, and the ease with which one can examine limits and other mathematical properties of the solutions.

10.3 SIMPLIFIED EQUATIONS WITH ANALYTIC SOLUTIONS

In this section, specific assumptions and simplifications are made to allow the derivation of analytic solutions for overland flow on a plane, and for interrill and rill erosion with overland flow. Analytic solutions to the runoff and erosion equations are used to illustrate field data needed for estimation of parameter values and for interpretation of processes controlling erosion.

10.3.1 The Basic Assumptions

In addition to the assumptions necessary for derivation of the one-dimensional shallow water equations and their approximating kinematic wave equations, specific assumptions are required for the erosion equations to have an analytic solution. In equation form, the assumptions are:

$$\frac{\partial(ch)}{\partial t} + \frac{\partial(cq)}{\partial x} = K_I R + K_R((B/K)q - cq) \tag{19}$$

with initial and boundary conditions as

$$c(0,x) = K_I \tag{20}$$

and

$$c(t,0) = K_I \tag{21}$$

We also assume a pulse input of the form

$$R(t) = \begin{cases} R \text{ for } 0 < t < t_* \\ 0 \text{ otherwise} \end{cases} \tag{22}$$

for the rainfall excess.

The first term on the right hand side of equation 19 means that we assume

$$E_I = K_I R \tag{23}$$

and the second term on the right hand side of equation 19 means that we assume

$$E_R = K_R(T_c - cq) \tag{24}$$

with $T_c = Bh^{3/2}$. Since $q = Kh^{3/2}$, we can write $T_c = (B/K)q$, and the result is the second term on the right hand side of equation 19.

Equations 20 and 21 mean that the initial concentration is K_I and, furthermore, that the concentration at the upstream boundary remains equal to K_I throughout the runoff hydrograph. These results can also be seen by taking limits of the equations presented in the Appendix. These assumptions and results are very significant in designing field experiments and in interpreting the resulting data.

10.3.2 Implications

The limit of the concentration as t approaches zero is:

$$C_0 = K_I \tag{25}$$

as the initial concentration. The limit of $c(t,x)$, for fixed x and as t approaches infinity, is C_f, and is given by

$$C_f = B/K + (K_I - B/K)\exp(-K_R x) \tag{26}$$

Finally, Shirley and Lane (1978) showed that the mean concentration, C_b, over the entire hydrograph is

$$C_b = Qs/Q = B/K + (K_I - B/K)(1 - \exp(-K_R x))/K_R x \tag{27}$$

If $B/K > K_I$, then $C_o < C_b < C_f$ and $c(t,x)$ for fixed x is a non-decreasing function of t. It can also be shown for fixed t that if $B/K > K_I$, then $c(t,x)$ is a non-decreasing function of x. These two non-decreasing functions mean (in the context of this particular model) that if $B/K > K_I$, then there is more transport capacity in the rills than is being satisfied by sediment input from the interrill areas. As a result, rill erosion occurs at all times and at all positions on the plane. In terms of sediment concentration graphs measured in the field, measured concentrations would tend to start at K_I near $t = 0$, and increase throughout the duration of runoff, assuming, of course, that the model is a good representation of reality.

If $B/K < K_I$, then the opposite is true. Under these conditions, $c(t,x)$ for fixed x would be non-increasing, or tend to decrease with increasing t. Also, $c(t,x)$ would be non-increasing with x and a fixed t. Again, if the model is correct, then measured concentrations would tend to start at K_I near $t = 0$, and decrease throughout the duration of runoff. If $B/K = K_I$, then transport capacity and existing sediment load are in equilibrium, so $C_o = C_f = C_b$, and, in fact, $c(t,x) = K_I$ for all x and t.

The implications of these results for plot and hillslope studies are that sediment concentration should be measured throughout the duration of runoff, and that analysis of data, using this model for parameter identification, should concentrate on events with nearly constant rainfall intensity and nearly saturated initial soil water content. The last two conditions will tend to make rainfall excess nearly constant, as assumed in the analysis. Fortunately, these conditions can nearly be met in rainfall simulator studies if data from runs where the initial soil water content is near saturation and the infiltration rate is nearly a constant are obtained for analysis.

Therefore, as a first approximation, one can examine the shape of the sediment concentration *vs.* time curve from a particular event on an experimental plot, and infer whether transport capacity in the rills ($B/K < K_I$) or detachment rate ($B/K > K_I$) in the rills is limiting sediment yield.

10.4 DISCUSSION

Although the Universal Soil Loss Equation remains the most often used model for predicting erosion on upland areas, more physically based models are emerging, and may become practical tools in the near future (i.e. see Rawls and Foster, 1986). As these new models emerge, they will probably be based upon unsteady and nonuniform overland flow modelled with the kinematic wave equations. Moreover, interrill and rill erosion processes will probably be explicitly represented in the partial differential equation used to describe erosion and overland flow.

The implications for plot and hillslope studies are that more, and more intensive, data need to be collected throughout the duration of runoff events, and at various positions on the slope. Only then can we begin to quantify unsteady and spatially varying overland flow and erosion processes.

APPENDIX

Summary of Solution Regions and Solutions for the Overland Flow Equations in the $t-x$ Plane

Recall that the kinematic wave equations are:

$$\frac{\partial h}{\partial t} + \frac{\partial q}{\partial x} = R \tag{A1}$$

and

$$q = Kh^m \tag{A2}$$

where the variables are as defined previously in the text.

1. Domains in the $t-x$ Plane for Solutions of the Kinematic Overland Flow Equations

Solutions for the overland flow equations require that the positive quadrant of the $t-x$ plane be divided into four regions. The regions listed below are also presented in Figure 10.4.

a. *Domain of Flow Establishment.* This region of the $t-x$ plane represents time from zero until cessation of rainfall excess at time t_* and distance down the plane such that steady state has not been reached:

$$0 \leq t \leq t_*$$
$$x \geq KR^{m-1}\, t^m \tag{A3}$$

b. *Domain of Established Flow.* This region of the plane represents time from zero until cessation of rainfall excess and distance down the plane such that steady state has been reached:

$$0 \leq t \leq t_*$$
$$0 \leq x \leq KR^{m-1}t^m \tag{A4}$$

c. *Domain of Prerecession.* This region of the plane represents time after cessation of rainfall excess and before depth of flow starts receding:

$$t \geq t_*$$
$$x \geq K(1-m)R^{m-1}\, t_*^m + Km(Rt_*)^{m-1}t \tag{A5}$$

d. *Domain of Recession.* This region of the plane represents time after the cessation of rainfall excess and depth of flow is receding:

$$t \geq t_*$$
$$0 \leq x \leq K(1-m)R^{m-1}\, t_*^m + Km(Rt_*)^{m-1}t \tag{A6}$$

2. Solutions in the Regions

a. *Domain of Flow Establishment.* In this region, the flow is unsteady but uniform:

$$h(t,x) = Rt \tag{A7}$$

b. *Domain of Established Flow.* In this region, the flow is steady but not uniform:

$$h(t,x) = (Rx/K)^{1/m} \tag{A8}$$

c. *Domain of Prerecession.* In this region, the flow is steady and uniform:

$$h(t,x) = Rt_* \tag{A9}$$

d. *Domain of Recession.* In this region, the flow is unsteady and not uniform:

$$h(t,x) = f_t^{-1} (Rx/K) \tag{A10}$$

where

$$f_t(u) = u^m + Rmu^{m-1} (t - t_*) \tag{A11}$$

The solutions described above are also shown in Figure 10.1.

Summary of Solution Regions and Solutions for the Sediment Concentration Equations in the $t - x$ Plane

Recall that the erosion equations are:

$$\frac{\partial(ch)}{\partial t} + \frac{\partial(cq)}{\partial x} = E_I + E_R \tag{A12}$$

with

$$E_I = K_I R \tag{A13}$$

and

$$E_R = K_R(Bh^n - cq) \tag{A14}$$

where the variables are as defined previously in the text.

1. Domains in the $t - x$ Plane for Solutions of the Sediment Concentration Equations

Solutions for the concentration equations require that the positive quadrant of the $t - x$ plane be divided into seven regions. The regions listed below are also shown in Figure 10.2.

a. *Domain 1.* This region of the plane represents time from zero until cessation of rainfall excess and distance down the plane such that concentration and flow have not reached steady state:

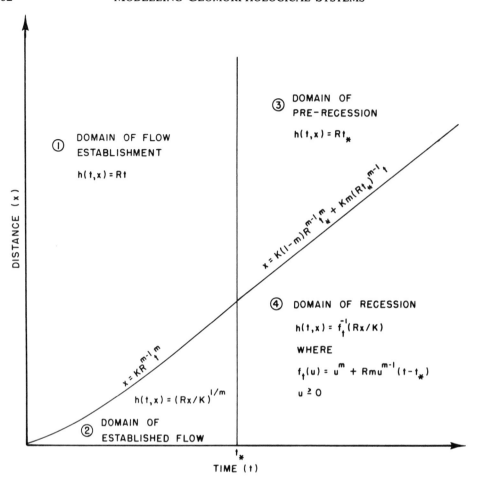

Figure 10.1. Domains in the $t-x$ plane for solutions of the kinematic overland flow equations for a constant and uniform rainfall excess rate of duration t_*

$$0 \leq t \leq t_*$$

and   (A15)

$$x \geq KR^{m-1} \, t^m$$

b. *Domain 2*. This region of the plane represents time from zero until cessation of rainfall excess and distance down the plane such that concentration has not reached steady state, but flow has:

$$0 \leq t \leq t_*$$
$$Km^{-m}R^{m-1} \, t^m \leq x \leq KR^{m-1} \, t^m \tag{A16}$$

c. *Domain 3*. This region of the plane represents time from zero until cessation of rainfall excess and distance down the plane such that concentration and flow have reached steady state:

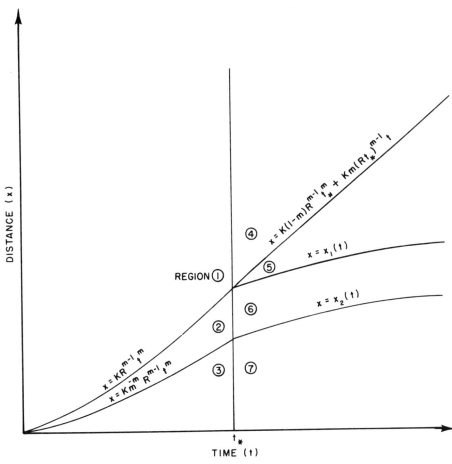

Figure 10.2. Domains in the $t-x$ plane for solutions of the overland flow erosion equations for a constant and uniform rainfall excess rate of duration t_*

$$0 \le t \le t_*$$
$$0 \le x \le Km^{-m}R^{m-1} \, t^m \tag{A17}$$

d. *Domain 4*. This region, corresponding to the domain of prerecession for flow, represents time after cessation of rainfall excess before depth of flow is receding, and before the arrival of the slower travelling concentration disturbance from the interaction of the water wave with cessation of rainfall excess:

$$t \ge t_*$$
$$x \ge K(1-m)R^{m-1} \, t_*^m + Km(Rt_*)^{m-1}t \tag{A18}$$

In Domains 5–7, let

$$a(u) = t_* + (K_0 u^m - mu^{-1}/(m+1))/R(m-1), \tag{A19}$$

and

$$b(u) = K(mK_0u - u^{-m}/(m+1))/R(m-1) \tag{A20}$$

e. *Domain 5*. This region represents that portion of the domain of recession before the arrival of the concentration disturbance propagating from the interaction of the water wave with cessation of rainfall excess. With the above definitions of a and b, let

$$x_1(t) = b(a^{-1}(t)) \tag{A21}$$

where

$$K_0 = m(Rt_*)^{m+1}/(m+1) \tag{A22}$$

Finally, the region is defined as:

$$t \leq t_*$$

and   (A23)

$$x_1(t) \leq x \leq K(1-m)R^{m-1} \, t_* m + Km(Rt_*)^{m-1}t$$

f. *Domain 6*. This region represents that portion of the domain of recession after the arrival of the concentration disturbance propagating from the interaction of the water wave with cessation of rainfall excess and before the arrival of the concentration disturbance propagating from the upper boundary. With the above functions a and b, let

$$x_2(t) = b(a^{-1}(t)) \tag{A24}$$

where

$$K_0 = m(Rt_*/m)^{m+1}/(m+1) \tag{A25}$$

With these definitions, the region is bounded by:

$$t \geq t_*$$

and   (A26)

$$x_2(t) \leq x \leq x_1(t)$$

g. *Domain 7*. This region represents that portion of the domain of recession after the arrival of the concentration disturbance propagating from the upper boundary:

$$t \geq t_*$$

and   (A27)

$$0 \leq x \leq x_2(t)$$

2. Solutions in the Regions

a. *Domain 1.* In this region

$$c(t,x) = K_I + K_R(B/K - K_I)uF(u) \tag{A28}$$

where

$$u = KR^{m-1} t^m/m \tag{A29}$$

and

$$F(u) = \int_0^1 v^{1/m} \exp(K_R u(v-1)) \, dv \tag{A30}$$

b. *Domain 2.* In this region

$$c(t,x) = B/K + (K_I - B/K)(1 - \exp(K_R(x_0 - x)) \\ ((K_R x_0)^2/mF(x_0/m) + 1 - K_r x_0))/(K_R x) \tag{A31}$$

where

$$x_0 = KR^{m-1} ((m(Rx/K)^{1/m} - Rt)/R(m-1))^m \tag{A32}$$

c. *Domain 3.* In this region

$$c(t,x) = B/K + (K_I - B/K)(1 - \exp(-K_R x))/K_R x \tag{A33}$$

d. *Domain 4.* In this region

$$c(t,x) = B/K + (K_I - B/K)(1 - K_R x_*/mF(x_*/m))\exp((K_R x_*/t_*)(t_* - t)) \tag{A34}$$

where

$$x_* = KR^{m-1} t_*^m \tag{A35}$$

e. *Domain 5.* In this region

$$c(t,x) = B/K + c_0\exp(-K_R x) \tag{A36}$$

where

$$K_0 = (R(m-1)(t - t_*) + mh(t,x)/(m+1))h^m(t,x) \tag{A37}$$

and

$$c_0 = (c(a(1/Rt_*), b(1/Rt_*)) - B/K)\exp(K_R b(1/Rt_*)) \tag{A38}$$

where c is computed using the formula from domain 4, equation A34.

f. *Domain 6.* In this region

$$c(t,x) = B/K + c_0\exp(-K_R(x - x_0)) \tag{A39}$$

where K_0 is defined above, $x_0 = K((m+1)K_0/m)^{m/(m+1)}/R$, and

$$c_0 = c(t_*,x_0) - B/K \tag{A40}$$

with $c(t_*,x_0)$ computed using the formula for domain 2, equation A31.

g. *Domain 7*. In this region

$$c(t,x) = B/K + c_0\exp(-K_R(x-x_0)) \tag{A41}$$

where K_0 and x_0 are as defined above and
$$c_0 = c(t_*,x_0) - B/K \tag{A42}$$

with $c(t_*,x_0)$ computed using the formula for domain 3, equation A33.

REFERENCES

Blau, J. B. (1986). *Parameter identifiability of an erosion simulation model*, M. S. Thesis, Dept. of Hydrology and Water Resources, Univ. of Arizona, Tucson, AZ.

Browning, G. M., Parish, C. L., and Glass, J. A. (1947). A method for determining the use and limitation of rotation and conservation practices in control of soil erosion in Iowa. *Soil Sci. Soc. of America Proc.*, 23, 246–249.

Campbell, F. B. and Bauder, H. A. (1940). A rating curve method for determining silt discharge of streams. *Trans. Am. Geophys. Union*, Part 2, Washington, 603–607.

Cook, H. L. (1936). The nature and controlling variables of the water erosion process. *Soil Sci. Soc. of America Proc.*, 1, 487–494.

Crawford, N. H. and Lindsley, R. K. (1962). *The synthesis of continuous streamflow hydrographs on a digital computer*, Tech. Report No. 12, Dept. of Civil Engineering, Stanford Univ., Stanford, California.

Dendy, F. E. and Boltan, G. C. (1976). Sediment yield-runoff drainage area relationships in the United States. *J. of Soil & Water Cons.*, 31(6), 264–266.

Ellison, W. D. (1947). Soil erosion studies. *Agric. Engr.*, 28, 145–146, 197–201, 245–248, 297–300, 349–351, 402–405, 442–450.

Flaxman, E. M. (1972). Predicting sediment in western United States. *J. Hydraul. Div., Proc. ASCE*, 98(HY12), 2073–2085.

Foster, G. R. (1982). Modeling the erosion process, in Haan, C. T., Johnson, H. P., and Brakensiek, D. L. (Eds.), *Hydrologic Modeling of Small Watersheds*. ASAE Monograph No. 5 American Soc. of Agric. Engr., St Joseph, Michigan, 295–380.

Foster, G. R. and Meyer, L. D. (1972). A closed-form soil erosion equation for upland areas, in Shen, H. W. (Ed.), *Sedimentation* (Einstein), Chapter 12, Colorado State Univ., Ft. Collins, Colorado.

Foster, G. R. and Wischmeier, W. H. (1974). Evaluating irregular slopes for soil loss prediction. *Trans. American Soc. of Agric. Engr.*, 17(2), 305–309.

Foster, G. R., Meyer, L. D., and Onstad, C. A. (1977). An erosion equation derived from basic erosion principles. *Trans. American Soc. of Agric. Engr.*, 20(4), 683–687.

Foster, G. R., Lombardi, F., and Moldenhauer, W. C. (1982). Evaluation of rainfall–runoff erosivity factors for individual storms. *Trans. American Soc. of Agric. Engr.*, 25(1), 124–129.

Henderson, F. M. and Wooding, R. A. (1964). Overland flow and groundwater flow from a steady rainfall of finite duration. *J. Geophys. Res.*, 69(8), 1531–1540.

Hjelmfelt, A. T., Piest, R. F., and Saxton, K. E. (1975). Mathematical modeling of erosion on upland areas. *Proc. 16th Congress, Int. Assoc. for Hydraulic Res., Sao Paulo, Brazil*, 2, 40–47.

Iwagaki, Y. (1955). Fundamental studies of the runoff analysis by characteristics. *Kyoto Univ. Disaster Prevention Res. Inst., Japan, Bull.*, 10, 1–25.

Johnson, Y. W. (1943). Distribution graphs of suspended-matter concentration. *Trans. of ASCE*, **108**, 941–964.

Kibler, D. F. and Woolhiser, D. A. (1970). The kinematic cascade as a hydrologic model. *Colorado State Univ. Hydrology Papers, No. 39*, Colorado State Univ., Ft. Collins, Colorado.

Lane, L. J. and Woolhiser, D. A. (1977). Simplifications of watershed geometry affecting simulation of surface runoff. *J. Hydrology, 35*, 173–190.

Lane, L. J. and Shirley, E. D. (1982). Modeling erosion in overland flow, in *Estimating Erosion and Sediment Rangelands*, Proc. Workshop, Tucson, Arizona, March 7–9, 1981, US Dept. Agric., Agric. Res. Serv., Agricultural Reviews and Manuals, ARM–W–26, June, 1982, 120–128.

Laws, J. O. (1940). Recent studies in raindrops and erosion. *Agric. Engr., 21*, 431–433.

Lighthill, M. J. and Whitham, C. B. (1955). On kinematic waves: flood movement in long rivers. *Proc. Royal Society (London),. Series A, 229*, 281–316.

Meyer, L. D. (1984). Evolution of the Universal Soil Loss Equation. *J. Soil and Water Conserv., 39*(2), 99–104.

Meyer, L. D. and Wischmeier, W. H. (1969). Mathematical simulation of the process of soil erosion by water. *Trans. American Soc. of Agric. Engr., 12*(6), 754–758, 762.

Musgrave, G. W. (1947). The quantitative evaluation of factors in water erosion, a first approximation. *J. Soil and Water Conserv., 2*(3), 133–138.

Negev, M. (1967). A Sediment model on a digital computer, *Tech. Report No. 76*, Dept. of Civil Engr., Stanford Univ., Stanford, California.

Nyhan, J. W. and Lane, L. J. (1986). *Erosion control technology: a user's guide to the use of the Universal Soil Loss Equation at waste burial facilities*, Manual LA–10262–M, UC–708, Los Alamos National Lab., Los Alamos, New Mexico.

Pacific Southwest Inter-Agency Committee (1968). *Factors Affecting Sediment Yield and Measures for the Reduction of Erosion and Sediment Yield*, 13 pp.

Prasad, S. N. and Singh, V. P. (1982). A hydrodynamic model of sediment transport in rill flows, *IAHS Publ. No., 137*, 293–301.

Rawls, W. J. and Foster, G. R. (1986). USDA Water Erosion Prediction Project (WEPP), EOS. *Trans. American Geophys. Union, 67*(16), 287.

Renard, K. G. and Laursen, E. M. (1975). Dynamic behavior model of ephemeral streams. *J. Hydraul. Div., Proc., ASCE, 101*(HY5), 511–526.

Rendon-Herrero, O. (1974). Estimation of washload produced by certain small watersheds. *J. Hydraul. Div., Proc. ASCE, 109*(HY7), 835–848.

Rendon-Herrero, O. (1978). Unit sediment graph. *Water Resourc. Res., 14*(5), 889–901.

Renfro, G. W. (1975). Use of erosion equations and sediment delivery ratios for predicting sediment yield. In *Present and Prospective Technology for Predicting Sediment Yields and Sources. Agric. Res. Serv., ARS–S–40*, 33–45. US Dept. Agric., Washington, D.C.

Rose, C. W., Williams, J. R., Sanders, G. C., and Barry, D. A. (1983a). A mathematical model of soil erosion and deposition processes: I. Theory for a plane land element. *Soil Sci. Soc. of America J., 47*(5), 991–995.

Rose, C. W., Williams, J. R., Sanders, G. C., and Barry, D. A. (1983b). A mathematical model of soil erosion and depositional processes: II. Application to data from an arid-zone catchment, *Soil Sci. Soc. of America J., 47*(5), 996–1000.

Sharma, T. C. and Dickinson, W. T. (1979). Discrete dynamic model of watershed sediment yield. *J. of Hydraulic Div., Proc. ASCE, 105*(HY5), 555–571.

Shirley, E. D. and Lane, L. J. (1978). A sediment yield equation from an erosion simulation model, in *Hydrology and Water Resources in Arizona and the Southwest, 8*, 90–96. Univ. of Arizona, Tucson, Arizona.

Singh, V. P. and Prasad, S. N. (1982). Explicit solutions to kinematic equations for erosion on an infiltrating plane, in Singh, V. P. (Ed.), *Modeling Components of Hydrologic Cycle*, Water Resour. Pub., Littleton, Colorado, 515–538.

Singh, V. P., Baniukiwicz, A., and Chen, V. J. (1982). An instantaneous unit sediment graph study for small upland watersheds, in Singh, V. P. (Ed.), *Modelling Components of Hydrologic Cycle*, Water Resour. Publ., Littleton, Colorado, 539–554.

Singh, V. P. (1983). Analytic solutions of kinematic equations for erosion on a plane: II. Rainfall of finite duration. *Advances in Water Resources,* **6**, 88–95.

Singh, V. P. and Regl, R. R. (1983). Analytical solutions of kinematic equations for erosion on a plane: I. Rainfall of indefinite duration. *Advances in Water Resour.,* **6**(1), 2–10.

Singh, V. P. and Quiroga, C. A. (1986). A dam breach erosion model: I. Formulation, *Water Resour. Manage.* (under review).

Smith, D. D. (1941). Interpretation of soil conservation data for field use. *Agric. Eng.,* **22**, 173–175.

Smith, D. D. and Whitt, D. M. (1947). Estimating soil losses from field areas of claypan soils. *Soil Sci. Soc. of America Proc.,* **12**, 485–490.

Smith, R. E. and Woolhiser, D. A. (1971). Overland flow on an infiltrating surface, *Water Resour. Res.,* **7**(4), 899–913.

Williams, J. R. (1975). Sediment yield prediction with universal soil loss equation using runoff energy factor, in *Present and Prospective Technology for Predicting Sediment Yields and Sources*, Agric. Res. Serv., US Dept. of Agric., Washington, D.C., **ARS-S-40**, 244–252.

Williams, J. R. (1978). A sediment graph model based on an instantaneous unit sediment graph, *Water Resour. Res.,* **14**(4), 659–664.

Williams, J. R. and Hann, R. W., Jr. (1978). Optimal operation of large agricultural watersheds with water quality constraints. *Texas Water Resour. Res. Inst.,* **TR-96**, 152 pp. Texas A&M Univ., College Station, Texas.

Wischmeier, W. H. and Smith, D. D. (1978). Predicting rainfall erosion losses—a guide to conservation planning. *Agric. Handbook No. 537*, US Dept. of Agric., Washington, D. C.

Woolhiser, D. A. and Liggett, J. A. (1967). Unsteady, one-dimensional flow over a plane—the rising hydrograph. *Water Resour. Res.,* **3**(3), 753–771.

Zingg, R. W. (1940). Degree and length of land slope as it affects soil loss in runoff. *Agric. Engr.,* **21**, 59–64.

Modelling Geomorphological Systems
Edited by M. G. Anderson
©1988 John Wiley & Sons Ltd.

Chapter 11

Hillslope solute modelling

STEPHEN TRUDGILL
Department of Geography, University of Sheffield

11.1 INTRODUCTION

Water flowing through soil transports both surface-applied and soil-derived solutes to streams and rivers; both vertical transport and lateral movement are involved (Figure 11.1). In addition, three dimensional considerations are important in relation to topography, contributing areas of water, and of solutes, and riparian transformations of solute dynamics. This chapter focuses on simple predictive models for soil leaching processes and considers how they might be applied to the understanding of linkages between fluvial and soil processes. The ethos of the chapter is not one of 'how complicated the processes are which have to be considered' (which is true) but one of 'what is the simplest way in which we can model the system and get a reasonable level of prediction?'

  The study of surface-applied solutes is important in the contexts of pollutant transfer and the leaching of agricultural chemicals such as fertilizers, pesticides and herbicides. The study of soil-derived solutes is important in geomorphological contexts, both in terms of weathering and pedogenesis and also in terms of hillslope evolution (Trudgill, 1986).

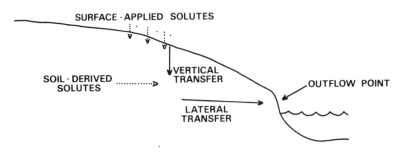

Figure 11.1. Links between hillslope soil systems and fluvial processes

The studies of surface-applied and soil-derived solutes are both important in the contexts of fluvial water quality modelling.

Deriving simple soil solute transport models rests upon combining water movement models with models for solute uptake and the behaviour of chemically reactive sorbed solutes. The models would therefore appear to demand detailed knowledge of both soil physics and soil chemical processes as well as adequate computational skill to cope with the variables involved. However, such an approach rapidly becomes cumbersome, although often possessing the virtue of describing many processes at a detailed level. In addition, the combination of soil physics and soil chemical processes is hampered by a lack of detailed knowledge of how the processes interact as well as by uncertainties as to when a particular type of model or specific consideration is appropriate. In particular, there are difficulties of knowing whether to apply equilibrium models or kinetic models to soil chemical processes when water input is sporadic and flows are unequal in the soil in time and space. In uniform, homogeneous, non-reactive media, progress can be rapidly made, but for heterogeneous soils, the choice is either to apply, and then calibrate, models of homogeneous media or to develop simple, but reasonably realistic, models for heterogeneous media. Both are attempted herein and particular attention is paid to the role of soil structure in field soils. Soil structure has the effect that some water flows in soil faster than others (Bouma *et al.*, 1981), especially when flow within structures, or peds, is compared with flow around peds (Beven and Germann, 1982). This differential flow gives differential opportunity for solute uptake and equilibrium (White, 1985a,b), with a contrast between the slower, pedal flow and the more rapid, interpedal flow. The models discussed below therefore refer to both rapid and slow flow and require a minimum of input data and computation. As such, they have more chance of application than more complex models, provided that they are reasonably faithful to actual processes. They also rest more on the evaluation of soil water flow than they do on the evaluation of soil chemical processes, which are more difficult to specify.

In addition, models for surface-applied solutes, where the solute is foreign to the soil, are often easier to evaluate because the passage of the solute through the soil can be more readily traced. However, surface-applied and soil-derived derived models are complementary in the sense that when talking of dilution of the output of soil-derived solutes by incoming rainfall, the dilution pattern may mimic that of a surface-applied solute because the rainfall behaves as a surface-applied solute would; similarly, dilution of surface-applied solute by pre-existing, non-labelled, soil water at an output may mimic the behaviour of a soil-derived solute because the non-appearance of a surface-applied solute tells us something of the way soil-derived solutes are displaced from the soil. Discussion of both sorts of models is therefore useful as one may illuminate the other; but the focus of attention is mainly on the more easily traced surface-applied solutes, especially nitrate fertilizer.

For a surface-applied solute, the crucial questions involve (a) the prediction of vertical, downward movement of the solute front on a soil profile scale and (b) on a hillslope scale, the spatial aggregation of (a) and the lateral transfer of the solute downslope to the stream. The study of solute penetration depth thus becomes an important focus and will form the main theme of the models discussed further, below. For a soil-derived solute, modelling follows on from this in the sense that while surface-derived quickflow water entrains surface-applied solutes, it will dilute soil-derived solutes in proportion to the amount of quickflow. The study of the dilution of near-equilibrium, base-flow waters by given amounts of quickflow thus appears to be a profitable line of thinking. Thus, while the potential factors influencing solute output from soils to streams are many and complex, one of the simplest approached involves modelling the amount of quickflow. This is not to say that the study of uniform, slower flow is unimportant, however. This will provide base-flow and provide a reference point against which to measure dilutions. Both uniform and quickflow models therefore have to be applied. It is an appropriate juncture in time to undertake this because the studies of stream hydrology, throughflow and soil water have progressed sufficiently to date to enable the linkages between soil water models and stream models to be assessed. The challenge is to provide simple, yet faithful, models, proposed on a sound basis, and then to test them in the light of their predictive power in actual field situations. A variety of models are therefore outlined below. These will then be tested against field data to evaluate their inadequacies and predictive powers before further discussion of how to improve simple, applicable and realistic models.

11.2 QUICKFLOW PRODUCTION

The recognition of quickflow in the stream hydrograph is well established. The sources of this quickflow production are many, ranging from direct channel precipitation, saturated contributing areas, man-made surfaces, overland flow, and rapidly flowing throughflow. The variety of sources has considered impact on the nature of the stream response, but from the work on flow around soil structures, largely in soil science rather than a hydrology or geomorphology (e.g. Bouma *et al.*, 1981; White, 1985a), it is evident that rapid flow within the soil and linkages with stream quickflow response are worthy of further attention.

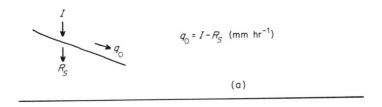

$$q_O = I - R_S \quad (\mathrm{mm\ hr}^{-1})$$

(a)

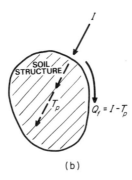

$$Q_f = I - T_p$$

(b)

Figure 11.2. (a) Initiation of infiltration excess overland flow; (b) Initiation of pedal infiltration excess bypassing flow (for notation, see text)

It is clear that as well as the partitioning of runoff waters into overland flow and soil throughflow, there is also a division of throughflow into rapid, preferential soil water flow and a slower flow. The preferential, or bypassing, flow travels round soil peds in structured soils in a manner analogous to the initiation of overland flow. In overland flow, infiltration excess overland flow occurs when rainfall intensity exceeds soil infiltration rate (Figure 11.2(a)):

$$q_o = I - R_s \qquad (1)$$

where:

q_o = volume of overland flow
I = rainfall intensity
R_s = infiltration rate of soil

The latter two in rates of mm hr^{-1} can be calculated for given areas to give q_o in volume terms.

In addition, return overland flow exists at slope foot sites where soil saturation leads to the return of soil water to the surface. The same kind of processes operate within the soil, only instead of the soil surface infiltration rate as the limiting factor, it is the rate of water entry into the soil structure surfaces which is critical (Figure 11.2(b)). Again, infiltration excess with respect to the ped leads to interpedal flow (mm hr^{-1}); (Trudgill et al., 1983b):

$$Q_f = I - T_p \qquad (2)$$

where:

Q_f = amount of preferential, interpedal flow
I = rainfall intensity
T_p = transmission rate of the ped

Again, these rates can be calculated as volumes of cm^3 water cm^{-3} soil.

In addition, flow may be returned to interpedal voids once the peds are saturated.

Such soil hydrological processes have clear implications for the uptake and transport of solutes. Just as the occurrence of infiltration excess overland flow limits the opportunity for the uptake of soil derived solutes, preferential flow, which bypasses soil peds, has similarly limited opportunity for soil-derived solute uptake. In both cases, only the displaced or return flow water will tend to be solute rich. Clearly, equation 2 predicts that the amount of preferential flow will increase as rainfall intensity increases. Thus, soil-derived solute dilution can be predicted to increases as rainfall intensity increases, that is to say that leaching efficiency will decrease. This partitioning means that preferential flow has to be taken into account in soil solute modelling. Only where it is absent can all input water be assumed to be available to mobilize soil-derived solutes; where it is present losses of soil-derived solute will be less than the input of water would predict. Conversely, the mobility of surface-applied solutes will be increased.

11.3   OUTFLOW AND SOIL WATER MOVEMENT

Quickflow models involving bypassing, preferential flow are in contrast to uniform displacement models (Figure 11.3). In simple displacement, output at any one time consists of displaced, long residence time water, the amount of displacement being proportional to the amount of input (Figure 11.3(a)). Quickflow, involving rapid preferential flow down macropores which bypasses soil structures, (Figure 11.3(b)) means that outflow involves both short residence time water and displaced long residence time water, often in sequence. For surface-applied solutes, (a) means that there is ample opportunity for reaction and attenuation of the solute and output concentrations will tend to be low, with mixing with soil-derived solutes; (b) means that concentrations will be high at some stage in the hydrograph, with little attenuation. In the latter situation contamination of water course with any surface-applied solute will tend to be much greater than in (a), at least transiently during the hydrograph. This will also apply to rainfall acidity, acid deposition being much more likely to penetrate to stream runoff under (b) than (a). For soil-derived solutes, output concentrations will be maximized under (a), with near equilibrium concentrations of solutes in displaced, long-residence time water; conversely, dilution of soil-derived solutes will occur under (b). In modelling it is therefore important to evaluate the proportion of quickflow in the soil in order to understand solute processes.

11.4   PENETRATION DEPTHS AND OUTFLOW RESPONSES

There is a relationship between the mode of soil water transport and the outflow response. For a surface-applied solute (or for a rainfall event diluting soil-derived solutes) the linkages between the surface and the outflow depends on the relationship between the penetration

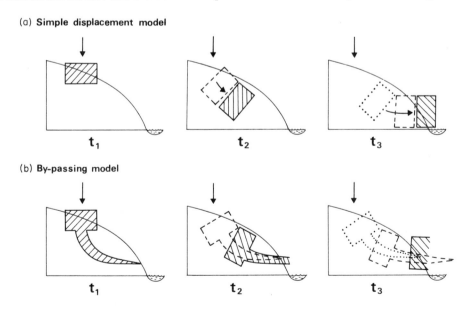

Figure 11.3. Soil water flow modes to streams: (a) Simple displacement — output involves long-residence time, displaced water; (b) Bypassing flow — output involves a proportion of current input water

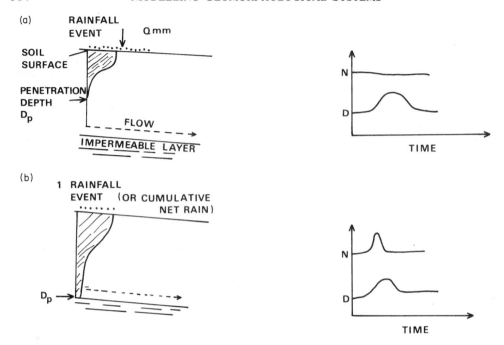

Figure 11.4. Soil profile penetration depths of surface-applied solute (dots) and outflow response. (i) No solute response, limited profile penetration under input rainfall (Q); (ii) Solute response, profile penetration to base. N = Nitrate concentration, D = discharge, D_p = solute penetration depth

depth of the solute from the surface and lateral transfer at some depth in the soil. If a simple situation is considered: that of a permeable soil profile and in impermeable bedrock sloping to a stream (Figure 11.4), the simplest statement about the system is that if a surface-applied solute (e.g. nitrate fertilizer) does not reach the zone of lateral flow at the base of the soil profile (Figure 11.4(a)) then there will be no response of the surface-applied solute at the soil water outflow. However, if the penetration front reaches the base of the soil profile (Figure 11.4(b)) a nitrate peak can be expected in the soil outflow water. Calculations of penetration depth (D_p) for a rainfall event, or for cumulative rainfall, is therefore important in predicting outflow response. Notice also that the solute pulse through the soil can be characterized by either the maximum concentration 'bulge', or model depth, of the solute or by the leading edge of maximum depth. In the case of overall outflow response, both are of interest, but for one solute application and first time of arrival at the outflow, it is the leading edge that is of primary importance, especially if bypassing flow occurs, as shown in Figure 11.5. Bypassing flow initially entrains the surface-applied solute but, subsequently, relocation of the soil within the peds forms the modal 'bulge' of the soil solute profile and the solute then behaves like a soil-derived solute. The bypassing flow therefore leads initially to solute peaks in the outflow but subsequently it leads to solute dilution in the outflow as the solute is preserved within the peds and bypassing flow leads to decreased leaching efficiency. The modelling endeavour is thus to predict outflow response in relation to soil water behaviour and solute disposition on and in the soil.

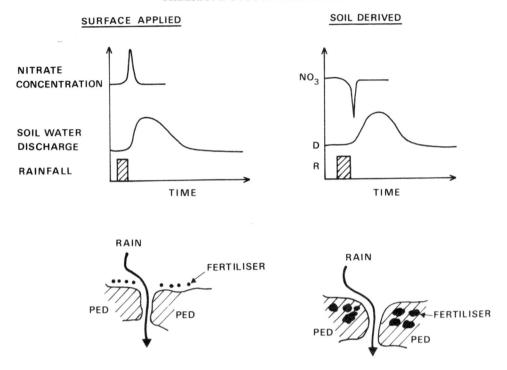

Figure 11.5. Bypassing flow and solute response in the outflow, (nitrate fertilizer, dots). Left — surface-applied solute entrained in bypassing flow: outflow solute peak. Right — soil derived solute protected from leaching by bypassing flow: outflow dilution. NO_3 = Nitrate, D = Discharge, R = Rainfall

11.5 PENETRATION DEPTHS AND OUTFLOW RESPONSES MODELS

11.5.1 Capacity Model

The simplest model to apply is one where the water-holding capacity of the soil profile is considered. Soil water outflow can be predicted to occur when this capacity is reached, much as a cup will overflow when filled to the brim. An immediately apparent estimate of this capacity is one of field capacity, beyond which excess water will drain off as gravitational water. Field capacity can be defined as the moisture status apparent when all gravitational water has drained off and this can be estimated (1) in the field as the moisture status 48 hours after heavy rain, though this may include an evaporation component and initial saturation is not assured; (2) climatically as those winter conditions where rainfall exceeds evaporation, leading to field soil moisture levels above field capacity; or (3) more precisely in the laboratory as the moisture status of a soil column after 48 hours free drainage from initial saturation. Operationally, the latter is the most standard, controllable estimate. Here, intact soil columns can be extracted from the field in plastic tubes of wide diameter (c 15–20 cm) lined with petroleum jelly ('Vaseline') to minimize edge flow. A tube is placed in a bucket and this is then gradually filled to displace entrapped air from below. After thorough soaking for some hours (i.e. overnight), the column can then be placed on a wire mesh grid to drain and weighed after 48 hours. Oven-drying can

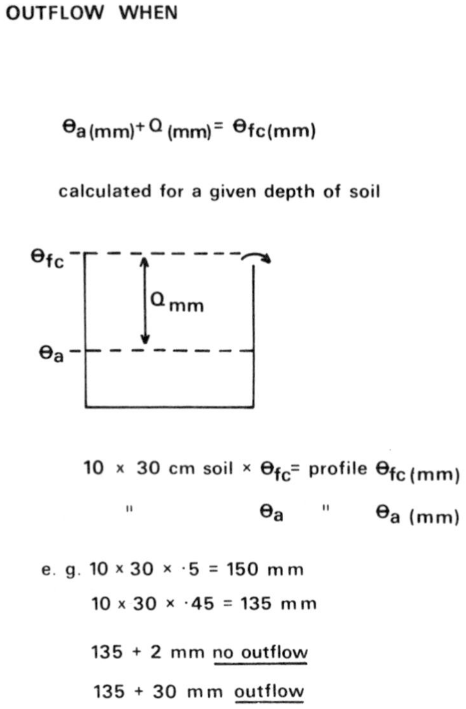

OUTFLOW WHEN

$$\theta_{a\,(mm)} + Q_{(mm)} = \theta_{fc(mm)}$$

calculated for a given depth of soil

10 × 30 cm soil × θ_{fc}= profile θ_{fc} (mm)

" θ_a " θ_a (mm)

e. g. 10 × 30 × ·5 = 150 mm

10 × 30 × ·45 = 135 mm

135 + 2 mm <u>no outflow</u>

135 + 30 mm <u>outflow</u>

Figure 11.6. Capacity model. For notation see text

be used to establish the dry weight of the soil so that field capacity can be expressed as a moisture content on a dry-weight basis and, by measurements of the volume of the soil column, converted to a θ value, i.e. one of cm^3 water per cm^3 soil and recorded as θ_{fc} cm^3 cm^{-3}. For a given depth of soil profile, θ_{fc} can be converted to cm water depth and to mm to make it comparable with rainfall data. Antecedent soil moisture in the field, assessed by the use of small (5 cm diameter) tins of known volume, wet weight and oven-dry weight, can also be recorded, as θ_a mm, and compared with θ_{fc} mm. Outflow is predicted to occur when input rainfall, Q mm, tops up θ_a to θ_{fc} (Figure 11.6):

OUTFLOW CONDITION IS WHEN $\theta_a + Q = \theta_{fc}$ (all in mm)   (3)

Thus, as shown in Figure 11.6, for a 30 cm soil profile at $\theta_{fc} = 0.5\,cm^3$ cm^{-3} and $\theta_a = 0.45\,cm^3$ cm^{-3} and ($\times 10$ to convert to mm), profile $\theta_{fc} = 150$ mm and profile $\theta_a = 135$ mm. Thus $150-135 = 25$ mm of rain needs to fall before outflow is predicted to occur. Clearly, this approach requires that no bypassing flow occurs. It also rests on estimates of θ_a and θ_{fc} from samples, assuming them to be representative of the field situation. The measurements are also of field capacity on *drainage* whereas the outflow situation is one of *wetting up*. Clearly, hysteresis may affect the moisture status, as shown in Figure 11.7, where field capacity on drying is a higher value than on drainage. Outflow is thus liable to occur somewhat before the moisture level predicted from θ_{fc} from drainage, as shown by Trudgill *et al.*, (1983a), in Figure 11.8. They also showed that some

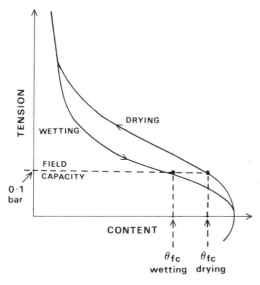

Figure 11.7. Soil water hysteresis and field capacity (θ_{fc}). Note that field capacity is more on drying than on wetting

data points for outflow occurred well below field capacity, implicating the occurrence of bypassing flow (in this work, an antecedent precipitation index was used instead of direct measurement of soil moisture, as explained in the figure caption).

11.5.2 Uniform Displacement Model for Predicting Modal D_p

Another simple model for uniform flow in the soil predicts the modal 'bulge' of D_p (Wild and Babiker, 1976):

$$D_p = \frac{Q}{\theta_a} \qquad (4)$$

where D_p and Q are expressed in cm and θ_a in cm^3 cm^{-3}

The workings of the model are illustrated in Figure 11.9. It assumes that all water entering the soil displaces antecedent soil moisture and that the volume operational in this displacement is equal to the volume of antecedent moisture. Thus, for smaller values of θ_a, higher values of D_p are seen for given rainfall amounts, Q: for given values of θ_a, higher Q also leads to higher values of D_p. Thus, displacement depth is greater for higher rainfall and lower soil moisture. This model may adequately predict modal D_p but does not predict the position of the leading edge. In addition, the assumptions about displacement and operational volume are liable not to hold in very dry conditions when wetting up would probably absorb much of the incoming water, nor in a very wet soil when gravitational water is liable to flow rapidly. Accordingly, further developments along the lines of this model involve predictions of the leading edge and also the processes of wetting up and rapid flow. The first of these is a wetting front model.

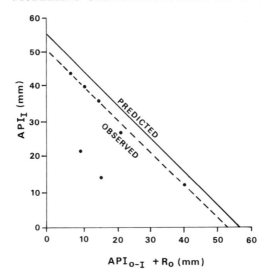

Figure 11.8. Outflow response (dots) against antecedent moisture (vertical axis), where API_I = Antecedent Precipitation Index on Input (mm) and input rainfall, (horizontal axis) where API_{0-I} = API increments from surface tracer placement to output and R_0 = rainfall at output, (i.e. increments from API_I to output). Predicted line is for soil profile water content (mm) at θ_{fc}. $API = API_n = (API_{n-1}.K) = P_{n-1}$, $K = 0.9$ (Gregory and Walling, 1973; p. 187) (After Trudgill *et al.*, 1983a)

11.5.3 Wetting Front Model

This can be used to predict maximum penetration depth (depth of the wetting front) but assumes that flow is uniform in the soil. It contrasts with the above model in that it assumes no displacement and instead assumes that incoming water is used up by filling pore space not antecedently occupied by soil moisture. It is thus more applicable to drier soils but the assumption is that the wetting front is equal to the same as D_p and also that no bypassing flow occurs. The model is described by Rose *et al.*, (1982) and is illustrated in Figure 11.10:

$$D_p = \frac{Q}{\theta_{fc} - \theta_a} \qquad (5)$$

For a given rainfall, D_p decreases as θ_a decreases because there are more pore spaces to fill in drier soils. For given θ_a values, D_p increases with increasing rainfall as higher Q values are distributed deeper in the same available pore space. Outflow can be predicted to occur when D_p is equal to or greater than soil profile depth, that is when the wetting front reaches the zone of lateral flow at the base of the soil. In order to do this, for each soil layer, any water in excess of θ_{fc} should flow down to the next layer to occupy the space between θ_a and θ_{fc} until the base of the soil is reached.

11.5.4 Mobile:Retained Water Model

Clearly, neither a displacement model nor an adsorption (wetting front) model can be entirely appropriate across the complete range of antecedent soil moisture values as

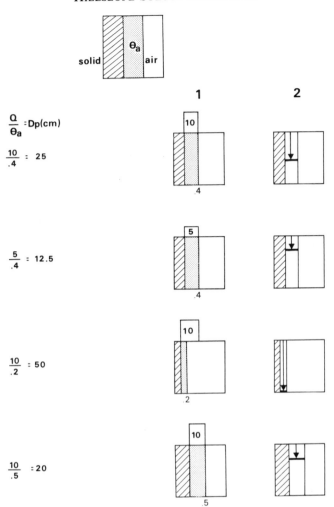

Figure 11.9. Uniform displacement model. Top: antecedent conditions, viewing antecedent soil moisture (θ_a) as a single component of operational volume, with four illustrative examples. Left: antecedent conditions, right: displacement depths. Top two: θ_a constant, Q varied; bottom two: Q constant, θ_a varied

absorption will be more prevalent in drier soils and displacement will be facilitated in moister soils. The mobile:retained boundary model specifies the domains of operation of the different processes. It thus combines some of the strengths of the previous two models but avoids some of their weaknesses. It assumes that there is a mobile:retained boundary of soil water, θ_r, set at 2 bars (Addiscott, 1977). In dry soils, where θ_a is less than θ_r, incoming water will occupy unfilled pore spaces to θ_r and follows the wetting front model, substituting θ_r for θ_{fc} (Figure 11.11):

$$D_p = \frac{Q}{\theta_r - \theta_a} \tag{6}$$

$$D_p = \frac{Q}{\theta_{fc} - \theta_a}$$

(Rose et al.
J. E. Q. 1982).

D_p = depth of penetration
of wetting front (cm)

Q = rainfall (cm)

θ_{fc} = volumetric moisture
content at field
capacity $(cm^3 cm^{-3})$

θ_a = antecedent volumetric
moisture content
$(cm^3 cm^{-3})$

— Assumes no displacement

— Distributes Q in pore space not
antecendently occupied

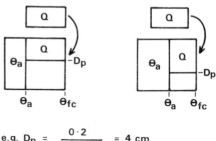

e.g. $D_p = \dfrac{0\cdot 2}{\cdot 5 - \cdot 45} = 4\ cm$

$D_p = \dfrac{3\cdot 0}{\cdot 5 - \cdot 45} = 60\ cm$

OUTFLOW WHEN D_p > PROFILE DEPTH

Figure 11.10. Wetting front model, with two examples (right) with constant Q and θ_a varied

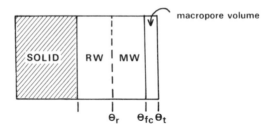

macropore volume

$\theta_r = \theta$ at 2 bars

$\theta_a < \theta_r$: "top up" $\theta_a \rightarrow \theta_r$
if $> \theta_r$ flows to next layer, i.e.

$$D_p = \frac{Q}{\theta_r - \theta_a}$$

$\theta_a > \theta_r$

$:D_p = \dfrac{Q}{\theta_a - \theta_r}$

Figure 11.11. Mobile:retained water model. RW = Retained water, MW = Mobile water, θ_r = mobile:retained
boundary, θ_t = total pore space

In dry soils, $(\theta_a < \theta_r)$, outflow will not occur until profile θ becomes greater than θ_r and for each soil layer Q in excess of $\theta_r - \theta_a$ will be mobile and will be distributed to the next layer, thus defining D_p with respect to Q.

In soils wetter than the mobile:retained boundary, where $\theta_a > \theta_r$, (but drier than θ_{fc}) displacement of mobile water $(\theta_a - \theta_r)$ will occur:

$$D_p = \frac{Q}{\theta_a - \theta_r} \tag{7}$$

A simpler model is to predict that if $\theta_a > \theta_r$, outflow will occur and if $\theta_a < \theta_r$ outflow will not occur.

However, when θ_a is between θ_{fc} and θ_r there is still an uncertainty as to whether incoming water will *displace* $\theta_a - \theta_r$ water or *occupy* the pore space between θ_{fc} and θ_a; this distinction is fundamental to solute dynamics as the former will displace solute-rich (soil derived solute) water and the latter will not. If the latter, the prediction for D_p becomes the same as equation 5 again; but equation 7 is of utility if it is assumed that capillary water held at tensions less than 2 bars is mobile and can be displaced. It is, however, probable that if rainfall is intense, then unoccupied pore space may become operational and/or water may pass into the macropores defined by $\theta_t - \theta_{fc}$ where θ_t is the total pore space where gravitational water moves. Thus, the displacement versus occupation (sorption) modes are difficult to resolve on a sound basis. The factor of intense rainfall, however, leads onto consideration of the bypassing model. Here, intense rainfall gives rise to the operationalization of flow in pores which would not necessarily flow according to considerations of amount of rainfall alone.

11.5.5 Bypassing (Preferential Flow) Model

The use of the general term 'preferential flow' implies that some water is flowing faster in the soil than other water, thus rendering the use of a uniform flow model inappropriate. The use of the term 'bypassing flow' is a specific term applying to preferential flow in well structured soils where flow occurs around the larger soil structures (Bouma *et al.*, 1981). Bypassing flow has been predicted to occur around soil structures if rainfall intensity, I, is greater than the transmission rate of the peds, T_p (equation 2; Figure 11.2(b)). This does not mean to say that uniform flow will not be occurring within the peds during intense rainfall as it can be assumed that the water flowing at T_p will be flowing uniformly in the peds. However, it is the excess of $I - T_p$ (Q_f) which is of interest since this will be flowing most rapidly towards the outflow. The problem lies in estimating the operational volume, θ_f, down which the excess flows:

$$D_p = \frac{Q_f}{\theta_f} \text{(where } Q_f = I - T_p) \tag{8}$$

where θ_f is in $cm^3\ cm^{-3}$.

Where $Q_f = 0$, any of the above models could be considered (equations 3–7) but where $Q_f > 0$, there are several possibilities (Figure 11.12). Occurrence of Q_f could simply lead to outflow, with water flowing down the sides of macropores to the outpoint point.

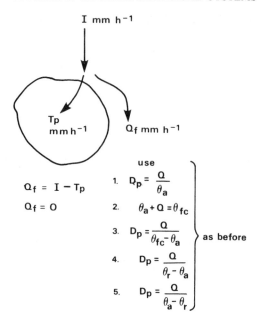

$$Q_f = I - Tp$$

$$Q_f = 0$$

use

1. $D_p = \dfrac{Q}{\theta_a}$

2. $\theta_a + Q = \theta_{fc}$

3. $D_p = \dfrac{Q}{\theta_{fc} - \theta_a}$ as before

4. $D_p = \dfrac{Q}{\theta_r - \theta_a}$

5. $D_p = \dfrac{Q}{\theta_a - \theta_r}$

$$Q_f > 0$$

(1) occurrence simply leads to output

(2) estimation of D_p of Q_f down θ_f

θ_f from $\theta_{sat} - \theta_{fc}$ (or θ drainage at drop of 1 cm head

Beven and Germann JSS 1981).

$D_p = \dfrac{Q_f}{\theta_f}$ calculated hourly.

Figure 11.12. Bypassing flow. For notation, see text

This, simplest, approach does not involve any calculation of θ_f but merely assumes that if Q_f exists, outflow will occur. This assumes no sorption of water in its travel down the pore. This may not be the case with intense rainfall on dry soil which would wet the surface, initiating Q_f, but which could then be sorbed lower down the profile; under winter rainfall conditions, with soils close to θ_{fc}, however, lack of sorption would be a more valid assumption. In addition, the duration of I in excess of T_p must also be important in order not only to initiate but also to sustain such flow, but this is difficult to speculate on without knowledge of the pore size distribution. Alternatively, therefore, specification of operational volume, θ_f could be useful. This is not the same thing as actual pore size or pore size distribution, merely a concept of the portion of the soil volume partaking in bypassing flow. It could involve complete flow down all the available macropore space, $\theta_t - \theta_{fc}$, or partial flow down the sides of such spaces. Thus either full pores of a range of sizes holding low tension water or thin ribbons of water running down the side of larger pores could

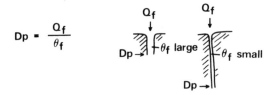

Figure 11.13. Influence of θ_f on D_p for given Q_f

be involved. θ_f is thus a conceptual, modelling device rather than a physical statement. Clearly, from equation 8, larger θ_f values predict lower values of D_p with given amounts of Q_f, whereas deeper penetration involves smaller values of θ_f and/or larger values of Q_f (Figure 11.13).

Two approaches to modelling θ_f are possible. Firstly, the estimation of θ_f from evidence independent of actual field behaviour, followed by calculations of D_p where predictions can be checked against tracer recovery depths (D_r) in the soil and/or tracer outflow patterns (for surface-applied tracers). Secondly, also for surface-applied tracers, observed values of D_r can be employed to calculate θ_f using calculated values of Q_f.

Independent estimates of θ_f can be calculated (Figure 11.13) by drainage of a soil column from saturation to field capacity (as above under *Capacity models*) but, this time actually measuring the amount of water drained (Figure 11.14a). This gives the maximum potential volume of θ_f.

Beven and Germann (1981) estimated macropore volume by saturating soil blocks and measuring the outflow volume which occurred consequent upon a drop in hydraulic head of 1 cm (this involves attaching the base of the block to one end of a 'U' tube and levelling the other end with the block; lowering the free end will release low-tension water held in macropores and the drainage can be measured as it flows from the free end of the tube, Figure 11.14(b)).

With these independent estimates of θ_f, $D_p = Q_f/\theta_f$ can be calculated hourly if $Q_f = I - T_p$ is calculated hourly in cm and D_p is in cm (Q_f is in cm for the hour and θ_f is in cm^3 cm^{-3}). If the calculation is for 1 cm^2 surface area of soil, D_p can be compared directly to profile depth for each hour of storm duration or for portions of hours. It can also be resolved for portions of hours to allow for variability of I during the storm. Alternatively, a simpler, but less realistic, approach can be made by using I_{max}, the maximum intensity observed during the storm or simply a general I value if intensity is more or less steady through the storm.

A further complication is that T_p could vary during the storm, especially if the soil is initially dry ($\theta_a < \theta_{fc}$) and sorption occurs initially until the peds are saturated. In saturated soils T_p will be equivalent to pedal saturated hydraulic conductivity (K_{sp}) but, in drier soils, T_p will be excess of K_{sp}, thereby overestimating Q_f if K_{sp} is applied as the value of T_p. K_{sp} values can be measured in a small permeameter (Burt, 1978) where small soil blocks are encased in small-bore tubes (1–2 cm diameter). Actual peds could be encased in paraffin blocks or, more easily, tubes filled with intact soil could be used, with a number of replicates (say, 10) and the lowest value obtained used.

θ_f can be back-calculated if tracers are used to measure D_p (Figure 11.14(c)). D_p in this case can be referred to as recovery depth, D_r, as it is an empirical observation dependent

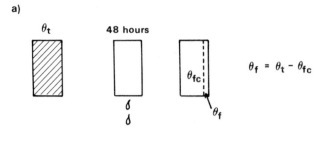

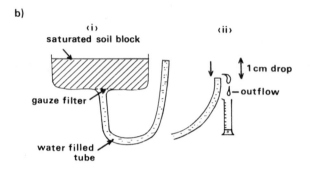

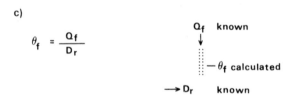

Figure 11.14. (a) Estimation of θ_f from saturation; (b) Estimation of θ_f from 1 cm head drop; (c) Estimation of θ_f from D_r

on the soil sampling scheme used and is therefore not necessarily the same as actual D_p. Thus:

$$\theta_f = \frac{Q_f}{D_r} \tag{9}$$

Using this procedure, values of θ_f are often much smaller than values gained using drainage, indicating very small operational volumes. However, these small volumes must often be transient if rainfall continues, leading to wetting up of the soil profile. If small values of θ_f did apply, very high velocities of flow are implied, far in excess of those which might be reasonably expected, if all the soil water flowed in very small volumes. Thus, initial flow only may be described. However, from observed deep values of D_r (Trudgill *et al.*, 1983b), such small initial volumes appear to be appropriate and the concept of $\theta_{f\,min}$ is necessary to predict D_r;

$$D_r = \frac{Q_f}{\theta_{f\,min}} \tag{10}$$

Thus, a range of models are available to predict surface-applied solute penetration depth and they make different assumptions about the mode of flow of water in the soil. It remains to test and refine these models by the matching of predicted, calculated depths with observed depths of penetration, together with observations of soil drainage water outflow behaviour. If there is a solute response in the outflow, it can be assumed that solute penetration has reached the base of the profile and moved laterally to the outflow. For soil-derived solutes, similar inferences can be made if the outflow shows a dilution pattern, assuming that input rainfall has penetrated to the outflow (Figure 11.5, right hand side).

11.6 TESTING AND REFINEMENT OF MODELS

All the models described above make simplifying assumptions and therefore may have virtue in ease of application but they are detracted from because of their simplifications. It is therefore necessary to test these models before refining them and making them more adequate. The models have been tested against data gained in the context of the study of the movement of nitrate fertilizer from the soil surface to soil drainage outflow at Slapton in South Devon, as described by Coles and Trudgill (1985). Here, a monitored plot on weakly structured silt–loam soil under grass was set up, with discharge response from a tile drain at 85 cm depth measured using an Ott R16 vertical stage recorder. Nitrate levels were measured using a Rock and Taylor 48 Interval Liquid Sampler, with analysis by hydrazine–copper method on a Chemlab Autoanalyser. Discharge response at the outflow was assessed for 40 rainfall events together with rainfall characteristics using an autographic rain gauge. Rainfall intensity was assessed as the maximum intensity for an individual storm, provided that more than 3 mm total rainfall was recorded at this maximum intensity (Kneale and White, 1984). Soil moisture was assessed gravimetrically on samples of constant volume (Kubiena tins, 221 cm^3) and expressed as the mean of five samples for the 5–30 cm layer in cm^3 cm^{-3} (θ values). Field capacity was assessed on laboratory drainage from saturation for 48 hours on nine columns, giving a mean of 0.46 cm^3 cm^{-3}. The mobile:retained boundary, θ_r, was assessed from a soil–moisture characteristic curve using pressure plate apparatus and, at 2 bars, was 0.29 cm^3 cm^{-3}. θ_f was defined on free drainage from saturation to field capacity, with a mean of 0.005 cm^3 cm^{-3} and a minimum of 0.001. In addition, tracer recovery (Coles and Trudgill, 1985) and the use of equation 9 gave a minimum value of 0.0001. T_p was set at 2.5 mm hr^{-1} from mean values obtained from field infiltration measurements (Coles and Trudgill, 1985).

The models described above are dealt with in turn to test their applicability. All models can be rejected as hypotheses for the universal prediction of outflow response as, for each model, there are cases where occurrences of outflow response do not match predictions. However, as shown below, if the jurisdiction of some models is restricted to specific conditions of antecedent soil moisture domains and input conditions (specifically θ_{fc}, θ_r, and I) then in specific domains, some hypotheses are not refuted and are left open for

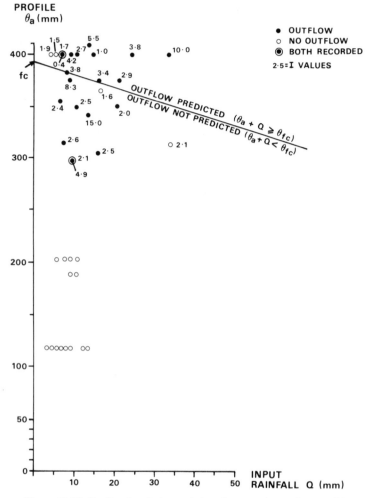

Figure 11.15. Predicted and observed data for capacity outflow model

further testing. Each model is considered in turn and the domains of application are discussed. In cases where D_p is predicted, if $D_p \geqslant$ profile depth, 85 cm, outflow is predicted, values > 85 merely implying outflow rather than having a physical meaning in soil depth terms.

11.6.1 Capacity Model

The data for predicted and actual behaviour are shown in Table 11.1. As displayed in Figure 11.15, outflow should occur when soil profile θ_a (vertical axis) is topped up by input rainfall, Q, (horizontal axis) to field capacity, θ_{fc}, calculated at 382.5 mm water depth for an 85 cm profile. Three cases show no actual outflow when outflow boundary conditions are apparently sufficient to lead to outflow ($\theta_a + Q \geqslant \theta_{fc}$). Eight cases of outflow occurred when conditions were apparently not sufficient to cause outflow ($\theta_a + Q < \theta_{fc}$), but in five

Table 11.1. Application of capacity model (equation 3)

θ_a	Profile θ_a (mm)	Deficit (mm)	Input (mm)	Predicted Outflow	Actual Outflow
0.48	408	0	13.2	YES	YES
0.47	399.5	0	3.7	YES	NO
0.47	399.5	0	6.5	YES	NO
0.47	399.5	0	6.0	YES	NO
0.47	399.5	0	33.3	YES	YES
0.47	399.5	0	14.5	YES	YES
0.47	399.5	0	10.4	YES	YES
0.47	399.5	0	24.3	YES	YES
0.47	399.5	0	6.5	YES	YES
0.47	399.5	0	9.0	YES	YES
θ_{fc}					
0.45	382.5	0	8.2	YES	YES
0.44	374	8.5	21.2	YES	YES
0.44	374	8.5	16.8	YES	YES
0.44	374	8.5	8.2	(YES)	YES
0.43	365.5	17.0	16.0	(NO)	NO
0.42	357	25.5	6.7	NO	YES
0.41	348.5	34	10.7	NO	YES
0.41	348.5	34	20.4	NO	YES
0.40	340	42.5	13.4	NO	YES
0.37	314.5	68	7.3	NO	NO
0.37	314.5	68	34.1	NO	YES
0.36	306	76.5	16.7	NO	YES
0.34	289	93.5	5.1	NO	NO
0.34	289	93.5	5.0	NO	YES
0.33	280.5	102	16.8	NO	YES
0.33	280.5	102	7.5	NO	NO
θ_r					
0.24	204	178.5	9.0	NO	NO
0.24	204	178.5	7.8	NO	NO
0.24	204	178.5	6.9	NO	NO
0.24	204	178.5	11.9	NO	NO
0.22	187	195.5	10.2	NO	NO
0.22	187	195.5	9.3	NO	NO
0.14	119	263.5	4.8	NO	NO
0.14	119	263.5	5.2	NO	NO
0.14	119	263.5	8.6	NO	NO
0.14	119	263.5	7.0	NO	NO
0.14	119	263.5	12.5	NO	NO
0.14	119	263.5	3.4	NO	NO
0.14	119	263.5	12.9	NO	NO
0.14	119	263.5	7.3	NO	NO

( ) = near prediction
Profile depth 85 cm
Field capacity, $\theta_{fc} = 0.46$ cm^3 cm^{-3}. For profile = 382.5
Mobile:retained boundary, $\theta_r = 0.29$ cm^3 cm^{-3}

Table 11.2. Application of $D_p = Q/\theta_a$ model (equation 4)

Q (cm)	θ	$D_p = Q/\theta$	Predicted outflow	Actual outflow
1.32	0.48	2.8	NO	YES
0.37	0.47	.8	NO	NO
0.65	0.47	1.4	NO	NO
0.60	0.47	1.3	NO	NO
3.33	0.47	7.1	NO	YES
1.45	0.47	3.1	NO	YES
1.04	0.47	2.2	NO	YES
2.45	0.47	5.2	NO	YES
0.65	0.47	1.4	NO	YES
0.90	0.47	1.9	NO	YES

θ_{fc}

Q (cm)	θ	$D_p = Q/\theta$	Predicted outflow	Actual outflow
0.82	0.45	1.8	NO	YES
2.13	0.44	4.8	NO	YES
1.68	0.44	3.8	NO	YES
0.82	0.44	1.9	NO	YES
1.60	0.43	3.7	NO	NO
0.67	0.42	1.6	NO	YES
1.07	0.41	2.6	NO	YES
2.04	0.41	5.0	NO	YES
1.34	0.40	3.3	NO	YES
0.73	0.37	2.0	NO	NO
3.41	0.37	9.2	NO	YES
1.67	0.36	4.6	NO	YES
0.51	0.34	1.5	NO	NO
0.50	0.34	1.5	NO	YES
1.68	0.33	5.1	NO	YES
0.75	0.33	2.3	NO	NO

θ_r

Q (cm)	θ	$D_p = Q/\theta$	Predicted outflow	Actual outflow
0.90	0.24	3.8	NO	NO
0.78	0.24	3.3	NO	NO
0.69	0.24	2.9	NO	NO
1.19	0.24	5.0	NO	NO
1.02	0.22	4.6	NO	NO
0.98	0.22	4.2	NO	NO
0.48	0.14	3.4	NO	NO
0.52	0.14	3.7	NO	NO
0.86	0.14	6.1	NO	NO
0.70	0.14	5.0	NO	NO
1.25	0.14	8.9	NO	NO
0.34	0.14	2.4	NO	NO
1.29	0.14	9.2	NO	NO
0.73	0.14	5.2	NO	NO

Profile depth = 85 cm

of these, $I \geqslant T_p$. This implicates the role of bypassing flow but for the three other cases $I < T_p$ ($I = 2.4$, 2.1, 2.0). Thus, while it might be suggested that $I > T_p$ may override capacity considerations, three cases refute this suggestion. Moreover, it is especially problematical to account for the three cases of no observed output when $\theta_a > \theta_{fc}$. Reference to Figure 11.7 (hysteresis loop) is unhelpful in this context as it suggests that θ_{fc} is *lower* for wetting than for drying, indicating that, if anything, outflow should be *before* the value of θ_{fc} determined on drying, rather than at a higher level. Lack of outflow is evident some 6.5 mm above mean θ_{fc}. This could suggest that spatial variability of θ_{fc} occurs, but there is no independent evidence to test this. It is, however, clear that this model is inadequate, both for the reasons of cases of no outflow above calculated θ_{fc} and for instances of outflow below θ_{fc}. These cannot be allowed for by a slight recalibration of θ_{fc} values. It should be noted, in addition, that from Table 11.1, outflow does not occur when $\theta_a < \theta_r$, as described further below under *Mobile:retained boundary model*.

11.6.2   Uniform Displacement Model

This does not predict the leading edge of the surface-applied solute but predicts the modal position of the solute 'bulge' in the soil profile. However, it is clear from Table 11.2 that predicted depths are all less than 10 cm. Even with dispersion ahead of the modal depth (Cameron and Wild, 1982, Figure 11.1) this is unlikely to reach the required depth of the base of the soil profile at 85 cm depth. This model is thus not a fruitful line of procedure.

11.6.3   Wetting Front Model

As shown in Table 11.3, for the first ten cases, $\theta_a > \theta_{fc}$ and the model is inapplicable—except insofar as if θ_a exceeds θ_{fc} outflow can be predicted to occur, as discussed under *Capacity model*, but this it does not for the three cases already discussed. For the remaining 12 instances of outflow response when $\theta_a < \theta_{fc}$, only two are predicted (only one a near prediction). This hypothesis is also thus clearly refuted. The basic mechanism proposed by the model is that water occupies the pore space not antecedently filled between θ_a and θ_{fc} and that it flows in a uniform manner. Thus, if the model is of no predictive value this implies that the envisaged process does not occur and, as outflow is, in fact, evident, that the flow is more preferential than the model assumes.

11.6.4   Mobile:retained Boundary Model

This model partitions soil water in that it is assumed that water held at tensions higher than θ_r is not mobile and thus it allows for a more preferential flow where all the soil water is not operational in movement. However, this is also refuted by observation (Table 11.4). For $\theta_a > \theta_r$, only two cases of outflow response are predicted by D_p calculations when 19 actually occurred. Application of $\theta_r - \theta_a$ is similarly unsuccessful, predicting outflow responses only when very small values of $\theta_r - \theta_a$ obtain, predicting that water in excess of θ_r will be mobile. In six cases there is predicted to be sufficient excess to lead to outflow, but there is no actual response.

Taking the simpler propositions, where: if $\theta_a > \theta_r$, then outflow is predicted to occur because mobile water is present and: if $\theta_a < \theta_r$, then outflow is predicted not to occur. The latter is not refuted and remains open for further testing; the former, however, is refuted as seven out of 26 cases above θ_r do not have outflow.

Table 11.3. Application of $D_p = Q/\theta_{fc} - \theta_a$ model (equation 5)

Q (cm)	$\theta_{fc} - \theta_a$	D_p	Predicted outflow	Actual outflow
1.32	0 $(\theta_a > \theta_{fc})$		YES	YES
0.37	0 $(\theta_a > \theta_{fc})$		YES	NO
0.65	0 $(\theta_a > \theta_{fc})$		YES	NO
0.60	0 $(\theta_a > \theta_{fc})$		YES	NO
3.33	0 $(\theta_a > \theta_{fc})$		YES	YES
1.45	0 $(\theta_a > \theta_{fc})$		YES	YES
1.04	0 $(\theta_a > \theta_{fc})$		YES	YES
2.43	0 $(\theta_a > \theta_{fc})$		YES	YES
0.65	0 $(\theta_a > \theta_{fc})$		YES	YES
0.90	0 $(\theta_a > \theta_{fc})$		YES	YES

————————————————————— θ_{fc}

Q (cm)	$\theta_{fc} - \theta_a$	D_p	Predicted outflow	Actual outflow
0.82	0.01	81	(NO)	YES
2.13	0.02	106.5	YES	YES
1.68	0.02	84	(YES)	YES
0.82	0.02	41	NO	YES
1.60	0.03	53	NO	NO
0.67	0.04	17	NO	YES
1.07	0.05	21	NO	YES
2.04	0.05	41	NO	YES
1.34	0.06	22	NO	YES
0.73	0.09	8	NO	NO
3.41	0.09	38	NO	YES
1.67	0.10	17	NO	YES
0.51	0.12	4	NO	NO
0.50	0.12	4	NO	YES
1.68	0.13	13	NO	YES
0.75	0.13	6	NO	NO

————————————————————— θ_r

Q (cm)	$\theta_{fc} - \theta_a$	D_p	Predicted outflow	Actual outflow
0.90	0.22	4	NO	NO
0.78	0.22	4	NO	NO
0.69	0.22	3	NO	NO
1.19	0.22	5	NO	NO
1.02	0.24	5	NO	NO
0.98	0.24	4	NO	NO
0.48	0.32	2	NO	NO
0.52	0.32	2	NO	NO
0.86	0.32	3	NO	NO
0.70	0.32	3	NO	NO
1.25	0.32	4	NO	NO
0.34	0.32	1	NO	NO
1.29	0.32	4	NO	NO
0.73	0.32	2	NO	NO

( ) = near prediction
Profile depth = 85 cm
$\theta_{fc} = 0.46$ cm^3 cm^{-3}

Table 11.4. Application of mobile:retained model, $D_p = Q/\theta_a - \theta_r$ or, if $\theta_a < \theta_r$, $D = \theta_r - \theta_a$ (equations 6 and 7)

Q (cm)	$\theta_a - \theta_r$	$\theta_r - \theta_a$	D_p	Predicted outflow	Actual outflow	
1.32	0.19	—	69.5	NO	YES	Outflow predicted as $\theta_a > \theta_r$
0.37	0.18	—	20.6	NO	NO	Outflow predicted as $\theta_a > \theta_r$
0.65	0.18	—	36.1	NO	NO	Outflow predicted as $\theta_a > \theta_r$
0.60	0.18	—	33.3	NO	NO	Outflow predicted as $\theta_a > \theta_r$
3.33	0.18	—	185.0	YES	YES	Outflow predicted as $\theta_a > \theta_r$
1.45	0.18	—	80.6	(NO)	YES	Outflow predicted as $\theta_a > \theta_r$
1.04	0.18	—	57.8	NO	YES	Outflow predicted as $\theta_a > \theta_r$
2.43	0.18	—	135.0	YES	YES	Outflow predicted as $\theta_a > \theta_r$
0.65	0.18	—	36.1	NO	YES	Outflow predicted as $\theta_a > \theta_r$
0.90	0.18	—	50.0	NO	YES	Outflow predicted as $\theta_a > \theta_r$
						θ_{fc}
0.82	0.16	—	5.1	NO	YES	Outflow predicted as $\theta_a > \theta_r$
2.13	0.15	—	14.5	NO	YES	Outflow predicted as $\theta_a > \theta_r$
1.68	0.15	—	11.2	NO	YES	Outflow predicted as $\theta_a > \theta_r$
0.82	0.15	—	5.5	NO	YES	Outflow predicted as $\theta_a > \theta_r$
1.60	0.14	—	11.4	NO	NO	Outflow predicted as $\theta_a > \theta_r$
0.67	0.13	—	5.2	NO	YES	Outflow predicted as $\theta_a > v_r$
1.07	0.12	—	8.9	NO	YES	Outflow predicted as $\theta_a > \theta_r$
2.04	0.12	—	17.0	NO	YES	Outflow predicted as $\theta_a > v_r$
1.34	0.11	—	12.2	NO	YES	Outflow predicted as $\theta_a > v_r$
0.73	0.08	—	9.1	NO	NO	Outflow predicted as $\theta_a > \theta_r$
3.41	0.08	—	42.6	NO	YES	Outflow predicted as $\theta_a > \theta_r$
1.67	0.07	—	23.9	NO	YES	Outflow predicted as $\theta_a > \theta_r$
0.51	0.05	—	10.2	NO	NO	Outflow predicted as $\theta_a > \theta_r$
0.50	0.05	—	10.0	NO	YES	Outflow predicted as $\theta_a > \theta_r$
1.68	0.04	—	42.0	NO	YES	Outflow predicted as $\theta_a > \theta_r$
0.75	0.04	—	18.75	NO	NO	Outflow predicted as $\theta_a > \theta_r$
						θ_r
0.90	—	0.05	180.0	YES	NO	Outflow not predicted as $\theta_a < v_r$
0.78	—	0.05	156.0	YES	NO	Outflow not predicted as $\theta_a < \theta_r$
0.69	—	0.05	138.0	YES	NO	Outflow not predicted as $\theta_a < \theta_r$
1.19	—	0.05	238.0	YES	NO	Outflow not predicted as $\theta_a < \theta_r$
1.02	—	0.07	145.7	YES	NO	Outflow not predicted as $\theta_a < \theta_r$
0.98	—	0.07	132.9	YES	NO	Outflow not predicted as $\theta_a < \theta_r$
0.48	—	0.15	32.0	NO	NO	Outflow not predicted as $\theta_a < \theta_r$
0.52	—	0.15	34.7	NO	NO	Outflow not predicted as $\theta_a < \theta_r$
0.86	—	0.15	57.3	NO	NO	Outflow not predicted as $\theta_a < \theta_r$
0.70	—	0.15	46.7	NO	NO	Outflow not predicted as $\theta_a < \theta_r$
1.25	—	1.5	83.3	(NO)	NO	Outflow not predicted as $\theta_a < \theta_r$
0.34	—	1.5	22.6		NO	Outflow not predicted as $\theta_a < \theta_r$
1.29	—	1.5	86.0		NO	Outflow not predicted as $\theta_a < \theta_r$
0.73	—	1.5	48.7		NO	Outflow not predicted as $\theta_a < \theta_r$

( ) = near prediction
Profile depth = 85 cm
$\theta_r = 0.29$ cm^3 cm^{-3}

Table 11.5. Application of bypassing model, outflow will occur if
$I > T_p$

Q (mm)	I (mm hr^{-1})	Outflow predicted	Actual outflow
13.2	5.5	YES	YES
3.7	1.9	NO	NO
6.5	1.7	NO	NO
6.0	1.5	NO	NO
33.3	10.0	YES	YES
14.5	1.0	NO	YES
10.4	2.7	YES	YES
24.3	3.8	YES	YES
6.5	0.4	NO	YES
9.0	4.2	YES	YES

————————————————————————————————————θ_{fc}

8.2	3.8	YES	YES
21.3	2.9	YES	YES
16.8	3.4	YES	YES
8.2	8.3	YES	YES
16.0	1.6	NO	NO
6.7	2.4	(NO)	YES
10.7	3.6	YES	YES
20.4	6.3	YES	YES
13.4	2.5	YES	YES
7.3	2.0	NO	NO
34.1	15.0	YES	YES
16.7	2.6	(YES)	YES
5.1	2.1	NO	NO
5.0	2.5	(YES)	YES
16.8	4.9	YES	YES
7.5	2.1	NO	NO

————————————————————————————————————θ_r

9.0	1.5	NO	NO
7.8	2.1	NO	NO
6.9	1.3	NO	NO
11.9	1.8	NO	NO
10.2	1.3	NO	NO
9.3	2.4	(NO)	NO
4.8	3.2	YES	NO
5.2	2.6	(YES)	NO
8.6	2.9	YES	NO
7.0	6.0	YES	NO
12.5	15.6	YES	NO
3.4	4.3	YES	NO
12.9	2.2	NO	NO
7.3	1.8	NO	NO

( ) = near prediction
$T_p = 2.5$ mm hr^{-1}

11.6.5 Bypassing Model

In some of the discussions above, preferential flow, with some portion of the soil water flowing faster than others, was implicated by the lack of application of models which assumed uniform flow. Uniform flow models generally underpredict the depth necessary for outflow. Indeed, where outflow responses are, in fact, recorded, then this implies that some water is flowing faster, and deeper, than uniform flow predicts. Thus, a model which includes preferential flow should be less likely to be refuted than one assuming uniform flow.

The simplest case to consider first is where it is predicted that if I exceeds T_p then outflow will occur. A more detailed case is where excess flow, Q_f, is routed down a given operational volume, θ_f (equation 8). The first of these is considered in Table 11.5. Above θ_{fc}, there is a reasonable match but the hypothesis is refuted by two cases out of 10 in which outflow actually occurs but is not predicted to occur. In the first case, I is 1.0 mm hr^{-1}, well below T_p (at 2.5 mm hr^{-1}) but rainfall amount is heavy at 14.5. In the second, I is only 0.4 mm hr^{-1} and Q is 6.5 mm. The latter is the most difficult to explain. Some possibility of combined effects of I and Q can be discussed, together with errors in measurement (see *Discussion*, below) but the discussion is liable to be inconclusive. Below θ_r, the hypothesis is also refuted in that $I > T_p$ occurs but there is no outflow. This implies that either pedal excess water could be initiated in the upper profile, only to be absorbed in the lower, or that T_p is an underestimate in dry soils. This latter is highly likely and it does imply, from the data, that I values of up to 15 mm hr^{-1} can be sorbed by the ped.

Between θ_{fc} and θ_r there are three cases of near prediction, two at or slightly above T_p (2.6 and 2.5) and one just lower at 2.4. If it is accepted that the 2.6 and 2.5 values are adequate to initiate preferential flow and if the 2.4 value is accepted as within experimental error, then the hypothesis is not refuted. The case is marginal, however, resting on the acceptance of the 2.4 value. It is, however, the best match of the entire data set and the suggestion is that this model is still open for further testing, rather than for rejection—the case for rejection being equally marginal and also resting on the 2.4 value.

It is therefore appropriate to proceed to Table 11.6 where equation 8 is considered. The three values of θ_f discussed earlier are used. Mathematically, D_p must increase as θ_f decreases and thus for $\theta_f = 0.005$ (mean value, drainage calculation), 10 cases out of 19 outflows are not predicted but for $\theta_f = 0.001$ (minimum drainage) this drops to three. For $\theta_{f\,min}$ at 0.0001, calculated from equation 10 using observed D_r values on a separate set of data, predictions match observed outflow in 13 out of 19 cases, but give the impression of unfeasible rates of flow. These high rates involve very small operational volumes, and while derived independently from observed tracer recovery, D_r, must be limited to initial flow as, if all the Q_f flowed down such a small volume, unreasonable velocities are implied. The model does not necessarily indicate actual physical processes, simply because decreasing θ_f must eventually give the right answer. In physical terms, however, as flow continues, θ_f must increase as the soil wets up. If the argument for Table 11.5 is accepted: that $I > T_p$ provides a reasonable model for D_p at values of θ_a between θ_{fc} and θ_r, then it is clear that very small operational volumes must be involved. This increases the predictive power of the model, as shown by the increasing match from column 3 to 4 and 5 on Table 11.6 and as supported by calculation of θ_f from independent observed D_r values. However, for $\theta_a < \theta_r$, six cases of $\theta_{f\,min}$ predict outflow when it does not occur and in four other cases, outflow response occurs without prediction. Two are between θ_{fc} and θ_r, one of these is

Table 11.6. Application of $D_p = Q_f/\theta_f$ model (equation 8)

Q (cm)	Q_f	D_p ($\theta_f = 0.0005$)	D_p ($\theta_f = 0.001$)	D_p ($\theta_{f\,min} = 0.0001$)	Actual outflow
1.32	0.30	60	300*	3000*	YES
0.37	0	—	—	—	NO
0.65	0	—	—	—	NO
0.60	0	—	—	—	NO
3.33	0.75	150*	750*	7500*	YES
1.45	0	—	—	—	YES
1.04	0.02	4	20	200*	YES
2.43	0.13	26	130	1300*	YES
0.65	0	—	—	—	YES
0.90	0.17	34	170*	1700*	YES

θ_{fc}

Q (cm)	Q_f	D_p ($\theta_f = 0.0005$)	D_p ($\theta_f = 0.001$)	D_p ($\theta_{f\,min} = 0.0001$)	Actual outflow
0.82	0.13	26	130*	1300*	YES
2.12	0.04	8	40	400*	YES
1.68	0.09	18	90*	900*	YES
0.82	0.58	116*	580*	5800*	YES
1.60	0	—	—	—	NO
0.67	0	—	—	—	YES
1.07	0.11	22	110*	1100*	YES
2.04	0.38	76	380*	3800*	YES
1.34	0	—	—	—	YES
0.73	0	—	—	—	NO
3.41	1.25	250*	1250*	12 500*	YES
1.67	0.01	2	10	100	YES
0.51	0	—	—	—	NO
0.50	0	—	—	—	YES
1.68	0.24	48	250*	2400*	YES
0.75	0	—	—	—	NO

θ_r

Q (cm)	Q_f	D_p ($\theta_f = 0.0005$)	D_p ($\theta_f = 0.001$)	D_p ($\theta_{f\,min} = 0.0001$)	Actual outflow
0.90	0	—	—	—	NO
0.78	0	—	—	—	NO
0.69	0	—	—	—	NO
1.19	0	—	—	—	NO
1.02	0	—	—	—	NO
0.93	0	—	—	—	NO
0.48	0.07	14	70	700*	NO
0.52	0.01	2	10	100*	NO
0.86	0.04	8	40	400*	NO
0.70	0.35	70	35	350*	NO
1.25	1.31	262*	131*	1310*	NO
0.34	0.18	36	180	1800*	NO
1.29	0	—	—	—	NO
0.73	0	—	—	—	NO

*Denotes outflow predicted.

$Q_f = I - T_p$

$\theta_f = 0.005$, 0.001 (drainage) or $\theta_{f\,min}$ 0.0001 ($\theta_f = Q_f/\theta_r$)

the 2.4 discussed above, one is at 2.5 (T_p) and therefore no excess is calculated. Two are above θ_{fc} and remain problematical to explain. Thus, if the 2.4 and 2.5 cases are accepted, the model is not refuted for conditions between θ_{fc} and θ_r. It now remains to discuss these results and possible combinations of models for different conditions.

11.7 DISCUSSION

No one model is universally applicable, but if θ_a domains are specified it can be seen that some possibilities are left open for further testing because they cannot be conclusively refuted.

Refuted are the application of:

Capacity model

Uniform flow model for modal depth

Wetting front model

Mobile:retained model, except for $\theta_a < \theta_r$

Bypassing flow model, except for $\theta_a < \theta_{fc} > \theta_r$ (accepting one marginal case)

These refutations are based on the assumption that the procedures are a fair test of the models, which they may not be. θ_{fc}, T_p, I and θ_f have all been measured by methods which are open to debate and to questions of spatial representativeness. They are all reasonable estimates with defensible procedures but they all have limitations as representations of reality. These limitations are, for field capacity, that the measure is on drying and the field process is one of wetting and the 48 hour period is reasonable but arbitrary. Both it and T_p have problems of spatial variability, which are unknown and thus of sample representativeness; the same is true of θ_f. Consistent standardized measures of I are obtained by the method used, but such an index does not allow for all the internal variations in intensity. Thus, the parameters used are approximations or simplified modelling indices which can be readily evaluated. Given these points, it is perhaps surprising that some of them are of utility, rather than that many of them are not! In order to assess the situation more clearly, some sensitivity analysis is necessary in order to describe the variations in answers obtained by small variations in inputs. In addition, further work is necessary on those models which have not been refuted, and even some that have, since, under natural rainfall conditions at one site, it is difficult to obtain adequate representation of all possible combinations of conditions.

In terms of arguing back from the observed data on Table 11.5, between θ_{fc} and θ_r, no outflow is seen at I values of 1.6, 2.0, 2.1 and 2.1. The lowest I value for observed outflow is 2.4. While there are only 16 examples studied, it can be proposed that if T_p is set to lie between 2.1 and 2.4 then the model will not be refuted. Although tautalogical, the suggestion is that the model—and the processes envisaged—actually apply, it is merely that T_p was inadequately calibrated for the model. This leaves the model open for further testing but stresses the importance of adequate characterization of the variables in order to give a fair test of the model. To be of predictive value, applicable to a wide number of field sites, T_p has to be evaluated independently of the outflow observations and, moreover, it is likely to be more reliable at soil moistures close to θ_{fc}; at lower moisture contents, sorption will lead to overestimation of Q_f and therefore of D_p. Thus, spatially and temporarily representative samples of T_p should be worth further effort, since the bypassing model cannot be clearly refuted.

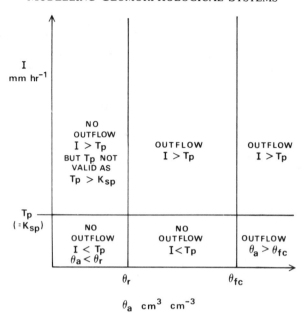

Figure 11.16. Proposed model of outflow response based on I, T_p and θ_a domains. T_p is not valid for dry soils as it is evaluated for pedal saturated hydraulic conductivity (K_{sp}) and sorption in dry soils gives rise to much higher T_p values that K_{sp}

The interest in the model testing, however, must lie in the way in which the test procedure has illuminated understanding of soil water flow processes. The clear underpredictions of the uniform flow models implicate the importance of the role of soil structure and rainfall intensity. If the approach is one of looking for verification, evidence amasses for the applicability of the bypassing model, together with some role of field capacity (Figure 11.16). If a stricter, falsification of hypotheses, approach is adopted, only the bypassing model for soils between θ_{fc} and θ_r remains open for further testing. In addition, if it is proposed that outflow will occur if both $\theta_a > \theta_{fc}$ and $I > T_p$, only in two cases in Table 11.5 is there an exception, indicating that this may be worthy of further consideration. Moreover, the question remains of how soil specific these models are. For instance, Kneale and White (1984) found that bypassing flow occurred in dry cracking clay soils, whereas for the Devon silt loam soils studies here there was no outflow, even at intensities as high as 15.6 mm hr^{-1}. To modify the bypassing hypothesis to allow for this is difficult as it is problematical to obtain a single measure of soil structure which could be built in as a boundary condition. Clearly, if peds are 'small' and pathways 'diffuse' more sorption would be evident than if peds are 'large' and pathways 'well-defined'; defining such terms precisely is less clear. Rose (1973) used mean aggregate radius as a parameter in predicting leaching in artificial porous aggregates, but this is difficult to apply to field soils *in situ*.

A major assumption in the model testing work is that water flows vertically for 85 cm before flowing laterally to the outflow. Dye testing at other sites has shown this to be substantially true (Trudgill, 1987) but this does not necessarily mean that it is also true for this site. In addition, water table fluctuations in the soil may mean that travel depths have to be less to reach the water table and thus zone of lateral flow in wet conditions.

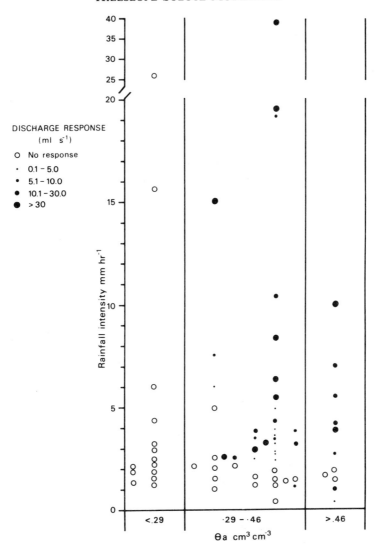

Figure 11.17. Outflow responses in relation to I and θ_a. $\theta_r = 0.29$, $\theta_{fc} = 0.46$

However, this would only explain the NO predictions with YES outflow occurrences in the tables above θ_{fc}. In Table 11.4 this occurs four times, Table 11.5 twice and Table 11.6, twice for no Q_f, and three times of $\theta_f = 0.005$. Observation of water levels in piezometer tubes shows that in general water levels are around 85 cm but that on occasions transient levels can reach 70 cm from the surface during periods of high rainfall. For Table 11.6, D_p data for $\theta_a > \theta_{fc}$ and at $\theta_f = 0.005$ only reach 60 cm but in Table 11.4 one of the four cases reaches 69.5 cm, the others being 57.8, 50.1 and 36.1. Thus, while depths in many cases still appear not to reach the water level, the few cases that do reach it show that this possibility is worthy of further investigation, especially in relation to models involved when $\theta_a > \theta_{fc}$.

A final consideration is whether D_p calculations are necessary in themselves. If θ_a domains appear to be important and if I values have a predictive value (Table 11.5), and bearing in mind the difficulties of specifying T_p, outflow responses can be plotted in relation to I and θ_a alone (Figure 11.17). This clearly shows a pattern of no outflow at $\theta_a < \theta_r$, only three cases out of 11 of no outflow at $\theta_a > \theta_{fc}$ (a 73 per cent predictive success) and emphasizes a threshold of around 2.5 mm hr^{-1} for outflow between θ_r and θ_{fc}, with only one case of outflow below (a 92.9 per cent predictive success) and two of no outflow above (a 93.3 per cent predictive success). This pattern thus coincides with that of Figure 11.16 within the percentage terms specified and thus forms a reasonable approximation to the model of Figure 11.16 in a verification sense, though strictly speaking the few mispredictions refute the model in a falsification sense. The data shown in Figure 11.17 are empirical, however, and cannot be derived independently unless T_p is specified as well as θ domains. It does provide a working approximation which can be tested in further situations.

11.7.1   Observations From Other Investigations

The data reported above refer only to one site and the combinations of θ, I and Q that happened to occur in a two year study period. Reference to other work will also be useful for a fuller model evaluation. For example, the applicability of the bypassing model was established by Trudgill *et al.* (1983a,b) for a calcareous loam soil under forest cover, with a refutation of the capacity model and, again, support for a low value of θ_f. For a soil profile of 44 cm depth, and a θ_f of 0.01, 100 per cent prediction was achieved for outflow response. Similarly, on a clay loam soil under arable cropping, Smettem *et al.* (1983) successfully predicted outflow response with θ_f values of 0.029–0.032 and $I - T_p$ at 2.406 giving D_p values of 82.96–75.19. Soil profile depth was 80 cm where outflow occurred under the event studied, the predicted D_p value of 82.96 cm, derived from the smaller θ_f, forecasting that outflow should occur. The work reported here parallels somewhat different approaches to similar problems by other workers. These include the work of White (1985a,b); Wild (1981) and Cameron and Wild (1982) together with those already reviewed by Trudgill and Briggs (1979, 1980, 1981). However, the models discussed above are the simplest models, with the least input parameters and therefore have merit in that sense. In common with the work of White (1985a,b) the conclusion is that the bypassing models probably have the greatest potential for further work on predicting the leading edge of surface-applied solute penetration, and therefore of outflow to streams.

ACKNOWLEDGEMENT

I would like to thank Nigel Coles for providing the test data (NERC Studentship GT4/81/AAPS/42).

REFERENCES

Addiscott, T. M. (1977). A simple computer model for leaching in structured soils. *Journal of Soil Science*, **28**, 544–563.

Beven, K. and Germann, P. (1981). Water flow in soil macropores. II. A combined flow model. *Journal of Soil Science*, **32**, 15–30.

Beven, K. and Germann, P. (1982). Macropores and water flow in soils. *Water Resources Research*, **18**, 1131–1325.

Bouma, J., Dekker, L. W., and Muilwijk, C. J. (1981). A field method for measuring short-circuiting in clay soils. *Journal of Hydrology*, **52**, 347–354.

Burt, T. P. (1978). *Three simple and low-cost instruments for the measurement of soil water properties*, Huddersfield Polytechnic Department of Geography and Geology, Occasional papers, 6.

Cameron, K. C. and Wild, A. (1982). Prediction of solute leaching under field conditions: an appraisal of three models. *Journal of Soil Science*, **33**, 659–669.

Coles, N. and Trudgill, S. T. (1985). The movement of nitrate fertiliser from the soil surface to drainage waters by preferential flow in weakly structured soils, Slapton, S. Devon. *Agriculture, Ecosystems and Environment*, **13**, 241–259.

Kneale, W. R. and White, R. E. (1984). The movement of water through cores of dry (cracked) clay loam grassland topsoil. *Journal of Hydrology*, **67**, 361–365.

Gregory, K. J. and Walling, D. E. (1973). *Drainage Basin Form and Process*, Edward Arnold, London, 458 pp.

Rose, C. W., Chichester, F. M., Williams, J. R., and Ritchie, W. (1982). A contribution to simplified models of field solute transport. *Journal of Environmental Quality*, **11**, 146–150.

Rose, D. A. (1973). Some aspects of the hydrodynamic dispersion of solutes in porous materials. *Journal of Soil Sciences*, **24**, 284–295.

Smettem, K. R. J., Trudgill, S. T., and Pickles, A. M. (1983). Nitrate loss in soil drainage waters in relation to by-passing flow and discharge on an arable site. *Journal of Soil Science*, **34**, 499–509.

Trudgill, S. T. (Ed.) 1986. *Solute Processes*, Wiley.

Trudgill, S. T. (1987). Soil water dye tracing, with special reference to the use of Rhodamine WT, Lissamine FF and Amino G Acid. *Hydrological Processes*, **1**, 149–170.

Trudgill, S. T. and Briggs, D. J. (1979, 1980, 1981). Soil and land potential. *Progress in Physical Geography*, **3**, 283–299; **4**, 262–275; **5**, 274–285.

Trudgill, S. T., Pickles, A. M., Smettem, K. R. J., and Crabtree, R. W. (1983a). Soil water residence time and solute uptake: 1. Dye tracing and rainfall events. *Journal of Hydrology*, **60**, 257–279.

Trudgill, S. T., Pickles, A. M., and Smettem, K. R. J. (1983b). Soil water residence time and solute uptake: 2. Dye tracing and preferential flow predictions. *Journal of Hydrology*, **62**, 279–285.

White, R. E. (1985a). The influence of macropores on the transport of dissolved and suspended matter through soil. *Advances in Soil Science*, **3**, 95–120. Springer-Verlag.

White, R. E. (1985b). A model for nitrate leaching in undisturbed structured clay soil during unsteady flow. *Journal of Hydrology*, **79**, 37–51.

Wild, A. (1981). Mass flow and diffusion. In Greenland, D. J. and Hayes, M. H. B. (Eds.), *The Chemistry of Soil Processes*, John Wiley, Chichester.

Wild, A. and Babiker, I. A. (1976). The asymmetric leaching pattern of nitrate and chloride in a loamy sand under field conditions. *Journal of Soil Science*, **27**, 460–466.

Modelling Geomorphological Systems
Edited by M. G. Anderson
©1988 John Wiley & Sons Ltd.

Chapter 12

Slopes on strong rock masses: modelling and influences of stress distributions and geomechanical properties

M. J. SELBY, P. AUGUSTINUS, V. G. MOON and R. J. STEVENSON
Department of Earth Sciences, University of Waikato, Hamilton, New Zealand

12.1 INTRODUCTION

Rock slopes are taken, in this chapter, to be hillslopes which are essentially free of soil or talus covers and which, consequently, are continuous, or nearly continuous, bodies of exposed rock owing their form to the intact strength of the rock and the spacing, continuity, dip and strike of their partings.

Little interest has been taken by geomorphologists in those properties of rock masses which influence hillslope form and development, and the study of processes which influence hillslope form has largely been limited to the development of talus deposits. The experiences of engineers and geologists, however, indicate that the formation and opening of joints and other discontinuities on a rock mass are not only the initial stages of weathering, but also major processes which influence, and sometimes control, the nature and extent of the dominant processes in rock slope evolution.

It has been shown elsewhere (Selby, 1980, 1982a,b,c; Moon and Selby, 1983) that many rock slopes have forms which are in equilibrium with the total strength of their rock masses; some, such as Richter denudation slopes have forms controlled by the angle at which their debris mantles rest (Bakker and Le Heux, 1952); others exhibit structural controls or are undercut by rivers, glaciers or waves; but, after rock mass strength, it is probable that rock slope development is most commonly controlled by the orientation of joints critical for stability (Terzaghi, 1962; Selby, 1982a).

The stability of rock slopes against mass wasting is of such economic importance that most attempts at modelling have been promoted by the necessity for controlling rock slope stability or designing safe cut slopes. A brief review will be given of the better known types of model, and this will be followed by a discussion of finite-element methods of assessing the stresses and strains in large rock bodies. In a subsequent section, an account will be given of current attempts to relate models of the genesis of species of volcanic rocks with models of their physical properties and of their resulting landforms.

12.1.1 Slope Stability and Modelling

Much published knowledge of the geological controls on the stability of rock slopes has been derived from efforts by engineers to develop larger and deeper open pit mines and deeper road cuts. Increases in sizes of mines have been due to the availability of large and efficient earth moving machines. In both mines and road cuts, the lower the angle of slope the more stable it is likely to be, but there is an increase in the volume of material which must be excavated, and hence in the cost.

The design of a safe, but economic, cut slope requires the development of a model which incorporates appropriate elements of (Piteau and Associates, 1979):
1. Geological structure;
2. Groundwater;
3. Stresses and deformations in the slope;
4. Rock mass strength;
5. Slope geometry;
6. Vibrations;
7. Climatic conditions;
8. Time.
In practice, investigations are usually reduced to studies of:
1. Rock mass strength with particular emphasis on structural discontinuities;
2. Recognition and characterization of potential failure surfaces which will follow segments of pre-existing structural discontinuities;
3. The production of a physical or mathematical model which can be used to analyse the stability of a slope.
In addition to these three components of an investigation there is increasing recognition of the importance for slope stability of the effects of stresses within the rock mass, both during pit construction and subsequently.

12.1.2 Types of Model

Existing models of rock mass strength are semiquantitative and derived from field observations and measurements. The method developed for geomorphic purposes has been described elsewhere (Selby, 1980). The importance of failure surfaces, and particularly of the frictional resistance available along them, has also been reviewed in an earlier essay (Selby, 1987) and neither of these topics will be discussed further.

The limit equilibrium approach to slope stability is the best known and most widely used method of analysis. In all limit equilibrium methods, it is assumed that a coherent rock mass will move over a failure surface which may be plane, curved, or a combination of these two shapes. All limit equilibrium methods make use of the Coulomb–Navier failure criterion either in total stress, or in effective stress, form. The result of a limit equilibrium analysis is usually expressed as a factor of safety for the slope. Many of the well known methods were developed in soil mechanics and have limited application to a jointed rock mass unless it contains a single large planar surface along which failure is likely to occur. Analytical methods appropriate for stepped surfaces and wedge failures have been developed for rock slopes and are treated in standard texts (e.g. Chowdhury, 1978; Goodman, 1980; Hoek and Bray, 1981).

Physical models, usually made of wood, cement or plaster blocks representing joint blocks, are commonly used where it is necessary to consider conditions in which slopes will fail not as a coherent mass above a failure plane, but by the progressive movement of many discrete units of rock along multiple surfaces. A disadvantage of physical models for the representation of reality is that the strength of the materials and the stresses operating have to be scaled in proportion to the reduction in size between the real feature (i.e. the prototype) and the model.

For many rock slopes the gravitational body forces are the only active forces inducing stresses. Gravitational body forces can be simulated in models by centrifugal loading which may greatly exceed the forces produced by Earth's gravitational acceleration. It has been shown by Hoek (1965) that the centrifugal acceleration which must be applied to the model is a multiple of Earth's gravitational acceleration. The value of this multiple is the ratio of prototype size to model size. Small scale models which are subjected to large centrifugal accelerations must use materials which are weaker than the prototype material. Large scale models stressed only by Earth's gravitational field must use very weak materials.

Most published accounts of centrifugal model testing indicate the difficulties which are encountered in attempts to model the complexities of natural jointing in rocks and the necessity of using three-dimensional models. These difficulties, which are recognized from such studies as those by Stacey (1973), have caused many engineers to turn to stress analysis as a method of modelling slope stability.

Photoelastic models have been used by several workers (Long, 1964; Stacey, 1973) to study the stress distribution induced around excavated slopes. The models are made from a birefringent photoelastic material such as an epoxy resin. The model is subjected to a load either in a press or a centrifuge, and observed in polarized light in a polariscope. Two sets of light interference fringes are produced and give directions and values of the stresses induced in the model. In practice, however, there are usually considerable difficulties encountered in calibrating the model and in evaluating the stress trajectories (Gyenge and Coates, 1964). As a result of the problems inherent in physical modelling of rock slopes increasing use is being made of the method based on finite-elements. Before a discussion is presented of this method a brief account is given of the nature and significance of stresses in rock masses.

12.2 STRESSES AND JOINT DEVELOPMENT

Rock bodies in uniform states of all round stress either will not develop joints, or preexisting joints in them will not open. Unbalanced compressive stresses deep within a rock body also will not cause joint development because of the confining effect of surrounding rock, but unbalanced compressive stresses near the surface of a rock body may cause zones of tension. Such zones, and those caused by tectonic, thermal, or dehydration processes, may be subject to tensile stresses which exceed the rock strength, then fracture propagation will lead to joint formation.

(a) Geostatic stress, σ_z (MN/m^2)

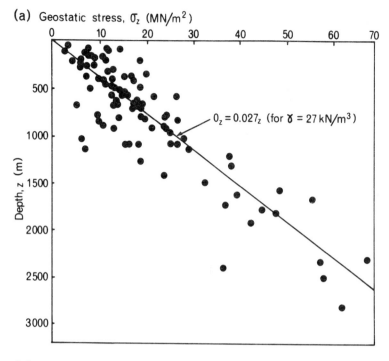

$0_z = 0.027_z$ (for $\gamma = 27\,\mathrm{kN/m^3}$)

(b) Coefficient of geostatic stress, K_0

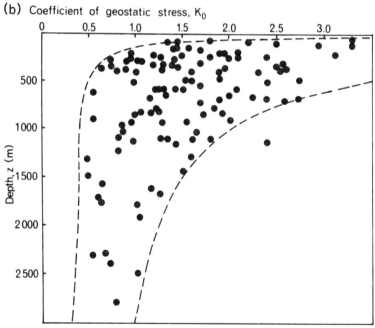

Figure 12.1 a & b (*caption opposite*)

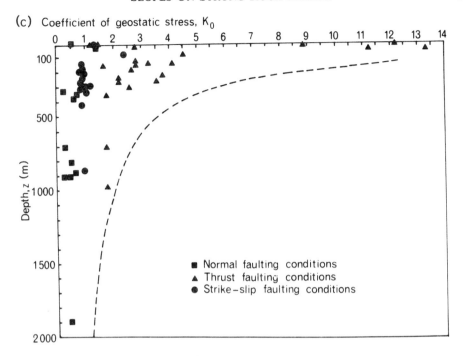

Figure 12.1. (a) Relationship between geostatic (total) stress and depth. Data from many sites. (b) Relationship between coefficient of geostatic stress and depth. (c) Effect of tectonic processes on the coefficient of geostatic stress. (a, b after Brown and Hoek, 1978 reprinted with permission from Pergamon Books Ltd; c, after Farmer (1983), reproduced with permission of Chapman and Hall Ltd, from an unpublished paper by Jamison and Cook)

Prediction of the distribution of stresses, and particularly of tensile stresses, can thus lead to an understanding of the location and mechanisms of joint opening and landform development on rock slopes.

12.2.1 Stress Modes

Four major sources of stress may be active in rock bodies at shallow depths: residual, gravitational, tectonic, and thermal.

1. Residual stresses are induced in a rock body by plastic deformation resulting from original crystallization, overburden weight, thermal gradients, tectonic pressures, or other processes which create a metastable stress condition in the absence of external stresses (Haxby and Turcotte, 1976). Removal of overburden or confining stresses disturbs the metastable state and the residual stresses may then be relieved by strains which may cause fracture propagation, buckling, slab failure, and other forms of instability.
2. Gravitational stresses at a point within a rock body are induced by the weight of the column of rock above that point: thus the total vertical stress (σ_z), which is also known as the lithostatic stress, is given by:

$$\sigma_z = \gamma z \tag{1}$$

where γ is the rock unit weight, and z is the depth of the point below the surface of the ground. Over a wide range of measurements this relationship has been found to be approximately true (Figure 12.1(a)). Scatter in the data is caused by insensitivity in the measuring techniques and by proximity to major geological and topographic features.

In saturated porous or fissured rock, the relationship should be expressed in terms of effective or geostatic stresses ($\sigma_z{}'$):

$$\sigma_z{}' = \gamma \ z_1 + \gamma_{sub} \ z_2 + \gamma_w \ h_w \tag{2}$$

where γ_{sub} is the submerged unit weight of rock equal to $\gamma - \gamma_w$,
  γ_w is the unit weight of water,
  h_w is the depth of water below the phreatic surface (in this case $z_2 = h_w$),
  z_1 and z_2 are the thicknesses of dry and saturated rock bodies respectively.

A similar convention should be adopted where horizontal and vertical stresses are being related, because the hydrostatic pressures will be equal in all directions and the phreatic and ground surfaces may not coincide. The convention assumes that horizontal stresses are equal in the whole horizontal plane and are related to the vertical stresses in the form:

$$\sigma_h{}' = K_0 \ \sigma_z{}', \text{ and } K_0 = \sigma_h{}' / \sigma_z{}' \tag{3}$$

where K_0 is the coefficient of lateral earth pressure at rest, or coefficient of geostatic stress,
  $\sigma_h{}'$ is the effective horizontal stress.

In the condition in which tectonic stresses have relaxed, and where the rock is dry, then $K_0 = 1$ and total stresses apply.

Gravitationally induced vertical stresses are related to horizontal stresses because rocks tend to expand horizontally in response to the vertical load. Elastic solids, such as rocks, tend to expand in directions perpendicular or transverse to the applied compressive stress.

The transverse expansion is described by Poisson's ratio (ν) which is the ratio of the transverse strain (ϵ_t) to the longitudinal strain (ϵ_ℓ), when the rock is free to expand transversely, thus:

$$\nu = \epsilon_t / \epsilon_\ell \tag{4}$$

If a rock is not free to expand transversely ($\epsilon_t = 0$), a transverse stress (σ_t) is created, it is the magnitude of this transverse stress which describes the coefficient of lateral earth pressure, thus:

$$K_0 = \frac{\nu}{1 - \nu} \tag{5}$$

and,

$$\sigma_t = \frac{\nu}{1 - \nu} \sigma_z{}' \tag{6}$$

This transverse component (σ_t) of the total horizontal stress (σ_h) has a value of about one-third of the vertical stress (σ_z), if the value of ν is 0.25, which it is for many rocks.

Weak rocks are less effective at preventing transverse expansion than strong rocks, consequently they have higher Poisson's ratios than strong rocks, and the magnitude of σ_t is greater in weak rocks for any value of the vertical stress. Weak rocks therefore tend to fracture readily with closely spaced joints, when confining rock bodies are removed by erosion, and strong rocks fracture less readily with more widely spaced joints in response to removal of confining rock.

3. Tectonic stresses result from earth movements and are especially important in areas of converging and diverging lithospheric plates. The compressive forces commonly active in zones of convergence induce stress fields in which horizontal stresses exceed vertical stresses. Very high horizontal stresses are common in zones of thrust faulting and give rise to high values for the coefficient of geostatic stress. Most K_0 values are <4 and are rarely >7 (McGarr and Gay, 1978) (Figure 12.1b, c). Zones of normal faulting are associated with crustal extension and values of K_0 are usually <1, if all horizontal stresses are relieved. Where strike-slip faulting is present, the stress field may be weakly compressional, but values of K_0 approximate to unity. Field evidence for tectonic stresses exists not only in the presence of faulting but also of jointing. Tectonically induced joints result from brittle yielding and are often displayed as conjugate joint pairs, with common strike directions but opposite dips, within considerable volumes of rock (Gerber and Scheidegger, 1973).

In all stress fields the values of horizontal and vertical stresses become increasingly similar at depth and K_0 approximates to unity. It has been suggested by Turcotte (1974) that only the upper 25 km of the lithosphere at temperatures below 300°C behaves elastically. At greater depths or temperatures, stresses are relieved by plastic flow.

4. Thermal stresses result from the prevention of expansion or contraction of a solid during heating or cooling. In a homogeneous rock the relationship between strain (ϵ) and temperature change (ΔT) is

$$\epsilon = \alpha \, \Delta T \qquad (7)$$

where α is the linear coefficient of thermal expansion of the rock.

If the rock is not free to expand or contract, stress is generated in the rock with a magnitude described by the elastic stress–strain relationship:

$$\epsilon_t = \frac{1}{E}\left[\sigma_t - \nu(\sigma_t + \sigma)\right] + \alpha \, \Delta T \qquad (8)$$

where E is Young's modulus of elasticity.

If a confined rock ($\epsilon_t = 0$) is cooled through a temperature interval ΔT, the induced thermal stress, ignoring the applied longitudinal stress (σ_l), is given by (Voight and St. Pierre, 1974):

$$\alpha_t = \frac{\alpha E \, \Delta T}{1 - \nu} \qquad (9)$$

Table 12.1. Some characteristic rock properties determined in uniaxial compression and indirect tensile tests (after Wuerker, 1955, and D'Andrea et al., 1965)

Rock type	$E(10^5$ MPa)	ν	σ_c(MPa)	Indirect tensile strength (MPa)
Granite	0.55–0.9	0.21–0.28	210	9–13
Gabbro	1.05	0.34	200	22
Basalt	1	0.28	290	16
Marble	1.1	0.28	250	15
Limestone	0.2–0.9	0.23–0.20	30–180	2.1–10
Sandstone	0.02–0.7	0.3–0.4	10–42	0.3–1.1
Chalk	0.05–0.7	0.4	15–19	0.5–0.9

Consider, for example, a confined rock that cools 100°C but may not contract ($\alpha = 10^{-6}/°C$, $E = 10^5$ MPa, $\nu = 0.25$). The induced tensile stress is then -13 MPa, which is similar in magnitude to the tensile strength of the rock (Table 12.1), and joints are likely to form.

Joints created by thermal contraction are most readily formed in heterogeneous rocks composed of minerals and grains with variable values for α, ν and E, and variable rates of cooling. Thermal stresses created at the scale of individual and interlocking crystals are major contributors to residual stresses.

12.2.2 Finite-element Method

The finite-element method was developed to analyse the structures of aircraft frames (Turner et al., 1956). The method has subsequently been applied to geological problems and is widely used in rock and soil mechanics (Zienkiewicz, 1971). Its application to geomorphological situations are few, but Sturgul and Grinshpan (1975) used it to study isostatic rebound of a formation of the Grand Canyon, Arizona, Sturgul et al. (1976) developed a finite-element model of the Hochkönig massif in Austria, and several studies have used the method to study the distribution of stresses in slopes (Yu and Coates, 1970; Kohlbeck et al., 1979; Valliapan and Evans, 1980; McTigue and Mei, 1981; Savage et al., 1985).

The finite-element method uses the assumption that the response of a body to loading can be approximated by considering a cross-section of the body to be made up of deformable elements, taken as either triangular or quadrilateral in shape, each connected at the corners or nodes. The amount of deformation will depend on the amount of load and how close the element is to the applied load. The chosen elements must be small enough for the straight lines which make up the sides of the elements to remain straight after deformation. The mathematical formulation of the method is based upon the principle that when loads are applied to a body, the total potential energy of the system is at a minimum. Possible loads are body forces (B), external nodal forces (N), and surface tractions (S). These forces will change the internal strain energy (U) in such a way that total potential energy (Π is given by:

$$(\Pi) = (U) - (U_B) - (U_S) - (U_N) \tag{10}$$

where (U_B), (U_S) and (U_N) are the potential energies due to body forces, surface tractions and nodal loads, respectively. The calculation involves determining the nodal displacement

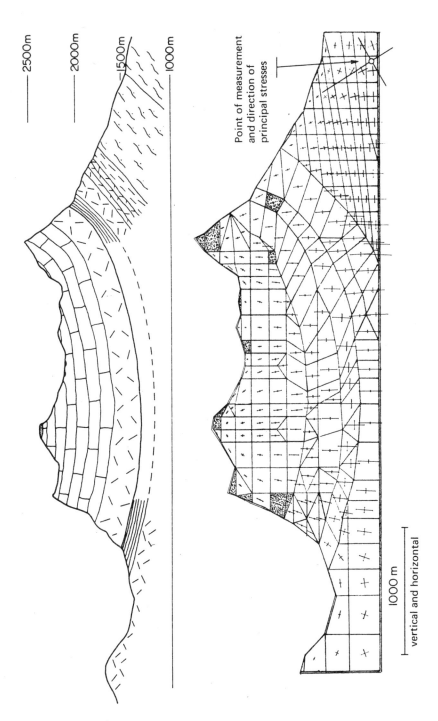

S

N

—— 2500m

—— 2000m

—— 1500m

—— 1000m

Point of measurement
and direction of
principal stresses

1000 m

vertical and horizontal

Figure 12.2. A simplified geological cross section through the Hochkönig massif, Austria, (top) and a finite-element grid (bottom). The calculated magnitudes and orientations of the principal stresses are shown by crosses. Zones of tensional stress are stippled. The measured stress is shown for a point in the lower right of the figure (after Sturgul *et al.*, 1976, reproduced with permission of the authors)

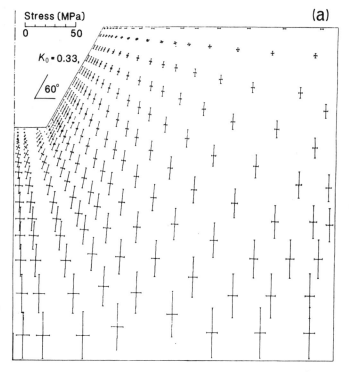

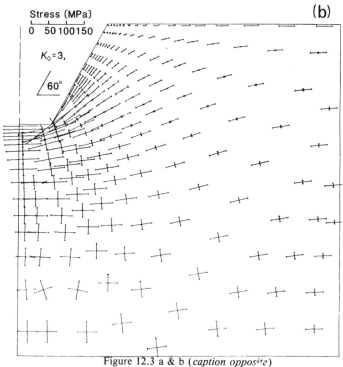

Figure 12.3 a & b (*caption opposite*)

(c)

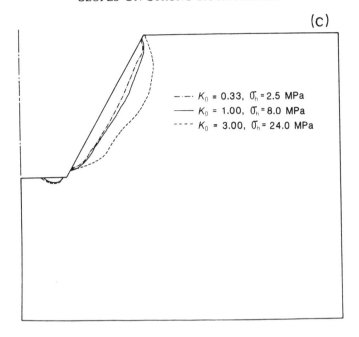

(c)

$K_0 = 0.33, \ \sigma_h = 2.5$ MPa
$K_0 = 1.00, \ \sigma_h = 8.0$ MPa
$K_0 = 3.00, \ \sigma_h = 24.0$ MPa

Figure 12.3. (a,b) Direction and magnitudes of principal stresses in an open pit with 60° slope and indicated values of the coefficient of geostatic stress. (c) Development of tensile zones around a 60° slope under compressive horizontal field stress (after Yu and Coates, 1970, reproduced with permission of the Department of Supply and Services, Canada)

resulting from the total loading of the system, then the strains induced and hence the stresses within each element.

In a study of a rock mass, a true scale cross-section through it is drawn and the input data needed are the distribution of rock densities and the elastic parameters, v and E. Consequently, the stratigraphy and structure must be known for a realistic determination of the stresses. It is common practice to plot the data showing the resulting direction of action and magnitude of the maximum and minimum principal stresses. The length of each bar indicates magnitude of the stress: compressive stresses are indicated by plain bars or converging arrows; tensile stresses are indicated by diverging arrows (see Figure 12.2 as an example).

12.2.3 Examples

1. In their study of the Hochkönig massif, Sturgul *et al.* (1976) set out to test the hypothesis that the stresses in the massif rocks are due to gravitational loading and not to tectonic forces. Measured stress orientations were available from the Mitterberg copper mine situated in the Palaeozoic basement beneath the calcareous rocks of the massif (Figure 12.2). The finite-element analysis confirmed that the measured and calculated stresses have a similar orientation and hence that the stress pattern is that which would develop from gravitational loading alone.

The figure also indicates that below the mountain the largest, or compressive stresses, are aligned close to the vertical, but near the flanks of the massif they trend parallel to surface of the ground. The significance of the latter observation is that where the stresses become tensional and of such magnitude that they exceed the tensile strength of the rock, then joints will open parallel with the rock face and give rise to slab failures. The figure also shows that tensional stresses are concentrated near the peaks of the mountain: tensile failures in such areas have been described in several papers as being responsible for opening of joints, the formation of elongated miniature graben known as 'sackung', and small scarps aligned parallel with mountain ridges (e.g. Gerber and Scheidegger, 1969).

The dominant vertical stress direction in the basement beneath the mountain is progressively transformed into a dominant horizontal stress pattern below the lowlands to the north and south of the massif. If deep and rapid incision by streams or ice were to occur in such localities then it is probable that the stresses imposed by the gravitational body forces of the massif would control joint opening and the details of valley landform development. This study shows also an example of the effect of a particular rock unit on stress patterns. The strata shown on the geological cross section by parallel-line shading are of greywacke with a low value of Young's modulus and a Poisson's ratio of 0.34. These rocks have a tendency to swell transversely to a higher degree than those above and below them, and show this in their stress pattern.

2. Studies by Yu and Coates (1970) and Stacey (1973) of stress induced by excavating open pit mines with a range of depths, shapes and volumes and with rocks with a range of physical properties, indicate the value of the finite-element model when used as a simulator of a large range of slope conditions. In Figure 12.3, the effect of varying the coefficient of lateral earth pressure alone is shown to be very great. The dominant compressive stresses are nearly vertical in (a), although parallel to the hillslope; but in (b) are more nearly parallel to the ground surface at shallow depths, and are of large magnitude at the toe of the slope. In (c) it is shown that a tensile zone has formed near the face of the slope. The depth of this zone into the rock is proportional to the value of K_0. The stress pattern and magnitude has a major effect, through joint propagation and opening, on the stability of rock slopes.

In a separate study Savage et al. (1985) have shown that large horizontal stresses develop under valleys, but decrease and become compressive at low values of Poisson's ratio. It is improbable that the bedrock could withstand large tensile stresses and would heave upwards and fracture, so increasing the effectiveness of erosion processes.

3. It has been postulated by Gerber (1980) that many features of glaciated mountains owe much of their form to stress patterns in their rocks, rather than to the erosional processes which are mostly effective as removers of prefractured rock. Testing of this hypothesis will involve extensive measurement of in situ stress and elastic properties of rocks by geomorphologists.

In studies of glaciated mountains of New Zealand it has been found (Augustinus, in preparation), from finite-element modelling, that high horizontal stresses are generated at the bases of valley-walls of glacial troughs. The magnitude of these stresses increases with the angle of the walls. Consequently, where the tensile strengths of the rocks forming the walls are exceeded, and pre-existing tectonic joint sets and other weaknesses exist, joints are formed, extended and opened. The rock mass is thus subdivided by joints into blocks which can be quarried by glaciers. The characteristic overdeepened form of glacial troughs owes much to the stresses within the rock mass. The high strength of the rocks forming the trough walls prevents non-glacial processes from reducing their slope angle and so contributes to the maintenance of the trough form.

12.3 GEOMECHANICAL PROPERTIES OF VOLCANIC ROCKS

In the central North Island of New Zealand, an active volcanic zone, late Quaternary volcanic rock units have been formed without modification by tectonism or burial. Their original extrusive forms such as lava sheets, domes and cones, have been modified by fluvial erosion and slope processes to leave the rock exposed in forms which express the properties of the rock. In the following examples, a brief account is given of the currently accepted mechanisms of the emplacement of the volcanic rocks, followed by a summary of the measurements we have made of their geomechanical properties and our interpretation of the influence that these mechanisms and properties have on rock slopes. We regard this work as a first step in the construction of a model, or series of models, which will relate emplacement mechanisms, rock properties and landforms.

12.3.1 Formation of Ignimbrites

Ignimbrite has recently come to be defined as all those deposits, whether welded or not, which are composed primarily of pumiceous material and show evidence of having been emplaced from a hot and concentrated particulate flow (a pyroclastic flow) (Walker, 1983). This genetic definition means that the term 'ignimbrite' can be applied to a wide range of geological materials ranging from loose, clast-supported pumice deposits through to fine-grained, densely welded rock. Such a wide diversity of materials leads to a great variety of geotechnical properties and hence to a range of geomorphic expressions.

A pyroclastic flow is composed of discrete pieces, or clasts, of juvenile magmatic material plus surrounding country rock. These clasts are buoyed up by gases derived from ingested air or exsolved from the clasts themselves. Such a flow can travel very readily over large distances as its movement is virtually frictionless. Two models of the initiation of a pyroclastic flow have been developed (Figure 12.4) and are discussed below.

Model A is a high eruptive energy version in which the initiation of the pyroclastic flow is preceded by a Plinian eruption column. Such a column is believed to occur in two stages (Sparks and Wilson, 1976); a lower gas thrust phase in which the energy of the eruption lifts the pyroclasts to heights of 1.5–4.5 km, and an upper convective phase which can transport the material to heights of up to 55 km (Wilson *et al.*, 1978). In the gas thrust phase air is ingested into the column, lowering the density of the suspension. If at the top of this phase the suspension is denser than the atmosphere then convection cannot occur and the entire mass collapses back to the ground surface under gravity. The great height of the eruption column provides the energy for rapid radial dispersion away from the eruption site as a pyroclastic flow.

Such a collapse will result in the development of a very high energy, highly fluidized pyroclastic flow capable of covering vast areas and largely unconstrained by the topography. The product of such a flow will be a thin, unwelded deposit which blankets the entire land surface but which is thickest in the valleys and thinnest on the ridges. The Taupo Ignimbrite of New Zealand is an example of such a deposit which consists largely of unwelded pumice clasts and glass shards and shows considerable sorting of the deposit (Wilson and Walker, 1985).

Model B applies to much more voluminous ignimbrites. The model is essentially similar to that above, but a much lower energy eruption and greater rate of material discharge

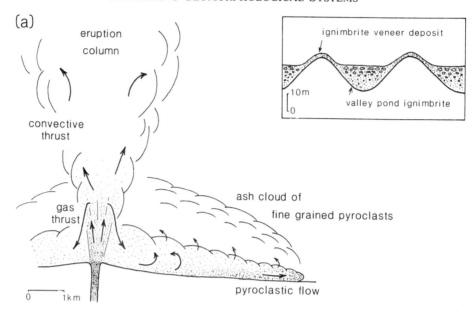

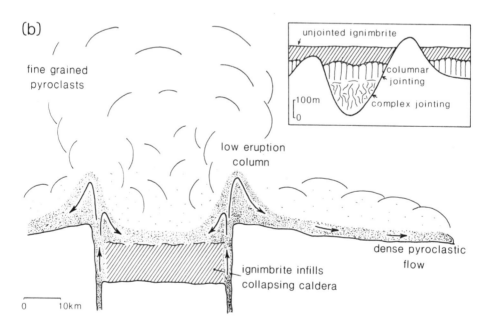

Figure 12.4. Models for the initiation of pyroclastic flows. (a) The collapse of a Plinian eruption column produces a fast moving, highly fluidized pyroclastic flow and a relatively thin, unwelded ignimbrite. (b) Low, continuously collapsing eruption columns produce hot, poorly fluidized pyroclastic flows and thick, welded plateau ignimbrites

may prevent the development of a Plinian eruption column at the start of the eruption (Lipman, 1986). The eruption will involve a relatively low, continuously collapsing pyroclastic fountain (McPhie, 1986). Such eruptions are frequently associated with caldera development and therefore may not have a unique vent, but may form along a series of vents or arcuate fissures marking the caldera ring faults (Hildreth and Mahood, 1986).

The pyroclastic flows produced will be poorly fluidized and relatively slow moving. They will thus be controlled by the topography and will conserve heat much more efficiently than the flows of model A (Walker, 1983; McPhie, 1986). Although these hot, poorly fluidized pyroclastic flows are largely restricted to the valleys, the quantity of material produced is such that they inundate the land surface, leaving a plateau which covers all but the highest peaks.

In practice these two models are not separate mechanisms, they define a spectrum of ignimbrite types which lie between the two end members described. The eruption and flow conditions, and hence the resulting ignimbrite, cover a wide range of types which are dependent upon the initial magmatic and eruption conditions.

12.3.2 Geomechanical Properties of Ignimbrites

Corresponding to the wide range of eruptive mechanisms responsible for the production of ignimbrites is an equally wide variety of physical properties. A number of geomechanical properties have been measured from various ignimbrite sheets from the central North Island, New Zealand. These are summarized in Table 12.2. In this table the relative position of each sample within the ignimbrite sheet is given as 'lower', 'middle', or 'upper'. This gives merely a relative height and does not necessarily imply separate flow units, as the material was sampled on the basis of physical properties rather than inferred stratigraphy.

The welding of the ignimbrites in Table 12.2 refers to the extent to which the glassy components have coalesced after deposition of the material. This depends primarily upon the heat of the flow and the amount of compaction the material is subjected to. In this table welding is assessed qualitatively; 'poor' refers to material which can readily be dug with a spade or pick-axe, 'slight' to material into which the point of a geological hammer will be deeply imbedded with a sharp blow, 'moderate' to material easily broken with a geological hammer and 'dense' to material harder than this. From the table it can be seen that the welding within one ignimbrite may vary from dense to poor. This variation commonly occurs vertically within a sheet, and may also occur with lateral variations in the flow.

Three common types of jointing are recognized: complex jointing, columnar jointing and no jointing. Typically ignimbrites are said to exhibit columnar jointing with regularly spaced vertical joints (Marshall, 1935). These joints are generally zones of several closely spaced joints rather than a single, clearly defined, joint plane and tend to define wide (3 to 5 m) curved columns with domed tops. However, the width is variable and may be as little as 30 cm in places (Figure 12.5). Complex jointing is also very common. In this case the vertical joints become distorted and closely spaced. They tend to have variable strike directions and thus form complex joint blocks (Figure 12.6). Ignimbrite for which the joint spacing is so great (generally greater than 10 m) that the outcrops are too small to give a reliable estimate of the joint spacing are classified as having no jointing. An example of such an unjointed ignimbrite is shown in Figure 12.7.

Table 12.2. Summary of selected geotechnical properties for a number of ignimbrites from the Central North Island, New Zealand. Data from Hind (1986) and Idral (1986)

Ignimbrite sheet	Relative position	Welding	Jointing	Porosity (%)	Slake (%)	Durability classif.	Moisture content	Compressive σ_c (MPa)	Strength softening factor	$c(c')$ (MPa)	Shear strength softening factor	$\phi(\phi')$ (°)
Whakamaru	Upper	Poor	None	44±1	30±1	Very low	Saturated	0.23±0.01	3.0	0.07	1.7	29
							Oven-dry	0.7±0.1		0.12		30
Whakamaru	Lower	Dense	Complex	17±0.1	99±1	Very high	Saturated	26±3	1.7	9	1.4	34
							Oven-dry	44±5		13		35
Ongatiti	Upper	Slight	Columnar	43±1	73±2	Medium	Saturated	0.8±0.1	6.3	0.24	6.6	34
							Oven-dry	5±1		1.59		33
Ongatiti	Middle	Moderate	Columnar	34±1	90±1	Med/High	Saturated	3.8±0.5	3.7	1.95	3.4	35
							Oven-dry	14±2		6.55		32
Ongatiti	Lower	Moderate	Columnar	20±1	98±1	Very high	Saturated	19±3	2.2	3.37	2.3	34
							Oven-dry	42±4		7.79		33
Owharoa	Middle	Moderate		26±1			Saturated	14±3	1.8	3.7	2.1	34
							Oven-dry	25±3		7.8		28
Owharoa	Lower	Dense	Complex	13±1			Saturated	32±5	1.8	9.0	1.5	33
							Oven-dry	56±6		13.3		34
Waimakariri	Upper	Poor	Columnar (Case-hardened)		15±2	Very low	Saturated	2±1				
							Oven-dry					
Waimakariri	Lower	Moderate	Columnar	50±5	93±2	Med/High	Saturated	3.1±0.1	1.6	1.2		26
							Oven-dry	5±1				
Waimakariri	Lower	Dense	Complex/ Columnar	10±1	98±1	Very high	Saturated	32±2	2.5	8.8		35
							Oven-dry	79±3				
Grey	Upper	Poor	Columnar	49±2	19±2	Very low	Saturated	1	4.5			
							Oven-dry	4.5±0.2				
Grey	Lower	Moderate	Complex	31±1	95±1	High	Saturated	12±1	2.5	4.6	1.4	26
							Oven-dry	30±3		6.3		34
Waiteariki	Lower	Dense	Complex	17±1	97±1	High	Saturated	14.5±0.5	2.8	4.8	2.7	32
							Oven-dry	41±1		12.9		33

Figure 12.5. Well developed columnar jointing in the Whakamaru Ignimbrite. The columnar joints are typically curved with rounded tops. In this example the joints are very closely spaced (~ 30 cm), but more commonly the joint spacing may be up to 3 to 5 m

Figure 12.6. Complex jointing in densely welded ignimbrite. The predominantly vertical jointing is strongly curved and has variable strike directions forming complex joint blocks

Figure 12.7. A cutting in poorly welded Waiteariki Ignimbrite. Joints in this cutting are so widely spaced that the material is classed as 'unjointed'

The porosity of ignimbrites is related to the degree of welding. The material is generally very porous compared with crystalline rocks and spans a wide range of actual porosities. The densely welded Whakamaru Ignimbrite, for example, has a porosity of 17 per cent, whilst its poorly welded counterpart has a porosity as high as 44 per cent. This compares with granite which typically may have a porosity of <1 per cent, sandstone for which the porosity may range from 5 per cent to 25 per cent or granular soils which may range from 20 per cent to 50 per cent (Farmer, 1968, 1983).

Slake durability is a measure of the resistance of the material to disintegration when subjected to repeated cycles of wetting and drying plus mechanical abrasion. The slake durability index (I_d) refers to the percentage of the original mass remaining after a standard slaking cycle. In Table 12.2 the second-cycle (I_{d2}) index is given which is the percentage of sample remaining after two complete cycles of wetting and drying. Striking results are noted for these ignimbrites in that there are two distinct forms of behaviour; most are extremely resistant to abrasion with 90 per cent or more of the material remaining after two cycles (generally high to very high (Brown, 1981)), however a number show an almost complete breakdown of the rock fabric ($I_{d2} < 30$ per cent, very low (Brown, 1981)). Only the slightly welded Ongatiti Ignimbrite falls between these two extremes ($I_{d2} = 73$ per cent).

The uniaxial compressive strength (σ_c) refers to the peak compressive stress which can be applied to a cylinder of rock before failure is induced. The ignimbrites tested span a range of two orders of magnitude from 0.23 MPa (Whakamaru, upper) to 79 MPa (Waimakariri, lower). These values range from a very low strength rock to a medium strength rock in the classification of Deere and Miller (1966), and compare with granite, $\sigma_c = 100$–300 MPa, sandstone, $\sigma_c = 20$–200 MPa, or coal, $\sigma_c = 5$–50 MPa (Farmer, 1968, 1983). Typically however, ignimbrites are low to very low strength rocks with $\sigma_c < 40$ MPa. Also of note in these values is that it is very common for one ignimbrite sheet to have units for which the strength varies by one order of magnitude (Ongatiti, Waimakariri and Grey Ignimbrites) or even two orders of magnitude (Whakamaru Ignimbrite). Likewise, the ignimbrites show marked variations in strength between oven-dry and saturated samples. A softening factor ($\sigma_{c,dry}/\sigma_{c,sat}$) of approximately 2 is typical, and in very porous samples the values reach 4 or even 6. This suggests that effective stresses due to water filled pores are very significant.

The values of cohesion (c, c') and angle of internal friction (ϕ, ϕ') refer to the strength of the material under a shear stress. Values of the shear strength (τ) were measured for various values of applied normal stress (σ_n). Plotting these on a graph of τ versus σ_n gives the cohesion as the y-intercept and friction angle as the slope of the best-fit straight line. The values given in Table 12.2 show consistent values of the friction angle of 30–35° for all the ignimbrites tested; these values do not appear to be significantly affected by the moisture content of the rock. The value of cohesion varies from 0.07 MPa (Whakamaru, upper) to 13 MPa (Whakamaru/Owharoa, lower). The cohesion increases with increased welding and the saturated effective cohesion is less than the dry cohesion (average softening factor = 2.7). These values compare with granite $c = 9.5$–40 MPa, $\phi = 51$–58°, sandstone $c = 4.1$–34 MPa, $\phi = 48$–50° (Gardiner and Dackombe, 1983), granular soils $\phi = 30$–40°, or drained clays $c = 0.01$–0.03 MPa (Lambe and Whitman, 1979). The ignimbrites tested have low friction angles compared with other rocks and friction angles comparable to those of soils; and low values of cohesion compared with other rocks, yet high values of cohesion compared with soils.

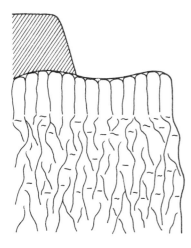

upper zone of poorly welded ignimbrite; unjointed; low strength and very low resistance to abrasion

central zone of moderately welded ignimbrite; vertical columnar jointing; medium strength and high resistance to abrasion

lower zone of densely welded ignimbrite; complex, curved, predominantly vertical joints with a few horizontal joints; high strength material with a high resistance to abrasion

Figure 12.8. Vertical zones of welding and jointing within an idealized, thick ignimbrite sheet

12.3.3 Hillslopes on Ignimbrite

A model for slope formation within an ignimbrite may be developed by considering an ideal sheet of uniform thickness. At the base of the sheet, heat storage is at a maximum, the material cools very slowly and becomes densely welded. Higher in the sheet the degree of welding decreases progressively until near the top the material may be totally unwelded. Taking a profile through such a sheet gives a series of zones as shown in Figure 12.8.

In the lower portion of the sheet the ignimbrite is densely welded and shows complex jointing. This jointing may develop as the ignimbrite cools and plastic deformation of the compressed material causes distortion of the vertical cooling joints, or may partially represent the release of tensile stresses as the rock is exposed by erosion. Due to the dense welding the rock has low porosity and high strength, together with a high resistance to abrasion. This high strength means that the jointing has a major control on the cliff form with most erosion occurring through the failure of joint blocks. This results in steep, almost vertical cliff faces with curved outcrops (Figure 12.6) which frequently retreat by exfoliation of platy joint blocks. The boundary between the lower and central zones is generally gradational.

In the central zone the columnar jointing of the ignimbrite becomes apparent. The material is of medium strength and porosity, but is still highly resistant to abrasion of the intact rock. Again predominantly vertical joint blocks control the form of the outcrop, with vertical cliff faces predominant (Figure 12.5). However, in this case the jointed zones defining the columns are zones of lowered resistance to abrasion and the columnar jointing pattern becomes accentuated by water percolating down the joint planes which become widened. Secondary precipitation may also occur along joint planes, such as the case-hardening from silica precipitation observed in the Waimakariri Ignimbrite. This hardening also serves to accentuate the columnar jointing pattern.

The boundary of this columnar jointed zone and the overlying poorly welded zone tends generally to be quite marked, with the welding grading from moderate to poor over a vertical distance of several metres or less. However, the boundary forms a very uneven surface ranging over vertical distances of 20 m or more. The columnar joints appear to coalesce into dome forms at their upper surface. Such domes may be from less than 30 m to greater than 100 m across, with areas of poorly welded material of similar extent between them.

(a)

(b)

(c)

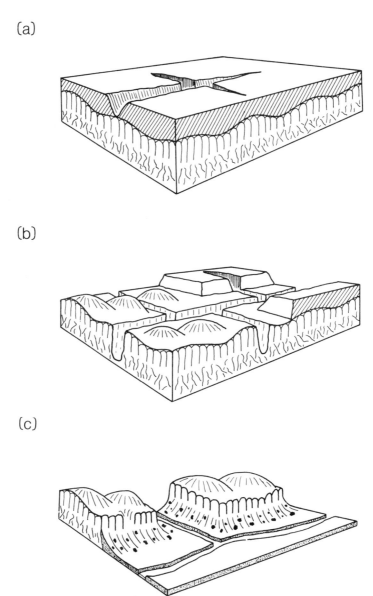

Figure 12.9. Stages in the erosion of an ignimbrite sheet. (a) Rapid erosion of steep gullies in the upper poorly welded zone. (b) Stripping of poorly welded material exhumes a rounded topography on the columnar jointed zone. Narrow gorges develop in the welded ignimbrite. (c) Valley widening leaves rounded hills, steep, jointed faces and blocky talus slopes

At the top of the profile is a zone of poorly welded material which is characterized by very few joints, high porosity, low strength and the tendency to completely break down under wetting and drying and mechanical abrasion. This zone is prone to physical weathering and erodes very easily, often showing extensive gully development. However, the material is capable of maintaining surprisingly steep slopes (Figure 12.7). With its low compressive strength and lack of joints the material comes to behave more like a soil than a hard rock. Under these conditions the shear strength has a major control on the stable slope angle. The relatively high cohesion of the poorly welded ignimbrite (compared with soils) allows the material to maintain steep slopes; a simple infinite slope analysis (Lambe and Whitman, 1979) for the poorly welded Whakamaru Ignimbrite in Table 12.2 shows that the dry material is able to maintain heights of approximately 35 m as vertical slopes whilst the saturated ignimbrite can support approximately 10 m high vertical cliffs.

An overall model of landform development for an ignimbrite sheet is shown schematically in Figure 12.9. Initially the upper surface is rapidly dissected by steep gullies resulting from the tendency of the superficial material to erode readily but maintain steep faces (Figure 12.9(a)). As this material is stripped away an erosional bench representing the boundary between the poorly and moderately welded ignimbrite is exposed (Figure 12.9(b)). This bench has an uneven surface reflecting the variation in the depth to welding as the columnar joints coalesce into dome forms at their upper surface (Figure 12.10). Below this the high strength and predominantly vertical jointing of the welded ignimbrite leads to very steep, precipitous cliffs and narrow gorges. These cliffs retreat by the collapse of joint blocks and so maintain vertical faces as the valleys widen. Talus slopes develop at the base of these cliffs; the talus consists of fallen joint blocks in a grus of sandy fragments produced by granular disintegration of the rock. The resulting landform units are: an upper, rounded eroded surface; vertical, jointed face; and blocky lower slope (Figure 12.9(c)). Continued degradation of these slopes may result in the complete masking of the jointed rock faces, producing a typical rounded ignimbrite topography as shown in Figure 12.11.

This idealized model assumes that a thick flow was erupted onto a flat surface. As discussed in section 12.3.1 there is a wide variety of ignimbrite emplacement conditions. However, each ignimbrite will contain one or more of the ideal units described above and will display the landforms associated with the units present. Where the ignimbrite is erupted onto a steep underlying topography the vertical variations will have complex lateral variations superimposed. This results in more complicated landforms in which erosional benches develop at various levels representing lateral changes in welding.

12.3.4 Rhyolite Emplacement Modes

Rhyolites are defined as silica-rich volcanic lavas ($SiO_2\% = 70$–75), that are emplaced in a variety of forms which are dependent on viscosity, temperature, water content, crystal and vesicle contents of lavas (Mueller and Saxena, 1977; Best, 1982; Fink, 1983; 1984).

Two models of emplacement of rhyolite lavas, are illustrated in Figure 12.12:

(a) Lava flow-units;
(b) Domes.

The characteristic stratigraphy of model (a) for each lava flow-unit comprises: a pre-flow basal tuff; basal breccia made up of angular chilled clasts in a vesiculated fine-grained matrix; a foliated interior; and surficial upper breccia (Christiansen and Lipman, 1966;

Figure 12.10. Erosional surfaces on the Whakamaru Ignimbrite. The bench in the middle distance represents the boundary between the lower, complexly jointed zone and the columnar jointed zone. Above this are vertical outcrops of columnar jointed ignimbrite with talus slopes at their base. These outcrops typically have domed upper surfaces representing the welding boundary between the moderately and poorly welded zones. The trees in the background are on the poorly welded ignimbrite

Figure 12.11. This rounded topography, reflecting variations in the degree of welding, is typical of eroded ignimbrite. The welding differences are the result of both vertical and lateral variations in the emplacement conditions

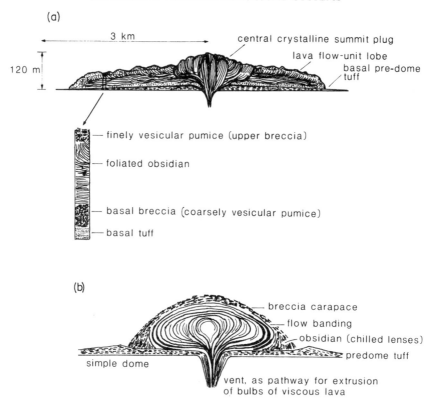

(a)

3 km

120 m

central crystalline summit plug
lava flow-unit lobe
basal pre-dome
tuff

— finely vesicular pumice (upper breccia)

— foliated obsidian

— basal breccia (coarsely vesicular pumice)
— basal tuff

(b)

breccia carapace
flow banding
obsidian (chilled lenses)
predome tuff

simple dome

vent, as pathway for extrusion
of bulbs of viscous lava

Figure 12.12. (a) Domes composed of lava flow-units (based on Fink, 1983); (b) Domes composed of viscous rhyolite lavas

Fink, 1980a, 1983). The interior of the flow is insulated by the enveloping chilled breccia which is a zone of rapid heat loss. Lava flow-unit lobes may comprise domes such as Little Glass Mountain and Inyo dome, northern California, U.S.A. (Fink, 1983; Fink and Manley, in press) (Figure 12.12).

Alternatively rhyolite may be extruded as viscous bulbs (Figure 12.12(b)) which build complex domes (cumulo domes) (Cole, 1970), simple domes (Eichelberger *et al.*, 1986) or plugs (Williams, 1932; Williams and McBirney, 1979).

Two examples from Pauanui and Onemana, Eastern Coromandel Peninsula, New Zealand, have been chosen as representative of the above emplacement models. The textural diversity of the lithologies represented in models (a) and (b) produce a great variety of geotechnical properties, and hence have a range of geomorphic expressions.

12.3.5   Geomechanical Properties of Rhyolites

A number of geomechanical properties have been measured for the two rhyolites at Pauanui and Onemana. Pauanui comprises a succession of lava flow-units and lobes of rhyolite and dense glassy rhyolite (known as pitchstone) forming a dome 387 m high and ~2 km

Table 12.3. Comparisons and contrasts between Pauanui and Onemana

Locality	Pauanui	Onemana
Landform	Large dome with coulees (flow lobes), made up of a succession of lava flow units	Eroded cumulo dome, bulbs of extrusive lava
Dimensions	Height 387 m Width ~2 km	*Estimated* ~150 m 800 m
Rock type/texture	Dense flow banded rhyolites (also brecciated)—mainly flow banded and glassy pitchstone textures	Vesicular rhyolite with large voids (av. ~10 mm diameter— up to 100 mm)
Joint spacing	Medium	Wide
Block geometry	Diamond shaped columns or angular plates	Irregular wide rectangular columns
No. joints m^{-3}	7–20	0.8–1.8

diameter (Stevenson, 1986). Onemana comprises eroded dome lavas. Both localities have exposures of rock in coastal cliff and wave-cut platform sections.

The differences between the geomechanical properties of rocks of the two areas are summarized in Table 12.3 and properties are listed in Table 12.4. Under uniaxial compression, Pauanui pitchstones suffer violent cataclastic failure (Hawkes and Mellor, 1970), which is illustrated in Figure 12.13, because their strength is controlled by a fine-grained glassy and devitrified groundmass of high density and low porosity. Microcracks radiate from scattered crystals and promote explosive brittle failure. These dense pitchstones ($\rho_{dry} = 2.50$ Mg m^{-3}) form columnar cliffs with diamond-shaped cross-sections cut by sets of parallel intracolumn fractures of possible cooling origin. Lower strength Pauanui rocks fail by partial shear, parallel to flow-band partings.

Onemana rocks, typically of low density ($\rho_{dry} = 1.66$–2.14 Mg m^{-3}) and high porosity (4.7–8.9 per cent) relative to the Pauanui rhyolites, have low peak strengths of 8.3–23 MPa. Under compressive stress, the core specimens fail by partial shear and slow crumbling caused by cracking between large voids.

12.3.6 Hillslopes in Rhyolite

Fresh domes comprising lava flow lobes have a mushroom-shaped profile with steep avalanche breccia edges. Eroded dome profiles form in response to joint spacing and rock intact strength.

A slope formation model may be conceived by considering an ideal flow-unit of uniform thickness. A basal breccia cools rapidly, compared with an insulated centre that remains plastic for a longer period of time. The upper part of the flow-unit, the flow carapace, breaks in response to the still moving lava below, becoming finely brecciated and vesiculated. Columnar joints may either form throughout the flow-unit or in the region of greatest heat loss (the basal breccia); they develop perpendicular to the cooling lava surface. Joint surfaces are smooth and annealed by escaping heat. An eroded slope profile develops in response to rock mass strength, joint spacing and intact strength (Selby, 1980). For example, rock of unit F (Figure 12.14) near the base has the lowest porosity, highest density and

Table 12.4. Geotechnical properties

	Density (Mg m^{-3})		Porosity %	Voids %	Glass content %	σ_c (MPa)
	ρ_{sat}	ρ_{dry}				
Pauanui rhyolites						
dense rhyolite	2.57 ± .01	2.55 ± .01	.93 ± .17	2.45	45.6	127.8
flow banded rhyolite	2.49 ± .01	2.44 ± .02	2.16 ± .44	5.16	20.8	124.8
red foliated and brecciated pitchstone	2.38 ± .02	2.33 ± .02	2.31 ± .25	5.38	53.7	62.1
brown foliated pitchstone	2.47 ± .02	2.46 ± .02	.57 ± .28	1.36	39.5	170.2
grey pitchstone	2.49 ± .01	2.49 ± .01	.35 ± .07	0.85	23.1	244.5
dark green pitchstone	2.50 ± .01	2.50 ± .01	.15 ± .04	0.35	39.6	240.3
Onemana rhyolites						
pink foliated vesicular rhyolite	2.16 ± .06	2.05 ± .06	5.07 ± .77	10.39	12.2	23.0
pink vesicular rhyolite	2.03 ± .04	1.85 ± .09	8.93 ± .39	16.75	15.9	8.3
grey void-rich rhyolite	1.19 ± .05	1.66 ± .06	7.85 ± .51	13.0	24.2	13.0

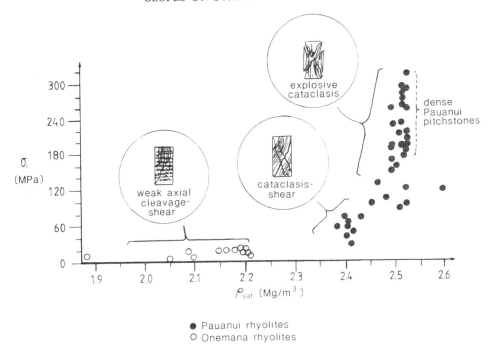

● Pauanui rhyolites
○ Onemana rhyolites

Figure 12.13. Uniaxial (unconfined) compressive strength versus saturated density for Pauanui and Onemana rhyolites with failure modes superimposed

strength. Both units E (a basal breccia with lenticular clasts and an annealed glassy surface rim) and F have steep slopes (76–82°) and medium joint spacing. The central unit G, characterized by tight flow folds, has some joints that appear to offset foliation, indicating that they formed while the lava was still partially mobile and cooling. Consequently the close joint spacing gives rise to a lower slope angle (48°). Above G, unit H is finely brecciated and vesiculated. Its relative lack of jointing ($\bar{J}_{sp} = 0.804$ m) enables steeper slope angles to be maintained, even though its density and intact strength are low relative to other units.

Similar properties and development can be attributed to upper flow unit I, which has a brecciated base and top, a foliated and dense interior, columnar joints extending throughout the flow and large breccia pods at its centre. The breccia pods may have occurred in response to gravitational instability and the rising of diapirs of less dense breccia, as described by Fink (1980b).

Onemana dome rhyolites comprise eroded steep-sided coastal headlands with irregular widely-spaced vertical columnar joints. The slope profile is largely controlled by the inclination of the major column-defining joints, flow-band inclination, and, to a lesser extent, secondary stress-release jointing and salt weathering.

Figure 12.15(a) illustrates an idealized section through a developing dome with bulbs of lava being extruded from opening vent fissures. As the bulbs come into contact with cooler lava, their surfaces chill, becoming glassy and trap large gas and liquid filled voids of up to 100 mm diameter. The voids form in this layer, prior to the cessation of

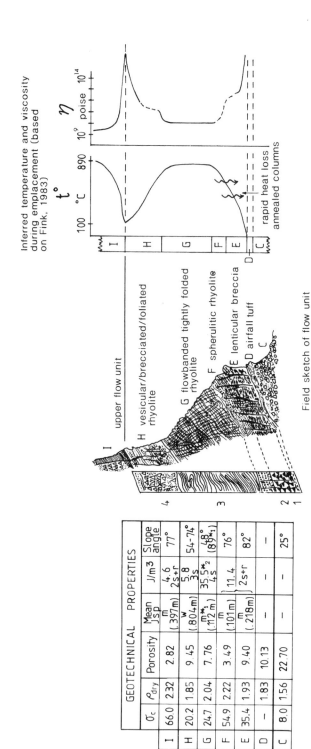

Inferred temperature and viscosity during emplacement (based on Fink, 1983)

rapid heat loss
annealed columns

Field sketch of flow unit

I   upper flow unit
H   vesicular/brecciated/foliated rhyolite
G   flowbanded tightly folded rhyolite
F   spherulitic rhyolite
E   lenticular breccia
D   airfall tuff

GEOTECHNICAL PROPERTIES					
	σ_c	ρ_{dry}	Porosity	Mean J$_{sp}$	J/m3	Slope angle
I	66.0	2.32	2.82	m (.397m)	4.6 2s+r	77°
H	20.2	1.85	9.45	w (.804m)	5.8 3s	54-74°
G	24.7	2.04	7.76	m*1 (.112m)	35.5*2 4s	48°(89°*1)
F	54.9	2.22	3.49	m (.101m)	11.4	76°
E	35.4	1.93	9.40	m (.218m)	2s+r	82°
D	–	1.83	10.13	–	–	–
C	8.0	1.56	22.70	–	–	25°

Figure 12.14. Model for slope formation within a lava flow-unit.

Key: Stratigraphic Interpretation
1 airfall tuff
2 basal breccia
3 central foliated rhyolite
4 upper vesicular breccia

Symbols: σ_c, uniaxial compressive strength (predicted from σ_c/point load strength correlations)
ρ, density
J_{sp}, joint spacing; m, medium, w, wide
J m^{-3}, number of joints per cubic metre (joint volumetric index); s, joint sets; r, random set
*1 basal cliff *2 central part of unit

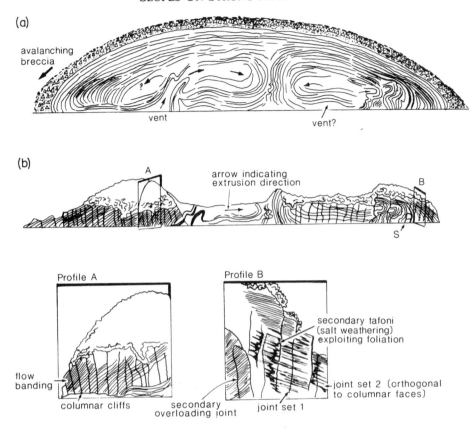

Figure 12.15. Slope model based on Onemana dome rhyolites: (a) Idealized slope profile of an emplacing dome (cross-section); (b) Eroded dome section and profiles, Onemana. Extrusion directions of lava bulbs occur parallel to fold axial planes. Chilled glassy surfaces (S) show striations and contain large oblate voids, providing evidence for rhythmic viscous extrusion of lava from within. Slope profile A is parallel to the seaward flow-band dip. Slope profile B is parallel to column-defining joint sets

extrusive flow, as is evident from their ovoid shape. Surface striations and tensional cracks on the chilled lava bulb foliations suggest viscous and perhaps rhythmic extrusion of lava bulbs during dome emplacement. However, the widely spaced random columnar joints that form in the cooling viscous lava and the flow-band dip appear to control the eroded-dome slope profiles (Figure 12.15(b)).

12.4 CONCLUSIONS

The models which have been found effective in analysing processes within rock bodies on which slopes develop, and for studying controls on slope form, include physical, numerical, geotechnical, and schematic descriptive types. Each type has its own properties, advantages and limitations. At present a full understanding of rock slopes can be achieved only by exploiting all of the available model types.

REFERENCES

Bakker, J. P. and Le Heux, J. W. N. (1952). A remarkable new geomorphological law. *Koninklijke Nederlandsche Akademie van Wetenschappen*, **B55**, 399–410 and 554–571.

Best, M. (1982). *Igneous and Metamorphic Petrology*, W. H. Freeman & Co., San Francisco.

Brown, E. T. (1981). *Rock Characterization, Testing and Monitoring. ISRM Suggested Methods*, Pergamon Press, Oxford.

Brown, E. T. and Hoek, E. (1978). Trends in relationships between measured in situ stresses and depth. *International Journal of Rock Mechanics and Mining Sciences and Geomechanics Abstracts*, **15**, 211–215.

Chowdhury, R. N. (1978). *Slope Analysis*, Elsevier, Amsterdam.

Christiansen, R. L. and Lipman, P. W. (1966). Emplacement and thermal history of a rhyolite lava flow near Fortymile Canyon, Southern Nevada. *Geol. Soc. Amer. Bull.*, **77**, 671–684.

Cole, J. W. (1970). Structure and eruptive history of the Tarawera Volcanic Complex. *N. Z. J. Geol. Geophys.*, **13**, 879–902.

D'Andrea, D. V., Fischer, R. L. and Fogelson, D. E. (1965). Prediction of compressive strength from other rock properties. *U.S. Bureau of Mines R.I.*, **6702**.

Deere, D. U. and Miller, R. P. (1966). Engineering classification and index properties for intact rock. *Technical Report No AFNL-TR-65-116, Air Force Weapons Laboratory, New Mexico.*

Eichelberger, J. C., Carrigau, C. R., Westrich, H. R., and Price, R. H. (1986). Non-explosive silicic volcanism. *Nature*, **323**, 598–602.

Farmer, I. W. (1968). *Engineering Properties of Rocks*, Butler and Tanner, London.

Farmer, I. W. (1983). *Engineering Behaviour of Rocks*, 2nd ed., Chapman and Hall, London.

Fink, J. H. (1980a). Surface folding and viscosity of rhyolite flows. *Geology*, **8**, 250–254.

Fink, J. H. (1980b). Gravity instability in Holocene Big and Little Glass Mountain rhyolitic obsidian flows, Northern California. *Tectonophysics*, **66**, 147–166.

Fink, J. H. (1983). Structure and emplacement of a rhyolitic obsidian flow: Little Glass Mountain Medicine Lake Highland, Northern California. *Geol. Soc. Amer. Bull.*, **94**, 362–380.

Fink, J. H. (1984). Structural geologic constraints on the rheology of rhyolitic obsidian. *Journal of Non-crystalline Solids*, **67**, 135–146.

Fink, J. H. and Manley, C. R. (in press). The origin of pumiceous and glassy textures in rhyolitic flows and domes. In Fink, J. (Ed.), The emplacement of silicic domes and lava flows. *Geol. Soc. Amer. Special Paper*, 212.

Gardiner, V. and Dackombe, R. V. (1983). *Geomorphological Field Manual*, George Allen and Unwin, London.

Gerber, E. (1980). Geomorphological problems in the Alps. *Rock Mechanics, Supplement*, **9**, 93–107.

Gerber, E. and Scheidegger, A. E. (1969). Stress-induced weathering of rock masses. *Ecologae Geologicae Helveticae*, **62**, 401–415.

Gerber, E. and Scheidegger, A. E. (1973). Erosional and stress-induced features on steep slopes. *Z. Geomorph. N. F., Suppl. Bd.*, **18**, 38–49.

Goodman, R. E. (1980). *Introduction to Rock Mechanics*, Wiley, New York.

Gyenge, M. and Coates, D. F. (1964). Stress distribution in slopes using photoelasticity-1. *Canada: Department of Mines and Technical Surveys Mines Branch, Fuels and Mining Practice Division, Divisional Report*, FMP 64/150–MRL, Ottawa.

Hawkes, I. and Mellor, M. (1970). Uniaxial testing in rock mechanics laboratories. *Eng. Geol.*, **4**, 179–286.

Haxby, W. F. and Turcotte, D. L. (1976). Stresses induced by the addition or removal of overburden and associated thermal effects. *Geology*, **4**, 181–184.

Hildreth, W. and Mahood, G. A. (1986). Ring-fracture eruption of the Bishop Tuff. *Geol. Soc. Amer. Bull.*, **97**, 396–403.

Hind, K. J. (1986). *A Geotechnical Investigation of Ignimbrite in the Ruahihi Area, Tauranga*, unpublished M.Sc. Thesis. University of Waikato.

Hoek, E. (1965). The design of a centrifuge for the simulation of gravitational force fields in mine models. *Journal of the South African Institute for Mining and Metallurgy*, **65**, 455–487.

Hoek, E. and Bray, J. W. (1981). *Rock Slope Engineering*, 3rd ed, The Institution of Mining and Metallurgy, London.

Idral, A. (1986). *Geotechnical Properties of Owharoa Ignimbrite*, Unpublished Dip. App. Sci. Report, University of Waikato.

Kohlbeck, F., Scheidegger, A. E., and Sturgul, J. R. (1979). Geomechanical model of an alpine valley. *Rock Mechanics*, **12**, 1–14.

Lambe, T. W. and Whitman, R. V. (1979). *Soil Mechanics, SI Version*, John Wiley and Sons, New York.

Lipman, P. W. (1986). Emplacement of large ash-flow sheets and relation to caldera collapse. *International Volcanological Congress, New Zealand, 1–9 February 1986. Abstracts*, p. 58.

Long, A. E. (1964). Problems in designing stable open pit mine slopes. *Canadian Mining and Metallurgy Bulletin*, **57**, 627.

Marshall, P. (1935). Acid rocks of the Taupo-Rotorua volcanic district. *Transactions of the Royal Society of New Zealand*, **64**, 323–366.

McGarr, A. and Gay, N. C. (1978). State of stress in the earth's crust. *Annual Review of Earth and Planetary Sciences*, **6**, 405–436.

McPhie, J. (1986). Eruption and emplacement of voluminous, very high aspect ratio ignimbrite: a late Permian example, northeastern N.S.W. *International Volcanological Congress, New Zealand, 1–9 February 1986. Abstracts*, p. 59.

McTigue, D. F. and Mei, C. C. (1981). Gravity-induced stresses near topography of small slope. *Journal of Geophysical Research*, **86**, B10, 9268–9278.

Moon, B. P. and Selby, M. J. (1983). Rock mass strength and scarp forms in southern Africa. *Geografiska Annaler*, **65A**, 135–145.

Mueller, R. F. and Saxena, S. K. (1977). *Chemical Petrology with Applications to terrestrial Planets and Meteorites*, Springer-Verlag, New York.

Piteau, D. R. and Associates (1979). *Rock Slope Engineering Reference Manual, Parts A–H*, Federal Highway Administration, Washington, D.C.

Savage, W. Z., Swolfs, H. S., and Powers, P. S. (1985). Gravitational stresses in long symmetric ridges and valleys. *International Journal of Rock Mechanics and Mining Science*, **22**, 291–302.

Selby, M. J. (1980). A rock mass strength classification for geomorphic purposes: with tests from Antarctica and New Zealand. *Z. Geomorph.N.F.*, **24**, 31–51.

Selby, M. J. (1982a). Controls on the stability and inclinations of hillslopes formed on hard rock. *Earth Surface Processes and Landforms*, **7**, 449–467.

Selby, M. J. (1982b). *Hillslope Materials and Processes*, Oxford University Press, Oxford.

Selby, M. J. (1982c). Rock mass strength and the form of some inselbergs in the central Namib Desert. *Earth Surface Processes and Landforms*, **7**, 489–497.

Selby, M. J. (1987). Rock slopes. In Anderson, M. G. and Richards, K. S. (Eds.), *Slope Stability: Geotechnical Engineering and Geomorphology*, Wiley, Chichester, 475–504.

Sparks, R. S. J. and Wilson, L. (1976). A model for the formation of ignimbrite by gravitational column collapse. *Journal of the Geological Society of London*. **132**, 441–451.

Stacey, T. R. (1973). Stability of rock slopes in mining and civil engineering situations. *National Mechanical Engineering Institute, CSIR Report*, **ME1202**, Pretoria.

Stevenson, R. J. (1986). *The Geotechnical Properties of Minden Rhyolite at Pauanui and Onemana, Eastern Coromandel*, unpubl. M.Sc. thesis, University of Waikato, Hamilton, New Zealand.

Sturgul, J. R. and Grinshpan, Z. (1975). Finite-element model for possible isostatic rebound in the Grand Canyon. *Geology*, **3**, 169–172.

Sturgul, J. R., Scheidegger, A. E., and Grinshpan, Z. (1976). Finite-element model of a mountain massif. *Geology*, **4**, 439–442.

Terzaghi, K. (1962). Stability of steep slopes on hard unweathered rock. *Geotechnique*, **12**, 251–270.

Turcotte, D. L. (1974). Are transform faults thermal contraction cracks? *Journal of Geophysical Research*, **79**, 2573–2577.

Turner, M. J., Clough, R. W., Martin, G. C., and Topp, L. J. (1956). Stiffness and deflection analysis of complex structures. *Journal of Aeronautical Science*, **23**, 805–824.

Valliapan, S. and Evans, R. S. (1980). Finite element of a slope at Illawarra Escarpment. *Proc. 3rd Australia–New Zealand Geomechanics Conference*, **2**, 241–246.

Voight, B. and St. Pierre, B. H. P. (1974). Stress history and rock stress. *Third Congress of the International Society for Rock Mechanics, Denver, Proceedings*, **2**, 580–582.

Walker, G. P. L. (1983). Ignimbrite types and ignimbrite problems. *Journal of Volcanology and Geothermal Research*, **17**, 65–88.

Williams, H. (1932). The history and character of volcanic domes. *Univ. Calif. Publ. Geol. Sci.*, **21**, 151–156.

Williams, H. and McBirney, A. F. (1979). *Volcanology*, Freeman Cooper and Co., San Francisco.

Wilson, C. J. N. and Walker, G. P. L. (1985). The Taupo Eruption, New Zealand. *Philosophical Transactions of the Royal Society of London*, **A**, 314.

Wilson, L., Sparks, R. S. J., Huang, T. C. and Watkins, N. D. (1978). The control of volcanic column heights by eruption energetics and dynamics. *Journal of Geophysical Research*, **83**, B4, 1829–1836.

Wuerker, R. G. (1955). Annotated tables of strength and elastic properties of rock. *Transactions of the Australian Institute of Mining Engineers*, **202**, 157.

Yu, Y. S. and Coates, D. F. (1970). Analysis of rock slopes using the finite element method. *Department of Energy, Mines and Resources Mines Branch, Mining Research Centre, Research Report*, **R229**, Ottawa.

Zienkiewicz, O. C. (1971). *The Finite Element Method in Engineering Science*, McGraw-Hill, New York.

Modelling Geomorphological Systems
Edited by M. G. Anderson
©1988 John Wiley & Sons Ltd.

Chapter 13

Modelling landform change

FRANK AHNERT

Geographisches Institut der RWTH, Aachen

13.1 TYPES OF LANDFORM CHANGE AND THEIR IMPLICATIONS FOR GEOMORPHOLOGICAL RESEARCH METHODS

Landforms range in size from the tiny impact craters of raindrops, which have diameters of just a few millimetres, to continental shields several thousand kilometres across. The larger the landform, the longer it usually remains in existence (Figure 13.1). The duration of its existence, in turn, will have considerable influence upon the kind and the number of changes to which it is subjected during its development.

There are two basically different kinds of landform changes. First, there are those that originate within the operating geomorphic system, without any change of the external factors of that system. These internal changes are termed *ensystemic* changes; examples would be the gradual change of a slope form by a particular combination of weathering and denudation processes or the progressive deepening of a stream valley during a long period of approximately uniform tectonic uplift. The second kind of landform changes is generated from outside the operating geomorphic system, by changes in climate or in the rate of crustal movement which alter the nature and/or the intensities of the processes of weathering, denudation, erosion, transport and deposition of rock materials. Those are termed *eksystemic* changes. Examples of major eksystemic changes are the alternations of glacial and interglacial periods during the Quaternary.

An explanation of ensystemic landform changes is based on observation of the existing forms, materials and processes and their interactions. If these components are well enough known, they may also permit predictions of future ensystemic changes. Past eksystemic changes may be explained from relief forms or materials and from a knowledge of past changes of climate or of crustal movements. However, one cannot predict future eksystemic changes by geomorphological methods because such changes are due to factors that lie outside the realm of geomorphology.

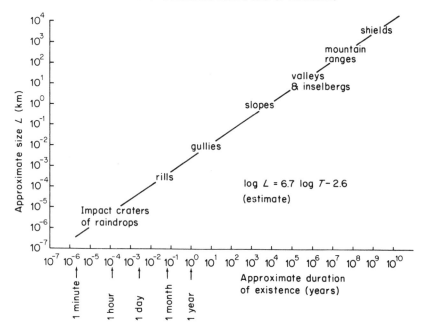

Figure 13.1. An estimate of the relationship between size and 'lifespan' of several types of landforms (modified after Ahnert 1981)

The longer a landform exists, the greater is the probability that it is subjected to eksystemic changes. The relationship between size and lifespan shown in Figure 13.1 is particularly important in this respect. Eksystemic changes have occurred during the long-term development of most medium-sized to large landforms. For a full explanation of their evolution, past tectonic and climatic changes normally will therefore have to be taken into account. On the other hand, it is possible to observe directly, over shorter time periods, the interaction between forms, materials and geomorphic processes of the present geomorphic system with regard only to the latter's ensystemic changes.

This difference expresses itself in the two main methodological branches of geomorphological research, functional geomorphology and historic-genetic geomorphology. Functional geomorphology concerns itself with the relationships between directly observable components of the geomorphic system. Its name indicates that it seeks to describe these relationships in equations that determine the existence or the variation of one component as a function of other components; also, this approach seeks to explain the present 'functioning' of the system.

'Functional' in this context is not the same as 'causal'. A causal relationship is unidirectional, from cause to effect; a functional relationship, however, can exist in both directions due to feedbacks in the system, or it can describe covariations (for example, between properties of form and properties of material) without direct reference to any process linkages. All causal relationships are also functional ones, but not all functional relationships are causal; furthermore, functional geomorphology is not synonymous with 'process geomorphology' (or, as this is sometimes called, 'morphodynamics'), but it contains process geomorphology as one of its parts.

Functional geomorphology and historic–genetic geomorphology both analyse landform evolution; however, functional geomorphology does so in terms of ensystemic changes while historic–genetic geomorphology places the main emphasis upon eksystemic changes. This difference is expressed also in the different use of the concept of time in these two branches. In functional geomorphology, time is physical time, the variable t which is an essential component of all process equations. As a factor, it indicates the duration of a process; as a divisor, it indicates the process rate. In history–genetic geomorphology, however, time is historical time, i.e. the coordinate axis on which the occurrences, the simultaneity and the succession of earth-historical events and periods are ordered. Obviously, an historic–genetic explanation of long-term landform evolution requires the aid of functional–geomorphological methods to interpret the effects of these earth-historical events upon geomorphological processes and, by way of these processes, upon landform development.

13.2 Models of Landform Change

Quantitative theoretical models of landform change are primarily functional–geomorphological. Once formulated, they may be expanded into historic–genetic models by the arbitrary insertion of eksystemic changes, i.e. by arbitrary changes of the constants, coefficients or exponents that control the operations of process equations in the model. This approach was present already in early nonquantitative models such as the equilibrium model by G. K. Gilbert (1877) and the cycle concept by W. M. Davis (1909). The Davisian cycle is a sequence of ensystemic changes and hence a functional–geomorphological model; it gets expanded into an historic-genetic model by the introduction of eksystemic changes through climatic or volcanic 'accidents' or through tectonic 'interruptions'. The semiquantitative slope development model by W. Penck (1924), too, was a functional–geomorphological one; although it included the effects of progressive increases or decreases in the rate of uplift to explain the 'waxing slope' and 'waning slope' forms, these changing rates of uplift were preset and were held constant during the development of the model slope. O. Lehmann (1933) devised a fully quantitative model of rock cliff development by parallel retreat of the slope, while Bakker and Le Heux (1947) simulated slope decline with their model of central rectilinear slope development. These early attempts were based more on geometrical assumptions than on observed process mechanics. They were followed by predominantly process-oriented models: qualitative ones by Jahn (1954) and Ahnert (1954), quantitative ones by Scheidegger (1961), Young (1963), Ahnert (1964), Hirano (1968), Gossmann (1970), Kirkby (1971) and others. Detailed reviews of the slope development models that existed up to 1970 have been given by Young (1972) and by Trofimov (1974). Later models are presented in two volumes edited by Ahnert (1976, 1987b) and by Woldenberg (1985).

Functional-geomorphological models can be designed to simulate *static systems* whose components are properties of forms and/or materials, *process systems* whose components are processes and process rates, or *process–response systems* within which static systems and process systems interact. Most functional–geomorphological models are process-response models. They may be used as research tools to examine several kinds of questions, for example:

1. What is the effect of one individual system component (e.g. the rate of denudation) upon one or several other components (e.g. the regolith thickness, the rate of weathering, or the steepness of the slope)?

2. How does a landform develop under specified process conditions of weathering, of denudation and stream incision?
3. Under what particular conditions of process and material can a given landform type (e.g. inselbergs) develop?
4. What changes of form, of regolith distribution and of other processes result if a particular process changes its intensity by a particular amount?
5. What influence have specified variations of rock resistance upon landform development?
6. Does a particular series of landforms (e.g. of slopes) observed in the field represent a developmental sequence which developed ensystemically under one set of environmental conditions?

The theoretical model approach has the advantage that it can isolate the effects of individual parts of the system, that it permits the extrapolation of observed processes over longer timespans and that it can serve as a means to test hypotheses. In these ways it can supplement the field observations in which the effect of an individual component often is difficult to isolate from the effects of other components; also, field observations cover in most instances too short a time span to determine rates of longer-term development directly.

The comprehensive FORTRAN program of landform development SLOP3D (Ahnert, 1976, 1977) was designed with these considerations in mind. It permits the generation of a great variety of process–response models and will therefore be used here.

13.3   Program Structure

SLOP3D is an expanded version of the earlier program COSLOP2 (Ahnert, 1973). It can be used for two-dimensional (profile) or for three-dimensional (land surface) models. Figure 13.2 is a simplified diagram of its overall structure. An initial profile or land surface is

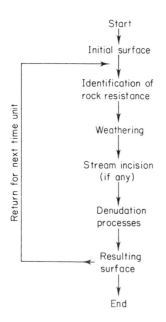

Figure 13.2. Simplified program structure of SLOP3D.

defined by the coordinates (x, y, z) of surface points. Bedrock strata of varying thickness and varying resistance to weathering can also be defined, with any strike and dip, as well as vertical zones of relative weakness or strength (dykes, shatterbelts or zones of densely spaced jointing). The initial surface is subjected to weathering, to stream incision (optional) and to denudation processes which may be selected and combined from the options available in the program; these include splash, suspended-load wash, bedload wash, slow mass movement and a rather primitive, probably inadequate solution routine; debris slides occur when the gradient exceeds a prescribed threshold value.

The denudation processes transfer regolith material from each individual surface point to the next point downslope; only in suspended-load wash is the material transferred directly beyond the slope foot without such point-to-point transfer. Denudation alters the local thickness of the regolith at the surface points and thereby the local surface elevation. This changed surface becomes the 'initial surface' for the next time unit of landform development in which it is subjected again to the work of weathering, stream incision and denudation. Each model time unit thus consists of one passage through the main loop of the SLOP3D program as shown in Figure 13.2.

13.4 Model Processes

Every process in SLOP3D is represented by a process equation which has been formulated on the basis of empirical knowledge and/or general physical principles. The process equations have common variables that link them in such a way that the dependent variable of one equation becomes an independent variable in another. These linkages form several feedback loops so that the components of the system are interdependent as shown in Figure 13.3. Bedrock weathering W produces waste material and thus tends to increase the regolith thickness C, which in turn, however, tends to have a negative influence upon bedrock weathering. Denudational waste removal R increases with increasing g. adient (measured in the downslope direction from the point in question) and has a tendency to diminish the local regolith thickness C. Denudational waste supply A increases with the gradient upslope and tends to add to the local regolith thickness. The latter thus becomes the

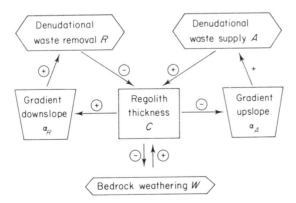

Figure 13.3. Functional interaction between weathering and denudation at a slope point in the SLOP3D process response system (Ahnert 1987b)

interaction centre of the three feedback loops in Figure 13.3. All three feedbacks are negative ones; they reverse the trend of initial changes and thus steer the system towards a state of dynamic equilibrium.

Of all the process options contained in the SLOP3D program, only some will be used here. Fluvial downcutting is programmed as the lowering of the lowest point of the landform at an arbitrarily prescribed uniform gross rate. It tends to increase the local relief and thus constitutes the input of potential energy into the system; it may be considered as an effect of tectonic uplift. However, only the gross rate is arbitrarily set. For most denudation processes in SLOP3D (an exception is suspended-load wash denudation) the net rate of fluvial downcutting equals the difference between this gross rate and the rate of waste supply from the slope. The latter may in some cases exceed the gross rate so that there is actually net aggradation despite continued gross downcutting.

Weathering in SLOP3D is essentially the preparation of bedrock material for denudational removal. It lowers the bedrock surface at the point in question by a specified amount W and adds that same amount to the thickness of the regolith cover. The process equation for mechanical weathering is

$$W_M = W_O \cdot e^{-k_1 C} \tag{1}$$

where W_O is the rate of weathering (amount per model time unit) on bare bedrock ($C = 0.0$), e is the base of natural logarithms, k_1 is a constant that represents properties of the affected rock material, and C is the overlying regolith thickness. The exponential decrease of the mechanical weathering rate W_M with increasing C corresponds to the exponential decrease of the temperature range with increasing depth below the surface (cf. Chang, 1958); it is worth noting that the frequency of freeze–thaw changes, which is a major factor in mechanical weathering, also decreases exponentially with increasing depth.

Chemical bedrock weathering, by contrast, seems to reach its maximum intensity under a certain optimum thickness of regolith which serves to store enough moisture in the contact zone with the bedrock to keep the chemical decomposition of the latter going even during the dry seasons in subhumid and semiarid regions. For regolith thicknesses that are smaller than this optimal value, the rate of chemical bedrock weathering W_{Ch} tends to increase with increasing C; this has been formulated in the program as

$$W_{Ch} = W_O \left(1.0 + k_2 \cdot \frac{C}{C_C} - \frac{C^2}{C_C{}^2}\right) \tag{2}$$

where C_C is a (prescribed) 'critical thickness' and k_2 is a constant.

By differentiating equation 2 and setting the differential equal to zero, one can determine the regolith thickness $C = C_{opt}$ at which the weathering rate W_{Ch} reaches its maximum, namely,

$$C_{opt} = k_2 \cdot \frac{C_C}{2} \tag{3}$$

The maximum rate can then be calculated by substituting this expression for C in equation 2:

$$W_{Ch\,(max)} = W_O \left(1.0 + \frac{1}{4} \cdot k_2{}^2\right) \tag{4}$$

If the regolith thickness is greater than the critical thickness C_C, chemical weathering continues at the rate

$$W_{Ch} = W_O \cdot e^{-k_3(C-C_C)} \tag{5}$$

W_M and W_{Ch} may be combined in any desired proportion according to

$$W = p \cdot W_M + (1.0-p) \cdot W_{Ch} \tag{6}$$

where W is the combined weathering rate and p is a prescribed positive proportionality coefficient smaller than 1.0. The chosen values of p, W_O, C_C, k_1, k_2 and k_3 can serve to simulate bedrock weathering under a wide variety of environmental and lithological conditions.

Of the denudation processes, slow mass movement is modelled as a point-to-point downslope transfer of waste material at a rate (amount per model time unit)

$$R_1 = k_4 \cdot (C \cdot \sin\alpha - k_5) \tag{7}$$

This equation corresponds to the expression used in soil mechanics for plastic flow (cf. Baver et al., 1972). k_4 and k_5 are prescribed input constants—k_4 a 'mobility coefficient' and k_5 a cohesion term that acts as a threshold; movement takes place only when the product $C \cdot \sin\alpha$ is greater than k_5. α is the local slope angle, measured between the point in question and the next point downslope.

The equation for wash denudation is somewhat more complex. It takes into account the results of field experiments on erosion rates, such as those by Musgrave (1935) and by Zingg (1940); according to these, the sediment yield by wash erosion from a test plot is a power function of the slope gradient and of the length of the test plot, measured at right angles to the contours. The exponents of the gradient and of the length in these experiments are commonly in the vicinity of 1.5. The SLOP3D version of wash denudation substitutes $\sin\alpha$ for the gradient, because this is a better measure for the tangential gravitational stress that acts upon a soil particle on a slope, and the runoff depth D for the slope length; of course D is a function of the length, but also of the rainfall intensity and of the infiltration rate, and it represents the relevant conditions more directly than does the length. Accordingly, the rate of wash denudation is

$$R_2 = (k_6 + k_7 \cdot C) \cdot D^{k_8} \cdot \sin^{k_9}\alpha \tag{8}$$

k_6 and k_7 are coefficients that characterize the erodibility of the material. k_7 and C have been included because a thick regolith on a slope may have weathered for a long time and therefore may be especially fine-grained and easily moved, in which case k_7 would be given a positive value, or there may have developed an erosion-resistant surface crust, in which case k_7 would be negative.

The SLOP3D program distinguishes two types of wash denudation. In bedload wash, the transported material is moved per time unit from each slope point to the next point downslope; the values of k_8 and k_9 are made greater than 1.0 (usually 1.5) in this case. In suspended-load the transported material is immediately removed from the model surface without redeposition at any other point. The essential difference between the two processes is that in bedload wash, the transported material that is accounted for at each point comes from the entire tributary area of that point, while in the case of suspended-load wash only that material is identified which is locally removed at the point in question. Consequently, for suspended-load wash the exponent k_8 is usually given the value 0.5 which it would have after differentiation of equation 8.

If the local regolith-covered slope exceeds a predetermined critical gradient (angle of repose), SLOP3D automatically activates debris slides, i.e. immediate downslope transfers of regolith material, until the angle of repose is restored. Bare bedrock slopes, of course, can be steeper; their weathered material then is transported away directly as 'rockfall' by the debris slide mechanism. The program at present does not contain any provision for bedrock landslides.

13.5 LANDFORM CHANGE UNDER UNIFORM ENVIRONMENTAL CONDITIONS (ENSYSTEMIC CHANGE)

If a low-relief denudational landscape is subjected to long-continuing uplift and if during this time the exogenic environment, in particular the climate, remains essentially the same, the landforms tend to change in an orderly manner which is governed by the rate of uplift and by the particular combination of stream erosion, weathering and denudation processes present. The sequence of forms that develop in this way can be simulated in process–response models, with the aim of identifying forms or form elements that are typical indicators of the specific processes or process combinations at work ('characteristic forms' of Kirkby, 1971). Such process–specific model forms can have great diagnostic value for the interpretation of landforms that have developed in the past. Also of interest is the obverse question whether possibly different sets of processes may produce the same landform shape. Such 'equifinality' or 'form convergence' would reduce the usefulness of the 'characteristic form' concept.

The comparison of several SLOP3D model runs with different denudation processes may serve as a theoretical model experiment to test these questions. In each case, the gross rate of fluvial downcutting remains constant for a sufficient time to see whether a 'characteristic form' develops; after that, the downcutting is stopped in order to observe the further form development under conditions of fixed baselevel.

13.5.1 With Slow Mass Movement

The first of these models is one with slow mass movement (*cf.* equation 7) as the denudation process (Figure 13.4). The initial profile is horizontal; stream incision takes place at the left end. In the early phases (until time $T = 101$) a remnant of the flat initial profile is preserved on the divide at the right end of the profile while the left part is strongly convex. This early phase is the phase of profile (or landform) differentiation: different parts of the profile are worn down at different rates so that a new profile shape develops; one might therefore call this also the phase of profile (or landform) transformation. It is a period of adaption of the landform to the new set of processes that acts upon it.

By time $T = 301$, the divide has been lowered too, no remnant of the initial (or palaeo-) form is left, and the entire form is 'recent' in the process-specific sense. At the foot of

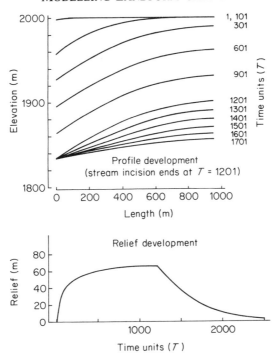

Figure 13.4. Slope development and relief development with slow mass movement. Stream incision takes place at the left end of the profile at a constant gross rate until time $T = 1201$ (Ahnert 1987b)

the slope, the angle has lessened somewhat between times $T = 101$ and $T = 301$. This is due to a change in the net rate of downcutting. At $T = 101$, only about two thirds of the entire profile length supply waste material to the slope foot—the uppermost third has not yet been reached by the new denudation. At $T = 301$, the entire slope supplies waste to the slope foot, and therefore more arrives there than at $T = 101$. Consequently, more of the stream's energy, represented by the constant gross downcutting rate, is used for waste removal from the slope foot at $T = 301$. The net downcutting rate is correspondingly smaller and the footslope less steep, because the downwearing of the slope points can keep up better with the lowering of the slope foot.

Between times $T = 601$ and $T = 1201$, the shape of the profile in Figure 13.4 has become virtually constant, which is merely the morphographic expression of the fact that all slope points are being worn down at the same rate at which the slope foot is lowered by net fluvial downcutting. This is the phase of dynamic equilibrium, or of 'equality of action' (Gilbert 1877) at all slope points. There also is no further change in the relief (or height) of the slope. The equilibrium profile is convex because the sine of the slope is the only independent variable in the denudational process equation (equation 7); in view of the downslope increase of the waste material that has to be transported, net denudation can only be equal at all slope points if the gradient increases also in the downslope direction.

The mechanism for the establishment of such an equilibrium was discussed above in connection with Figure 13.3. This state is sometimes also referred to as a 'steady state'. In order to avoid confusion it is useful to reserve the term dynamic equilibrium to the balance between process rates and the term steady state to the constant form and the constant regolith properties that accompany dynamic equilibrium.

The example of Figure 13.4 shows that a considerable time is needed, and a great amount of profile transformation must take place, before equilibrium is reached. The longer the slope, the more time is required. Since the principal condition for reaching equilibrium is that the gross rate of downcutting (i.e. of uplift) and the exogenic environmental conditions must remain constant for this long 'relaxation' time, there is not much chance that complete equilibrium could ever be reached except at a very local scale and on short slopes; for uplift rates and environmental conditions vary frequently.

However, Figure 13.4 also shows that the profile shape comes close to its steady-state form very early during its development, which means also that the system of interacting processes of stream erosion, weathering and denudation is very early close to equilibrium. In other words: after a change of the process system, the early pace of relief differentiation and of approaching equilibrium is fairly rapid. In practical terms, this means that one has good reason to assume that many landforms are in near-equilibrium condition and may be investigated with the equilibrium concept in mind. The near-equilibrium state of most of the world's landforms is also demonstrated by the predominance of a relatively thin soil cover on the land surface; it indicates that weathering and denudation must occur at very similar rates. If it were not so, there would either be a much larger share of bare bedrock surface (signifying an advantage of denudation over weathering) or a prevalence of very thick soils as a result of long-term excess of weathering over denudation.

The end of downcutting at time $T = 1201$ in Figure 13.4 means that an eksystemic change of process conditions has taken place: the input of endogenic energy that was necessary to maintain the steady state has stopped. Now the profile development enters into its third phase, namely, the phase of relief decline under conditions of fixed baselevel. During this phase the characteristic convex profile form is maintained but its height and steepness diminishes progressively. The three phases of form development are also expressed in the relief development diagram at the bottom of Figure 13.4: the phase of form differentiation with its at first rapid and then slower increase of relief, the phase of approximate dynamic equilibrium with nearly constant relief and the phase of relief decline which also is at first rapid and then asymptotically approaches zero.

13.5.2 With Wash Denudation and with the Combination of Wash Denudation and Slow Mass Movement

Wash denudation requires runoff; the runoff, in turn, is a function of the intensity and the duration of the rainfall that causes it. Rainfall intensity directly influences runoff intensity, while rainfall duration determines the length of slope over which the runoff discharge increases.

Runoff discharge can increase downslope only while the rain is falling. If a rainfall event of runoff-producing intensity lasts longer than the time in which the runoff can flow from the crest to the foot of the slope, the runoff discharge Q, represented in equation 8 by the runoff depth D, increase progressively from crest to foot (provided, of course, that in all segments of the slope the additions by rainfall are greater than the losses by infiltration). The characteristic slope form for wash denudation under these conditions of 'long-lasting' runoff-producing rainfall is concave—for suspended-load wash as well as for bedload wash.

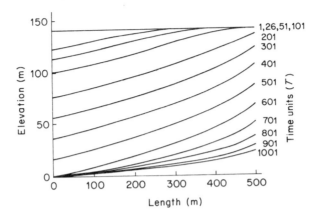

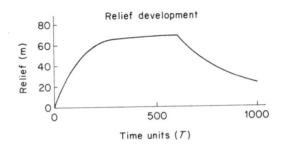

Figure 13.5. Slope development and relief development with bedload wash denudation. Stream incision ends at time $T = 601$

This is demonstrated in Figure 13.5 and Figure 13.6. During the early phase of profile differentiation the profile is usually convex (*cf*. Figure 13.5) because at first the fluvial downcutting is more intensive than the denudational lowering of the slope points. With the steepening of the slope the denudation rate increases until all slope points are being lowered at the same rate, i.e. until the entire slope is in dynamic equilibrium. This state is reached when the product of depth term times slope term in equation 8 is equal at all points of the profile. The concave shape of this equilibrium slope is the geometric expression of the fact that, since D in equation 8 increases from crest to foot, equal lowering at all points is only possibly if the slope term in equation 8 decreases accordingly.

If, however, the typical runoff-producing rainfall events are shorter than the crest-to-foot runoff time, the runoff will increase only part way down the slope, namely, as far as the water from the crest flows until the rain has stopped. Downslope from this point, the runoff discharge during the rain is constant at all points (barring differences of infiltration). After the end of the rain, much of the afterflow infiltrates and runoff ceases. For short rainfall events and on relatively long slopes, water from the slope crest may therefore not at all reach the slope foot and the stream bed that lies there. In such a case, the actual hydrological drainage area of that stream is smaller than its topographical drainage area; it receives water only from that portion of the latter that lies sufficiently close to the stream.

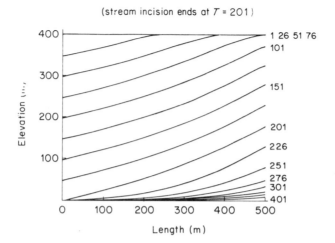

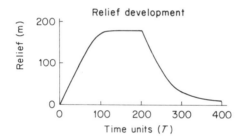

Figure 13.6. Slope development and relief development with suspended-load wash denudation. Stream incision ends at time $T = 201$

This situation, known also as 'partial area contribution of runoff' (*cf.* Yair, Sharon and Lavee, 1978; Richards, 1982, S. 39–41), has important effects upon the shape of equilibrium wash denudation slopes. The concave form described above reaches from the crest only as far downslope as the runoff discharge increases; below this point the equilibrium slope profile is convex in the case of bedload wash denudation (shown in Figure 13.7) because here the increased load to be moved requires—at constant runoff discharge—a corresponding increase in the value of the sine term in equation 8. The shorter the rain relative to the length of the slope (i.e. relative to the running time of the water over this length), the shorter is the upper concave segment and the longer is the lower convex segment. In the field, such concavo-convex profiles have been identified for ephemeral streams by Schumm (1961) and, on wash slopes on the Colorado Plateau, where the downslope runoff decrease was due to piping, by K.-H. Schmidt (1986). For suspended-load wash denudation from short rainfall, the characteristic profile form is concavo-rectilinear.

Another sequence of convex and concave segments in an equilibrium profile is produced by the combination of wash denudation and slow mass movement (Figure 13.8). The wash denudation (here with 'long-lasting' rainfall events) is most effective at the slope foot which is therefore concave; the mass movement dominates on the slope crest which has therefore

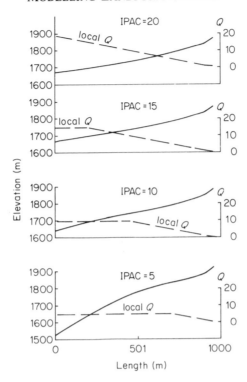

Figure 13.7. Equilibrium slope profiles developed by bedload wash denudation with different characteristic durations of the runoff-producing rainfall events. These profiles are 20 model length units long. Rain duration is controlled by the SLOP3D program constant IPAC. The runoff Q, here equivalent to the runoff depth, increases over the entire profile length for IPAC = 20, over three-quarters of the profile for IPAC = 15, over the upper half for IPAC = 10 and over the uppermost quarter for IPAC = 5 (Ahnert 1987b)

a convex shape. Of course, the slope form also depends on the overall relative intensity of these two processes.

13.5.3 Conclusions With Regard to Ensystemic Changes

The model examples of slope development under conditions of uniform downcutting and constant climatic environment suggest the following conclusions:

1. The ensystemic change of the slope form leads — after passing through an early phase of profile differentiation — to an equilibrium profile form that is characteristic for the particular process combination at work; in other words, there are process-specific characteristic slope forms which, by their existence, may serve as diagnostic indicators of the particular process system at work in a given landscape.
2. These characteristical forms are
   (a) convex for mass-movement slopes;
   (b) concave for wash slopes under conditions of long-lasting runoff-producing rainfall events;
   (c) concavo-rectilinear for suspended-load wash under conditions of short runoff-producing rainfall events;

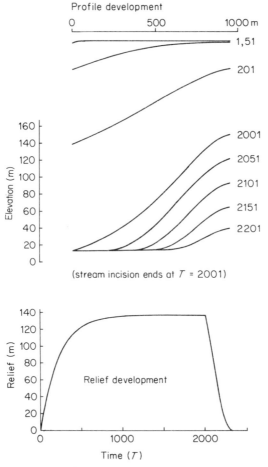

Figure 13.8. Slope development and relief development with a combination of slow mass movement and suspended-load wash denudation. Stream incision ends at time $T = 2001$

    (d) concavo-convex for bedload wash under conditions of short runoff-producing rainfall events.

3. If, as a single eksystemic change, fluvial downcutting ceases, the characteristic slope forms of the different denudation processes tend to be maintained, although with subdued emphasis, as long as the denudation processes themselves are active.

4. The attainment of complete dynamic equilibrium for the entire slope requires in most cases such a long time that, in nature, the eksystemic tectonic and/or climatic inputs have very probably changed before this time has passed; therefore, one should expect to find such complete equilibrium in nature only rarely. However, the entire ensystemic evolution of the slope form has the intrinsic tendency towards the establishment of the equilibrium with its characteristic slope shape, so that one may interpret the development of the slope in the light of this tendency. Such interpretation is both aided and justified by the fact that the slope form becomes similar to its characteristic process-specific equilibrium shape long before the state of equilibrium has actually been reached.

13.6 LANDFORM CHANGES DUE TO CHANGING ENVIRONMENTAL CONDITIONS (EKSYSTEMIC CHANGE)

13.6.1 Tectonic Changes

Perhaps the most common eksystemic changes to affect landform development are changes in the rate of tectonic uplift which in turn bring about changes in the gross rate of fluvial downcutting. Constant gross downcutting rates will, if they last for a sufficiently long time, lead to a state of dynamic equilibrium in which the slope form remains constant while all slope points are being lowered at the same rate, i.e. the net rate of downcutting at the slope foot and the rates of denudational lowering and of regolith production by bedrock weathering at all slope points are equal and constant.

Any change in the gross rate of downcutting will interrupt this previous tendency toward equilibrium and direct the entire landform development system towards a new, different equilibrium state. For example, an increase of gross downcutting will tend to steepen the valley sides — first at the slope foot, then gradually extending upslope by the process of 'regressive denudation' (Ahnert, 1954) that is akin to the headward erosion of streams. The steeper slope will conduct the waste material more rapidly to the stream, thus diminishing the net downcutting rate while increasing the denudation rate on the slope; the higher denudation rate in turn will cause the regolith cover to become thinner and regolith production by bedrock weathering to become more intensive. In other words, the initial increase of stream erosion generates an increase of denudation and of bedrock weathering on the slope, with a tendency towards a new equilibrium between waste supply and waste removal at a higher intensity level. Conversely, a decrease of fluvial erosion would mean formation of lag deposits (colluviation) at the slope foot, decrease of the denudation rate and hence thickening of the regolith. The latter would gradually extend upslope, with a concomitant decrease in the rate of bedrock weathering and with the overall tendency to establish an equilibrium at a lower intensity level.

Of particular geomorphological interest is not only the effect of the change as such, but also the effect of different lengths of the time intervals between successive changes. In Figure 13.9, gross downcutting remains constant at a rate of 4.0 m per time unit for the first 500 time units, then changes to 2.0 m per time unit for the next 500 time units and finally changes to 4.0 m again. The denudation process is bedload wash; the concave profile at time 501 shows the shape of the slope at the end of the 4.0 m per time unit period. It is very nearly the equilibrium shape for this intensity of downcutting.

With the onset of the change to the lesser intensity of gross downcutting (2.0 m per time unit), there is first a net accumulation at the slope foot (between times 501 and 526). It indicates that the reduced rate of downcutting is insufficient to remove all the waste that arrives from the slope; the waste production by weathering and the transport rate on the slope are still geared to the previous higher rate of downcutting. After time 526, however, the relative height of the slope decreases, its mean gradient is smaller, and therefore less waste is transported to the slope foot; the smaller gross rate of downcutting is now sufficient to cause net lowering. Soon afterwards the slope approaches the new equilibrium, as the great similarity of the profile shapes for times 701 to 1001 shows.

The change back to 4.0 m per time unit at time 1001 produces initially a high rate of net lowering of the slope foot until time 1026. From 1026 to 1076, however, the net lowering

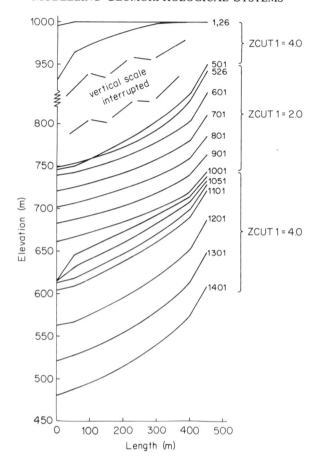

Figure 13.9. Effect of change of gross stream incision rate (ZCUT1) after every 500 time units upon profile development

of the slope foot is virtually zero, because with the steepening of the slope so much of the thick regolith cover (that had formed during the low-intensity time before 1001) is now moved to the slope foot that no net lowering can take place. Only after time 1076 has the evacuation of that 'inherited' regolith progressed enough to permit renewed net downcutting.

The relief development (Figure 13.10) reflects the development of the slope. The absences of relief change signify the near-equilibrium state of the model between times 400 and 500 at the more intensive gross rate of downcutting and between times 900 and 1000 at the less intensive rate. The short dip of the relief curve between times 1026 and 1076 indicates the short period of no net downcutting while at the same time the crest of the slope continues to be lowered.

If the change in the rate of downcutting takes place every 50 time units instead of every 500, the slope development never comes close to a state of dynamic equilibrium (Figure 13.11), and the relief oscillates widely (Figure 13.12). However, both the profile sequence and the relief development curve display a very strong periodicity with a recurrence of

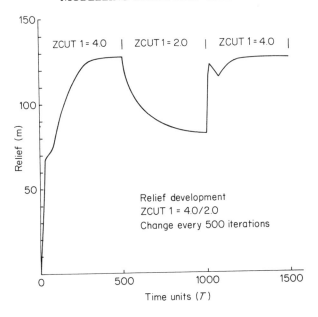

Figure 13.10. Relief development with change of gross stream incision rate (ZCUT1) after every 500 time units (*cf.* Figure 13.9)

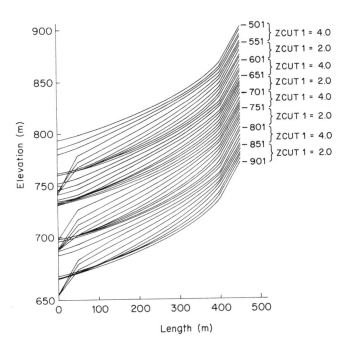

Figure 13.11. Effect of change of gross stream incision rate (ZCUT1) after every 50 time units upon profile development

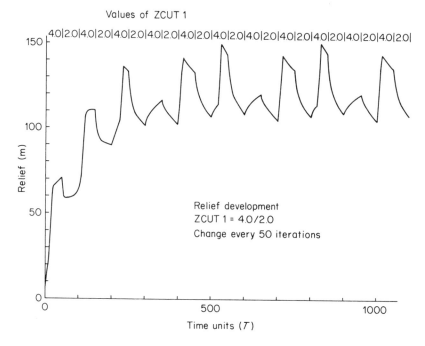

Figure 13.12. Relief development with change of gross stream incision rate (ZCUT1) every 50 time units (*cf.* Figure 13.11)

identical situations after about 300 time units—about six times as long as the interval between gross downcutting changes.

The following main phases may be distinguished in Figures 13.11 and 13.12, between times $T = 521$ and $T = 821$:

Phase 1: Excessive net downcutting and steepening of the footslope (strong relief increase); gross downcutting rate: 4.0.

Phase 2: Stop of net downcutting, stagnation of foot point yet lowering of slope crest (moderate relief decrease); gross downcutting rate: 4.0.

Phase 3: Net aggradation at slope foot while summit continues to be lowered (strong relief decrease); gross downcutting rate: 2.0.

Phase 4: Approximate stagnation at slope foot while summit is being lowered (moderate relief decrease); gross downcutting rate: 2.0.

Phase 5: Net downcutting of slope foot, but at a slower rate than in phase 1 and without excessive steepening of footslope (small relief increase); gross downcutting rate: 4.0.

Phase 6: Approximate stagnation at slope foot (similar to phase 4; moderate relief decrease); gross downcutting rate: 2.0.

Phase 7: Excessive net downcutting, footslope steepening and relief increase, but slightly less than in phase 1; gross downcutting rate: 4.0.

Phase 8: Stagnation at the slope foot and moderate relief decline (similar to phase 2, but for a longer time); gross downcutting rate: 4.0.

Phase 9: Aggradation at the slope foot—at first rapid, then slower—and accompanying relief decrease; gross downcutting rate: 2.0.

After phase 9 follows again phase 1. The entire sequence lasts for a succession of six time intervals of differing gross downcutting rates of 4.0 m and 2.0 m per time unit, respectively.

These model examples show that the geomorphological consequences of a particular change in the rate of downcutting are not the same every time but depend upon the slope shape, the regolith cover and the amount of colluvium that may be present at the onset of the change. The shorter the interval between changes, the greater is the probability that conditions of form and material inherited from one or several preceding periods influence the net result of the present process response system.

Gross downcutting changes that are frequent when compared with the time needed for a significant transformation of the slope also tend to affect mainly the lower part of the slope. Farther upslope the effect of each change gets progressively reduced and also modified by the effect of the next one. This is quite noticeable in Figure 13.11: while the slope form changes very dramatically in the lower half of the profile, the upper half is being lowered continuously parallel to itself at a nearly constant rate, despite all the changes in the rate of downcutting at the slope foot.

13.6.2 Change of the Type of Predominant Denudation Process

Climatic changes often bring about qualitative changes of the denudation processes in an area. It has been pointed out above that different denudation processes—in particular, slow mass movements and wash denudation—have their own characteristic slope form. A change from one process to the other would therefore tend to alter the shape of the profile. Again, the duration of the period between changes has an important influence upon the profile development and hence upon the profile shape. If that duration is 100 time units (Figure 13.13a), both the wash process (time 701–801) and the slow mass movement (time 801–901) persist long enough to develop their characteristic profile shapes—concave for wash, convex for mass movement. If the changes occur every 50 iterations (Figure 13.13b), the wash slope becomes concave only at the slope foot while the upper slope retains some of the convexity that was inherited from the previous period of mass movement.

If the intervals between changes are shorter still (Figure 13.14), the influence of the mass movement upon the slope form dominates even more strongly. Nevertheless, the wash-produced concavity at the slope foot is still discernible; because the runoff increases from crest to foot of the slope, wash denudation can be more effective on the lower part.

13.6.3 Changes in the Characteristic Duration of the Runoff-producing Rainfall

The effect of rainfall duration upon the shape of bedload wash slopes was shown above (Figure 13.7). Climatic change can alter the characteristic duration of rainfall events in an area and can lead, therefore, to a modification of slope forms. The amount of such modification depends not only upon the magnitude of the rainfall duration change but also upon the time that is available before a new change occurs. In the model examples of Figures 13.15 and 13.16 the intensity of the rain has been held constant; only the duration is varied and thereby also the length over which the runoff depth D (cf. equation 8) increases from the crest downslope. In Figure 13.15, with intervals of 500 and 250 time units between

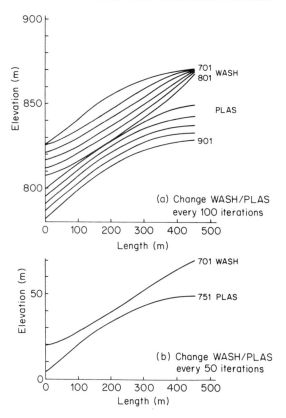

Figure 13.13. Effect of change between bedload wash denudation (WASH) and slow mass movement (PLAS) upon profile development (a) every 100 time units, (b) every 50 time units. In (b), only the slope profiles at the end of an interval (i.e. just before the process change) are shown

changes, the convex 'short-rain' (IPAC = 2) profile is fully developed and distinct from the concave 'long-rain' (IPAC = 10) profile; both are not far away from their equilibrium shape.

However, for progressively shorter intervals between changes (Figure 13.16), the 'short-rain' slope form is less and less pronounced. The shorter the intervals, the more does the middle part of the profile change, first from convex to straight and ultimately to concave. This means that the interval of 'long-rain' development dominates the shaping of the slope increasingly, simply because its greater characteristic runoff can do more denudational work than the runoff in the preceding 'short-rain' interval.

13.7 SLOP3D-SIMULATION OF A NATURAL SLOPE DEVELOPMENT SEQUENCE

The northern Eifel Mountains in Germany, close to the Belgian border, consist of high (5–600 m) peneplains of Tertiary age into which the streams have cut deep valleys during the Quaternary. One of these streams is the Kall, a 25 km long tributary of the Rur (not to be confused with the Ruhr River). The headwater of the Kall flows in a very shallow

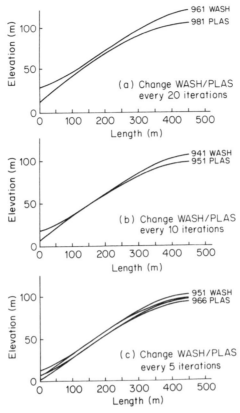

Figure 13.14. Effect of change between bedload wash denudation (WASH) and slow mass movement (PLAS) upon profile development (a) every 20 time units, (b) every 10 time units and (c) every 5 time units. Only the slope profiles at the end of an interval (i.e. just before the process change) are shown

valley on the peneplain. Downstream this changes gradually to a V-shaped valley of progressively increasing depth below the older erosion surface. The side slopes are very gentle and approximately straight in the headwater reach; down-valley they become steep, with a convex upper slope and a straight lower slope.

Since the Tertiary peneplain is older than the Quaternary valley incision and since the uppermost reach of the Kall still flows on the peneplain surface today, it is reasonable to assume that

1. Before incision, the river was flowing in its entire length on the peneplain;
2. The incision began at the mouth and progressed from there upstream by headward erosion;
3. The process of headward erosion is not yet completed and has yet to reach the uppermost part of the stream course.

From these considerations, one can derive the hypothesis that the valley side slopes in the uppermost section are similar to those that existed everywhere along the Kall before the incision started, and that—since the age of the incision is progressively greater downstream—the spatial succession of slope profiles from the headwater reach to the mouth represents a temporal sequence of stages of slope development.

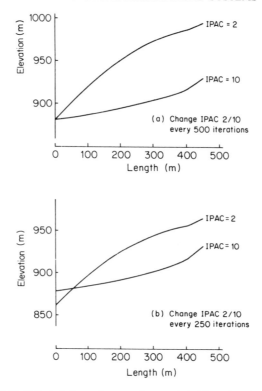

Figure 13.15. Effect of change between characteristic 'short-duration' rainfall (IPAC = 2) and 'long-duration' rainfall (IPAC = 10) upon profile development every 500 time and every 250 time units. The profile length is equivalent to 10 model length units

This notion was tested by means of SLOP3D. A simulation model of such a development sequence would have to begin with a natural 'initial slope' and to subject this slope to the processes that are likely to have occurred in the area during the Quaternary. If the hypothesis is reasonable, the successive slopes in the model should resemble the spatial sequence of slope shapes found in the Kall valley from head to mouth. A second criterion besides the shape of the slope was that the thickness of the regolith and its spatial distribution on the model slope profile should correspond with the regolith thickness and distribution on the field slopes.

As the initial slope, a profile 680 m in length with a relative height of only 28 m was taken from the uppermost part of the valley. This initial model slope was then subjected to fluvial incision, weathering, slow mass movement (solifluction and/or creep) in combination with a slight suspended-load wash denudation as it might occur in periglacial regions on tundra vegetation during the snowmelt season. Inclusion of the wash denudation became necessary because without it the regolith cover would have remained considerably thicker than it is in the field. Whenever the slope angle exceeded the critical preset value of 32 degrees, regolith material was transferred downslope by debris-slides.

The results are shown in Figures 13.17 and 13.18. The model profiles are numbered in their temporal, the field profiles in their spatial sequence. The elevations of the model slope

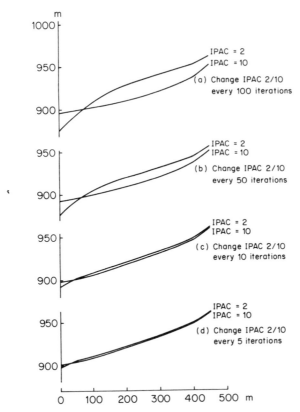

Figure 13.16. Effect of change between characteristic 'short-duration' rainfall (IPAC = 2) and 'long-duration' rainfall (IPAC = 10) upon profile development every 100 time units, every 50 time units, every 10 time units and every 5 time units. The profile length is equivalent to 10 model length units

summits generally are somewhat higher than those of the field slope summits, because all the model slopes developed from one and the same initial slope in the high divide region of the Tertiary peneplain while most of the field slopes are located in areas where the Tertiary peneplain was lower. In matching the two sets, the succession of the last two field profiles had to be reversed because field profile no. 7 lies already in the marginal zone of the Eifel where the Tertiary peneplain has been downwarped so that the overall relief there is smaller than at field profile no. 6. Apart from this anomaly, however, the model profiles and the field profiles match so well that the hypothesis that the side slopes along the Kall valley represent successive stages of one and the same slope development sequence appears justified. It seems that in relatively small valleys such as this one, the alternations of climates during the Pleistocene have not changed the mode of slope development enough to generate different slope forms during the different periglacial and interglacial climatic regimes. Instead, it seems that there may have been climate-caused fluctuations of process intensities, and thus accelerations and decelerations of development, but without any basic change in the direction of the development itself.

The lengths, the relative heights and the regolith thickness of the model slopes are approximately equal to those of the field slopes. The timescale of the model slope

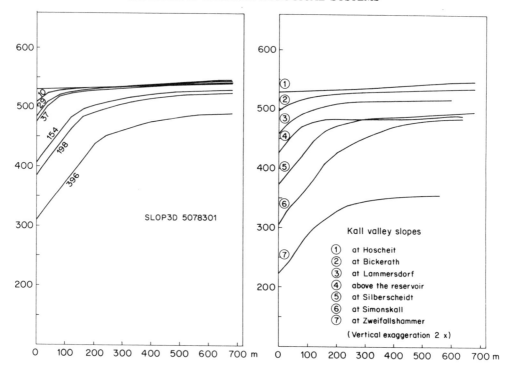

Figure 13.17. Model slope sequence (left) and field slope sequence (right) of the Kall valley, northern Eifel, Germany (Ahnert 1987a)

development can be calibrated by comparing the denudation rates of the model (per time unit) with known denudation rates (per 1000 years) of mid-latitude uplands of similar relief (*cf*. Ahnert, 1970) and by relating the two rates through simple proportion. In this case, the length of a model time unit equals approximately 2500 years of real time, therefore the duration of the entire slope development (396 time units) is close to one million years; this corresponds well with the generally known age of Quaternary valleys in the German uplands.

12.8   CONCLUSIONS

At the beginning of this chapter, the point was made that there are qualitatively different types of change possible in the development of landforms, namely, ensystemic and eksystemic changes, and that the modelling of landform development must take this difference into account. Most existing quantitative models, including those generated with the SLOP3D program, are of the functional–geomorphological type and therefore are primarily designed to simulate the mechanisms and the effects of ensystemic changes. However, the examples presented show that eksystemic changes may be introduced into such functional–geomorphological process–response models, with results that tend to improve the understanding of their effects upon landform evolution. The functional-geomorphological models thus can become, beyond their original purpose, also a useful tool for the historic–genetic explanation of long-term form development over geologic time spans.

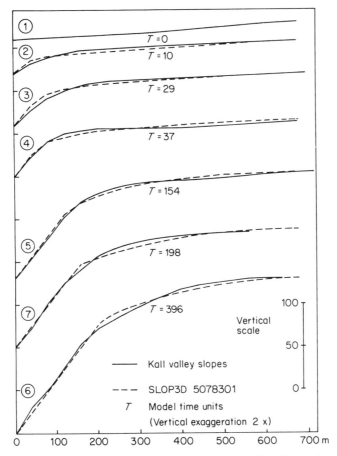

Figure 13.18. Match between model slopes and field slopes of the Kall valley. For the inversion of number 6 and 7, see text (Ahnert 1987a)

On the other hand, it is interesting to note that although the slope development in the Kall valley reached through several major Pleistocene climatic changes, it is modelled well as a purely ensystemic change. This could mean that the geomorphic effects of climatic changes upon slope development may have been occasionally overestimated. Compared to stream processes, shore processes or glacial processes, the energy consumption of slope processes per unit of area and time is rather small. Therefore, slopes seem to respond to climatic changes more slowly than those other process realms.

REFERENCES

Ahnert, F. (1954). Zur Frage der rückschreitenden Denudation und des dynamischen Gleichgewichts bei morphologischen Vorgängen. *Erdkunde*, **8**, 61–64.
Ahnert, F. (1964). *Quantitative models of slope development as a function of waste cover thickness.* Abstracts of Papers, 20th IGU Congress, London, 188 pp.
Ahnert, F. (1970). Functional relationship between denudation relief and uplift in large mid-latitude drainage basins. *Am. J. Sci.,* **268**, 243–263.

Ahnert, F. (1973). COSLOP2—a comprehensive model program for simulating slope profile development. *Geocom Programs,* **8,** 24 pp.

Ahnert, F. (Ed.) (1976). Quantitative slope models. *Z. Geomorph. N. F. Suppl.,* **25,** 168 pp. in this volume: Ahnert, F.: Brief description of a comprehensive three-dimensional process-response model of landform development, 29–49.

Ahnert, F. (1977). Some comments on the quantitative formulation of geomorphological processes in a theoretical model. *Earth Surface Processes,* **2,** 191–201.

Ahnert, F. (1981). Über die Beziehung zwischen quantitativen, semiquantitativen und qualitativen Methoden in der Geomorphologie. *Z. Geomorph. N. F. Suppl.,* **39,** 1–28.

Ahnert, F. (1987a). Approaches to dynamic equilibrium in theoretical simulations of slope development. *Earth Surface Processes and Landforms,* **12,** 3–15.

Ahnert, F. (1987b). Process-response models of denudation at different spatial scales. In Ahnert, F. (ed.): *Geomorphological models—theoretical and empirical aspects.* Catena Suppl. 10.

Bakker, P. and Le Heux, J. W. N. (1947). Theory of central rectilinear recession of slopes. *Kon. Ned. Ak. van Wetensch., Proc.,* **50,** no. 8 and 9; **54** (1953), no. 7 and 8.

Baver, L. D., Gardner, W. H. and Gardner, W. R. (1972). *Soil Physics* 4th ed., New York.

Chang, J.-H. (1958). *Ground Temperature* (2 vols.), Blue Hills Meteorol. Lab., Harvard University.

Davis, W. M. (1909). *Geographical Essays,* reprinted 1954 by Dover Books, New York.

Gilbert, G. K. (1877). *Report on the geology of the Henry mountains,* Washington.

Gossmann, H.(1970). Theorien zur Hangentwicklung in verschiedenen Klimazonen. *Würzburger Geogr. Arb.,* **31.**

Hirano, M. (1968). A mathematical model of slope development. *J. of Geosci., Osaka City Univ.,* **11** (2), 13–52.

Jahn, A. (1954). Denudacyjny bilans stoku. *Czas. Geogr.,* **25,** 38–64. Engl. translation: Denudational balance of slopes (1968). *Geogr. Polonica,* **13,** 9–29.

Kirkby, M. (1971). Hillslope process–response models based on the continuity equation. In Brunsden, D. (Ed.), *Slope Form and Process. IBG Spec. Publ.,* **3,** 15–30.

Lehmann, D. (1933). Morphologische Theorie der Verwitterung von Steinschlagwänden. *Vierteljschr. d. Naturf. Ges. Zürich,* **78,** 83–126.

Musgrave, G. W. (1935). Some relationships between slope-length, surface-runoff and the silt-load of surface-runoff. *Trans. Am. Geophys. Union,* **16,** 472–478.

Penck, W. (1924). *Die morphologische Analyse,* Stuttgart, 283 pp.

Richards, K. (1982). *Rivers—Form and Progress in Alluvial Channels,* London, 358 pp.

Scheidegger, A. (1961). *Theoretical Geomorphology,* 2nd ed. 1970, New York, 435 pp.

Schmidt, K.-H. (1986). Factors influencing the system of structural landform dynamics—empirical data as a basis for theoretical models. Lecture at Aachen Workshop on theoretical geomorphological models, April 1–4, 1986; to be published.

Schumm, S. A. (1961). Effect of sediment characteristics on erosion and deposition in ephemeral stream channels. *U.S.G.S. Prof. Paper,* **352 C,** 31–70.

Trofimov, A. M. (1974). *Osnovi analiticheskoi teorii rasvitiya sklonov* (Fundamentals of the analytical theory of slope development), Kasan, 212 pp.

Woldenberg, M. (Ed.) (1985). *Models in Geomorphology,* Boston, 434 pp.

Yair, A., Sharon, D., and Lavee, H. (1978). An instrumented water-shed for the study of partial area contribution of runoff in the arid zone. *Z. Geomorph. N. F. Suppl.,* **29,** 71–82.

Young, A. (1963). Deductive models of slope evolution. *Nachr. d. Ak. d. Wiss, Göttingen, II. Math.-Naturw. Kl.,* **5,** 45–66.

Young, A. (1972). *Slopes,* Edinburgh, 288 pp.

Zingg, A. W. (1940). Degree and length of land slope as it affects soil loss in runoff. *Agric. Engineering,* **21,** 59–64.

Modelling Geomorphological Systems
Edited by M. G. Anderson
©1988 John Wiley & Sons Ltd.

Chapter 14

Restrictions on hillslope modelling

IAN DOUGLAS

School of Geography, University of Manchester

14.1 INTRODUCTION

Hillslope processes and evolution form a core theme in geomorphology, with verbal and graphic models of hillslope development dominating large segments of geomorphological teaching for almost a century. Such models deal with extreme cases, unlikely to be widespread in nature, but typifying what might happen to landforms under constant conditions through long periods of time. Four main factors have changed the geomorphological view of hillslope models over the last two decades; the widespread availability of computers; the expanding engineering and agricultural demands for prediction of slope stability and soil erosion; the documentation by radiometric dating technique of many climatic changes during the Quaternary affecting nearly all parts of the earth; and the availability of the empirical field data from many, but not all, morphoclimatic environments for model calibration.

Today hillslope modelling is generally assumed to imply the mathematical modelling of earth surface processes and process-relationships. The spur to modelling is the belief that by a continual process of mathematical model building, model testing, and model redesign, better and better explanations of the forms and processes in geomorphic systems will emerge (Huggett, 1985). The possibility of modelling is brought about by the increasing technological sophistication, speed and complexity of computers to undertake the repetitive calculations. The inputs to modelling come from greater quantitative knowledge of hillslope systems. Modelling can be highly successful in handling systems in which all the individual components are well understood, and can be represented reliably and precisely in the analysis, and for which the input data are accurate (Philip, 1975). The restrictions on modelling stem from the incompleteness of our understanding of the components of hillslope systems and their interactions.

Modelling of hillslopes has been largely the product of tenuous interplay and exchange of ideas between agricultural engineers, geomorphologists, and soil mechanics specialists.

Little attention has been paid to the role of ecosystem dynamics in geomorphic processes, to the precise development of the energy relationships of erosive processes on slopes, despite some sound suggestions by Culling (1983). The importance of scale in hillslope models has not been adequately resolved, there being merely a general consensus that modelling of hillslope processes must be distinguished from the modelling of hillslope evolution; models of 'dynamic equilibrium' differing from models of the evolutionary development of landforms (Church and Mark, 1980). Up to 1972, most quantitative work on hillslopes had concentrated on the process or 'dynamic equilibrium' models (Carson and Kirkby, 1972), probably because more short-term field data were available and such models were relevant to predictions of soil loss and hillslope stability. Since then, despite new three-dimensional models of hillslope evolution (Ahnert, 1976; Armstrong, 1976, 1980), most effort has continued to be devoted to modelling of processes.

14.2   SCALES OF SIZE AND TIME

Process modelling is essentially short-term dealing with the prediction of the surficial form changes and sediment yields that will result from particular hydrometeorological inputs, usually calculated in terms of mass removed per unit of time of 10^0 to 10^6 s. Landform evolutionary models deal with the change in form, or shape, over time periods of 10^{10} to 10^{14} s, within which both ecosystem dynamics and the magnitude and frequency of geophysical events may change. Both broad categories of models may deal with either the slope profile as a line whose curvation and elevation change, or a surface that is altered over time. The failure to specify the lengths of hillslope profiles and the dimensions of the land surfaces in some modelling exercises is a restriction on their applicability and testing.

14.2.1   The Size and Dimensions of Hillslope Models

The hillslopes used to illustrate discussions of modelling usually range from 10^1 to 10^3 m in length, while the results of modelling are often depicted as dimensionless curves. The maximum area considered in most three-dimensional models is about 1 km^2, although again the relevant dimensions are not always clear. Slope length and hillslope size affect processes by offering more opportunities for diversity as they become larger. Slope length affects soil loss, being included in several empirical relations, such as the Universal Soil Loss Equation, in the general form:

$$X = kS^m L^n$$

where X is the soil loss in weight units, S is the slope gradient in per cent; and L is the planimetric slope length in feet. While clearly important if the Hortonian 'belt of no erosion' exists and to allow cross grading of rills down a long slope, length also influences the relationship between subsurface flow and surface saturation, particularly where translocation of material has occurred. Hillslope properties change downslope, thickness of surficial sediment and proportion of clay often increasing and amount of gravel and mean particle size decreasing with distance from the divide (Walker, 1966). Such changing properties affect plant and animal activity which in turn influence geomorphic processes and soil erodibility (Imeson, 1985).

In three-dimensional models, area is of greater significance, for the greater the size the less likely is uniformity of declivity, even in the unlikely event of the parent material being completely uniform. The most important size-related factor is the tendency to convergence and the development of hollows. Although affected by substrate and ecological conditions, such hollows are closely related to slope hydrology and sediment supply. Instability of slope profiles develops when such small hollows grow in size with a positive feedback, until they become valley heads into which flows of water and sediment are concentrated (Kirkby, 1980). Such stream head hollows are a key threshold in the hillslope system, tending to grow unstably if the inflowing water is able to transport more sediment than that carried into the hollow. As the hollow is fed by varying proportions of subsurface and overland flow, removal and deposition of sediment will alter with the extent of the seepage and saturation zones and the point of commencement of channel flow.

14.2.2 Timescales in Hillslope Models

Process models dealing with 10^0–10^6 s time spans are primarily concerned with describing how processes operate on unchanging slope forms. Landscape evolution models at 10^{10}–10^{14} s time spans endeavour to demonstrate form changes. Time however is a complex variable, implying a magnitude and frequency scale of events. Just as stratigraphy has tended to become more event based, and as neocatastrophism has spread from palaeontology (Dury, 1980) into both the earth sciences and biology, so the uneven importance of time in geomorphology has become emphasized and widely appreciated.

Within a time span of 10^{10} s, all the phenomena which occur at shorter scales, say of 10^3 or 10^4 s, will have occurred up to 10^7 times. In stratigraphy, distinction between abundant, rare, and unique sediments emphasizes that short duration high magnitude events may produce the same volume, and even type, of sediments as longer duration, smaller magnitude events. Some would argue that the phenomena varying at one timescale may be ignored in models concerned with another. Standard geomorphological practice appears to be to ignore short-term variation in models of landscape evolution and to assume certain processes to be constant in short-term process models (Phillips, 1986). However, an assumption that in geomorphic erosion models, processes acting on timescales of an order of magnitude different are effectively independent may prove to be a restriction on the applicability of models.

14.2.3 Seasonality in Hillslope Models

Many environments from close to the equator to the poles have seasonal rhythms of temperature or moisture or both. Much of the hillslope modelling so far conducted refers to processes in temperate environments (e.g. Armstrong, 1980) yet these environments are highly seasonal, with winter snow and, in many areas, deciduous vegetation without leaves in winter. This contrast between summer and winter affects the erosivity of precipitation and the protection of the soil. Snow is generally less erosive than rain, but large volumes of snow melt runoff can rapidly transport particles loosened by needle-ice formation and other freeze and thaw processes (Strömquist, 1985). A direct relationship between precipitation and transport of sediment would be inappropriate for winter conditions in a temperate catchment. Although in cooler, upland environments, the winter freeze and the spring melt dominate the hydrological cycle and sediment transport events. Yet relatively little attention has been paid to snow accumulation and melting phenomena in hillslope models. But the explanatory power of catchment water balance models for temperate lowland catchments is greatly improved if a term for snow is incorporated (Ward, 1985). Snowfalls, depth and persistence of lying snow, and number of freeze-thaw cycles vary considerably from year to year, changing the significance of seasonal extremes.

Autumnal leaf fall does not coincide with snowfall. Areas may be devoid of leaves when heavy rain occurs. If the fallen leaves do not provide protection to the soil, the bare ground will be subject to higher rainfall erosivity than at any other time of the year. However, a temperate forest floor has a biological rhythm with fallen litter decaying slowly during the winter and a resurgence of spring ground flora before the canopy becomes fully established again.

Winter frost, especially the number of daily cycles plays a major role in the preparation of material for transport. In some relatively low latitudes, around 30° N and S, in upland areas, such as the Atlas Mountains of North Africa (Raynal, 1960) and the highlands of New South Wales, nocturnal freeze-thaw cycles occur daily during the winter months. Higher latitudes and lower altitudes may experience fewer cycles because downal temperature ranges are far less. A whole range of slopes are thus affected by frost action yet such phenomena are seldom included in models.

In low latitudes, seasonality stems from moisture availability, with major contrasts between wet and dry seasons. Monsoonal regimes see great variations in moisture and the behaviour of regolith materials on slopes. Irregularly spaced wet season storms produce a succession of wetting fronts passing down through regolith profiles, causing montmorillonitic clays to swell, stimulating organic activity and translocating colloids and solutes. The rapid wet season growth of vegetation and the gradual build-up of soil moisture transforms the slope system from relative inactivity to high biological productivity and rapid biogeochemical processes. The precise effect of any storm or other hydrometeorological event in this seasonal ecological situation will depend on the stage of plant growth and soil moisture accumulation. Early wet season storms may block cracks in the soil with unwashed loose mineral and organic debris and produce more runoff than a theoretical calculation of infiltration capacity would suggest. Heavy late wet season storms when soil moisture levels are high will produce large volumes of runoff, but will have less surface material available for transport.

Even in the rainforests of equatorial regions, such seasonality of moisture availability is evident, particularly in southeast Asia. In south-central Java, for example, precipitation is less than evapotranspiration for four months from July to October. At this time litter fall increases, but soil moisture storage is not fully depleted until September or October (Bruijnzeel, 1982). Canopy, forest floor, and soil moisture conditions thus alter with a lag through the dry season. The environment in which slope processes operate is thus undergoing steady modification.

Despite these seasonal fluctuations, it may be argued that the biggest storms in low latitudes involve such high intensities and such large volumes of rainfall that ecological conditions on slopes are irrelevant. The problem then becomes that of understanding extreme events.

14.2.4 Extreme Events in Hillslope Models

Studies of other planets have emphasized that in other environments, surface features created by exceptional events persist for millions of years (Baker and Milton, 1974). On earth, the Missoula floods on the mega-scale, and the more moderate yet still extreme 1957 events in the Guil Valley in the French Alps (Tricart, 1974), and Hurricane Camille damage in Virginia, U.S.A. in 1969 (Williams and Guy, 1973), at a more moderate scale, provide examples of exceptional extreme events producing landforms which will dominate the local geomorphology for a long time.

Many areas have hillslopes dominated by infrequent landsliding or debris avalanches. In Japan, denudation by a single occurrence of densely spread debris avalanches is almost equivalent in volume to subsequent densely spread debris avalanches is almost equivalent in volume to subsequent denudation of their fears in about 100 years (Yoshikawa, 1985). Hillslopes of parts of Papua New Guinea reflect earthquake related landsliding (Simonett, 1967) while many parts of the Pennines and other Namurian sandstone and shale areas of Britain have hillslopes formed by post-glacial landslides (Tallis and Johnson, 1980).

Starkel (1976) distinguishes several regions on the basis of extreme events:

1. Regions with a frequency of extreme events of 5–10 per century, a total denudation caused by such events exceeding the mean 100-year denudation total and which are characterized by large-scale landscape and modifications (tropical monsoon and mediterranean climates, and farming lands of the temperate zone);
2. Regions in which extreme events are rare: when they occur they do not exceed the mean 100-year denudation total. In such regions the 'normal' processes include the effects of rapid downpours or snowmelts (semiarid areas);
3. Regions of rare extreme events that considerably surpass the 100-year denudation totals due to low mean precipitation (extremely dry areas and some mountains with boreal climates);
4. 'Stable' regions which show little deviation in denudation rates in particular years. Extreme events which do occur merely cause variations in the mean annual values (arctic, boreal–continental climates and lowlands of the temperate zone).

While it might be argued that most of the hillslope modelling attempted so far refers to stable (type 4) regions, the role of extreme events has to be seen in spatial scale perspective. If the hillslope model applies to a small area, the effects of a single event, such as a wind storm causing a tree fall and subsequent severe erosion, may dominate slope evolution for a long period. Other adjacent slopes may be unaffected for several decades by any such event. For a larger area, the fall of a single tree would represent, but a slight perturbation in the trend of slope evolution.

More pervasive may be global extreme events such as the major oscillations of climate now associated with the El Nino phenomenon. Drought in the western Pacific area and heavy rains and storms on the Pacific coast of South America related to the El Nino produce extreme geomorphic conditions. Large mass movements and severe flooding dominate slope evolution in the Peruvian Andes.

In Australia, persistent El Nino associated droughts reduce the density of vegetation and increase fire risk. Erosion following wildfires is often severe and may be a major force in slope evolution. Drought in equatorial regions often leads to rainforest fires such as those which affected over a quarter of Borneo in 1983. Canopy reduction in such areas greatly increased rainfall erosivity.

Not enough is known about extreme events, both localized and widespread, to say with certainty whether Starkel's subdivision into four regions is appropriate. Perhaps the short-lived phenomena, such as tornadoes, cyclones, and earthquakes are more clearly understood than the more pervasive drought–biota–fire linkages which can affect wide areas of continents at any one time. Hillslope evolution in many environments may be a combination of extreme events followed by adjustments, or short periods of large amounts of change, followed by slow obliteration of the traces of those changes.

14.3   UNDERSTANDING COMPONENTS

Models are frequently restricted by lack of data on particular components of the hillslope system. Weathering is a complex process involving a series of physical, chemical, and biological reactions. The nature and intensity of individual reactions is well known under laboratory conditions, but less understood in the field situation. In some hillslope models, chemical weathering is assumed to be a function of water supply and regolith depth (Cox, 1980), but regoliths range in character from the rapidly weathered slopes of young volcanic material, to the deep duricrusted weathering mantles of old land surfaces. The volcanic regoliths contain many highly soluble elements, whereas the ancient weathering crusts often consist of the least soluble aluminium and iron oxides which alter little further under present conditions.

Limitations on modelling due to lack of data (Freeze, 1978) and the shortage of suitable empirical data to test theoretical process response models (Ahnert, 1980) are often stressed. Few field experiments have been deliberately designed to provide the data necessary for modelling, a notable exception being that by Beven and Kirkby (1979). Such purposeful experiments and observations help elucidate the nature of system components, but, as always, field measurements will provide data only on the parameters under measurement, not on all the potentially significant reactions and processes.

14.3.1   Chemical Processes in Hillslope Modelling

A hillslope may be seen as a mass of particles, or minerals, some tightly bound together, others combined more weakly, some loose and some decomposed and circulating through the ecosystem as chemical elements. While the subatomic structure of these particles is unlikely to be relevant to hillslope evolution, their chemical composition influences weathering and erosion. The chemical characteristics of dispersibility, shrink and swell and solubility affect the rate of slope evolution. Soil chemistry and nutrient status contributes to plant growth which in turn has an impact on slope erosion.

While the catena and nine-unit land surface models of slopes suggest a systematic downslope variation in soil, physical and chemical properties, detailed investigations of the soil catena in the type area of Kware, northwest Tanzania show that such a systematic variation does not necessarily exist (Hathout and Bunting, 1975). In the Kware area considerable variation in soil properties occurs at a constant distance below the hill summit. Local variations in slope form, drainage, soil texture, soil stability, and soil moisture influence soil changes downslope. Such variations affect the ultimate products of weathering. Great variations in the geochemical environment in toposequences over limited area may give rise to diversity of soil profile development, micromorphology, and clay mineralogy (Remmelzwaal, 1978; Douglas, 1980). The chemical environment of slope evolution thus may vary down a hillslope and between adjacent hillslopes.

Such a problem of chemical environmental variation is severe for the short-term process modelling of hillslopes and a major difficulty for long-term evolutionary models. Weathering is usually incorporated into hillslope models as a single, unvarying process which may operate at different rates. Kirkby (1985) describes the degree of weathering at a given depth in a regolith in terms of the proportional substance remaining, as if there is a consistent gradual transition from bedrock to ground surface. Such a condition is unusual. Many hillslopes have weathering mantles derived from more than one set of climatic conditions. Weathering processes have to be envisaged as a combination of at least two stages, the contemporary soil development at the surface and the continuing geochemical evolution of the weathering mantle. For example, Radwanski and Ollier (1959) suggest that as a hillside is lowered by erosion, the ferricrete at depth in the profile may be dissolved at the top and redeposited at the base of the ferricrete. Elsewhere, weathering processes destroy old ferricretes high on slopes and produced new ferricretes further downslope (Ollier, 1976). In the majority of areas where the regolith was not modified or completely removed by glaciation, complex weathering mantles must be expected, with a discontinuity between what happens at or near the groundsurface in the soil profile proper and what happens at the basal surface of weathering where the mantle of decomposed material is in contact with the bedrock.

A two stage theory of weathering, advocated by both Büdel (1957) and Ollier (1959), is well portrayed in Ahnert's models of inselberg development (1982, 1987). Where the rate of weathering is less than the rate of removal of material downslope, inselbergs form and steep slopes retreat parallel to themselves (Ahnert, 1982; Armstrong, 1987). The rate of removal, however, is related to a number of thresholds, associated with the resistance of the ground surface and various subsurface layers to erosion.

A neglected area of geomorphological research, the interaction of chemical and physical properties and processes, may contribute to understanding the release of sediment on potentially active mudstones and shales slopes. The dispersibility of such material can be expressed through the sodium-absorption ratio (SAR). As the sodium content increases, so does dispersibility. A relatively small increase in electrolyte content may take material across a dispersibility threshold, which sees rapid disintegration of lumps of mudstone and the onset of rapid erosion. This threshold is often triggered in nature after the crossing of another threshold, the removal of the vegetation and soil cover protecting the mudrocks from direct raindrop impact. Once natural or people-made disturbance exposes these dispersible slope materials and the appropriate chemical conditions occur, rapid gullying and ravine development begin to dominate slope evolution. Lack of field research into these processes restricts their incorporation into hillslope models.

Another scale of environmental chemical changes affecting slope evolution is that of shifts in atmospheric chemistry. Such shifts may be natural, as a result of volcanic eruptions or the product of industrial growth and economic activity. The most recent phase of blanket peat erosion in the southern Pennines was probably set in motion by death of *Sphagnum* from air pollution (Tallis, 1985). The most serious consequence for terrestrial ecosystems of regional acidification at currently observed levels in Norway may be the increased rate of leaching of major elements and trace metals from forest soils and vegetation (Overrein *et al.*, 1981). Such gradual people-made changes contrast with the eruptive release of gases from volcanoes which affect varying areas, depending on the type of eruption and the weather at the time of the release. Given the observed impact of vulcanicity on weather in the last century and the evidence of tephra deposits, the role of such events in long-term slope evolution cannot be dismissed readily.

14.3.2   Ecological Processes and Hillslope Evolution

Despite the urgings of many geomorphologists with tropical experience (e.g. Tricart, 1978, 1979) hillslope studies still give insufficient attention to ecological processes. In forested areas, hillslope development is intimately interwoven with ecosystem development. Hillslope processes in such areas of high biomass and biological productivity must be set in the context of functioning ecosystem. Plants and animals alter the pathways and erosivity of precipitation and the protection and erodibility of hillslope materials. In the short and medium term, forests cannot be considered as a permanent slope cover because natural events such as tree fall, gap creation and plant recovery lead to changes in vegetation cover. Over the longer term, climatic change leads to shifts in vegetation, albeit often with a considerable lag after shifts in rainfall frequency and thermal patterns.

In addition to the seasonal variations in the influence of ecosystems on hillslopes discussed earlier, both individual species and plant communities affect water movement and on slopes. In the wettest parts of northeastern Australia, highly significant interspecific differences in interception storage capacity affect the routes and rates of throughfall and stemflow (Herwitz, 1986, 1987). In a mature lowland rainforest and a nearby one-year-old successional area in Costa Rica, throughfall varies greatly from day to day and from site to site in the same area, but averages 52 per cent of rainfall in the forest and 68 per cent in the young successional area (Reich and Borchert, 1984). Individual trees can funnel stemflow to the ground in prolonged heavy cyclonic storms in northeastern Australia at a rate sufficient to create a local saturated zone downslope of the trunk from which overland flow may begin. Just as the rainforest ecosystem is a mosaic of mature, building, and gap phase vegetation (Whitmore, 1978), so the rainforest floor during and immediately after rain is a mosaic of saturated, wet, and partially wet areas. Any throughflow or overland flow begins as a series of tree influenced points of concentration and convergence zones around stream-head hollows. Whether or not these zones coalesce or a uniform wetting front moves down through regolith depends on the depth and duration of precipitation and antecedent conditions. With such an irregular pattern of hydrologic response, slope erosion is uneven and discontinuous.

Superimposed on this pattern of hydrologic response due to species differences are the changes through time in forest structure. Gaps of $300\,m^2$ created by fall of individual trees to windflow gaps of regenerating plants cover the soil. Tree fall is an irregular occurrence, likely to affect a 1 ha area once or twice in a decade. The net result of the superimposed mosaics of interception characteristics and gap formation and recovery is spatial and temporal discontinuity in slope erosion. Data collection for modelling must take this variability into account.

These forest ecosystem variations are not confined to tropical rainforests. Temporary patches of saturation exist during and immediately after rain to the side of beach trees in English woodland where stemflow reaches the ground. Tree fall in pine plantations on the English Pennines is a major source of sediment and influences gully development. The uneven pattern of bare soil, litter accumulation, raindrop size and splash erosion in northern Luxembourg forests indicate that erosion depends on both climate and the dynamic properties of the ecosystem.

All ecosystem models, even for comprehensively studied areas like Hubbard Brook in New England, U.S.A. and El Verde Forest in Puerto Rico, are essentially lumped system models where variations over a few hectares are ignored. Slope models, however, are often concerned with smaller areas, or distances downslope of only a few hundred metres, and with prediction of short-term sediment yields. A major restriction on such process models is lack of understanding of ecosystem components. If the influence of plants is poorly understood; that of animals is even less well known.

14.3.3 Effects of Soil Fauna and Animals on Slope Processes

Many hillslope modellers recognize the role of burrowing animals, termites and earthworms in the movement downslope of soil material (Culling, 1983), but no formal attempt appear to have been made to incorporate such activity into hillslope models. Since 1882 when Charles Darwin showed that earthworms could move almost $25\,t\,ha^{-1}\,y^{-1}$, a variety of estimates of the geomorphic activity of insects and animals have been made (Table 14.1). Nowhere is the role of insects more evident than in the termite mounds of seasonally wet tropics of Africa, South America and Australia. A wide variety of features such as narrow, pillar-like mounds (Bonell et al., 1986), large circular dome-shaped structures (Komanda, 1978), and ferriginous crusts (Taltasse, 1957) result from termite activity. Each type of termite

Table 14.1. Estimates of the quantities of available sediment supplied by animal and insect activity in different environments

Organism	Locality	Sediment supplied $(t\,ha^{-1}\,y^{-1})$	Source
Semiarid regions			
Isopods and porcupines	Northern Negev, Israel	0–3	Yair and Rutin, 1981
Ants	Duchesne County, Utah	9.8	Yair and Rutin, 1981
Grasslands			
Prairie dog	Great plains, USA	56.25	Yair and Rutin, 1981
All animals	Great plains, USA	75–100	Yair and Rutin, 1981
Earthworms	Rothamsted, England	10–90	Pitty, 1971
Earthworms	Kent, England	25	Darwin (in Pitty, 1971)
Sub-Alpine/Arctic			
Cilleus undulatus	Alaska, USA	20	Yair and Rutin, 1981
All burrowing	Olympic Natl. Park	0.48	Yair and Rutin, 1981
Temperate Forest			
Burrowing animals	Ardennes, Luxembourg	0.903–0.916	Imeson and Kwaad, 1976
Worm cast	Jong Boesch, Luxembourg	15	Hazelhoff et al., 1981
Savanna			
Termites	Ouagoudougu, Upper Volta	1.2	Yair and Rutin, 1981
Termites	Shaba, Zaire	2–20	Komanda, 1978
Erosion of termite mounds	North Queensland, Australia	4–110	Bonell et al., 1986
Tropical Forest			
Earthworms	Abidjan, Ivory Coast	10–50	Yair and Rutin, 1981

mound has a differing geomorphic significance in terms of sediment accumulation by termites during the mound building phase and in sediment supply to slope processes in the mound destruction phase (Tricart, 1957). Mound destruction indeed may be viewed as an example of the lowering of topography through the type of diffusion process described by Scheidegger (1970). Not only are the mounds likely to become major sediment sources, but the work of termites thoroughly turns over the uppermost layers of the soil leaving it in a friable state with increased porosity (Millington, 1981). Such insects are a major contributor to the thorough biotic reworking of the upper 20 cm of soils in tropical rainforests in northeast Queensland, Australia, making them capable of admitting even the most intense and heaviest rains (Bonell et al., 1983).

Nevertheless the volumes of sediment reworked by biological activity on slopes equals or exceeds the volumes released from experimental catchments and plots (Yair and Rutin, 1981). The biologically prepared sediment is an addition to the supply side of the slope sediment system. Disturbance by animals probably has a marked seasonal rhythm. Most of the digging and burrowing activity of porcupines and isopods in the Northern Negev occurs in the hot and dry season. However, such activity is highly dependent on the rainfalls and runoff characteristics of the preceding wet season. Thus the rainfall and runoff characteristics of one winter could influence the amounts of sediment removed from the slopes in the next winter more significantly than such constant slope properties as slope length and gradient. Sediment removal on slopes in the Negev thus may be related to perodicity in the production of sediment by burrowing animals (Yair and Rutin, 1981).

In addition to the direct role of insects and animals in loosening and transporting soil material, they may alter hydrologic pathways, from their possible roles in the formation of soil pipes (Jones, 1981) to the concentration of overland flow in animal tracks and the damage to vegetation by trampling and overgrazing. Livestock and wild animals from cattle and sheep to deer, elephants and hippotami may create preferential pathways for rapid water movement and potential rilling and gullying.

The links between ecosystem dynamics, soil structure and geomorphological processes are essentially in terms of the influence of:

1. Soil structure, the size distribution of pores and aggregates and the stability of peds or voids on water acceptance, hydraulic conductivity, runoff, and the potential for erosion.
2. Changes in processes operating at the microscale on the hydrological response of areas from 1 m^2 to a hillslope or small catchment.
3. Vegetation cover and other organisms in creating localized extremes precipitation input to the ground which dominate the hydrogeomorphic response of an area (Imeson, 1985).

Clearly, ecosystem dynamics can only be ignored in modelling the most barren arid zone or artificially cut slopes. In well-vegetated areas lack of knowledge of the relevant ecosystem components is a real restriction on hillslope modelling.

14.4   UNDERSTANDING LINKAGES

In nature the linkages between biota, soil, weathering mantle, bedrock, slope gradient water and energy are complex. Slope process models simplify these linkages into such categories as water input, solute, sediment and water output often using water transfer as the key link. Simple budgets for individual points or transects on the slope can be seen in terms of material arriving from upslope, material made available by weathering and material transported further downslope. For complete slopes of some width, the linkages are more complex. As discussed earlier, the discontinuous spatial input of water to the soil and movement of water through and over the soil produces irregular detachment, transport, and deposition of slope material. Dispersion, aggregate breakdown, particle comminution may be followed by deposition, coagulation, cementation and surface crusting which may increase resistance to subsequent detachment and further transportation. Different linkages between components may exist at different places and different times. Plants that store nutrients and enhance slope stability while they are growing may accelerate erosion and release nutrients when they fall over and decay. Elsewhere the sediment released by the collapse of one plant may be trapped by the stems or exposed roots of the next. Essentially, the feedback in the system is such that the components of the system are not always coupled to each other in the same way. Although the transport of debris is basically downslope, biotic processes may lift material higher up the slope or the weathering profile. Understanding these linkages and the many temporary storages of material they involve is required before short-term changes in hillslopes can be modelled adequately.

14.4.1 Feedback Mechanisms on Hillslopes

A good example of the differences in linkages in geomorphic systems is provided by a consideration of how climate actually controls drainage basin morphometry in the U.S.A. (Patton and Baker, 1976). In regions such as the Appalachians, where rainfall is relatively evenly distributed in space and time, a feedback mechanism continually works to dampen basin response by aiding the development of thick soils and dense vegetation which increase infiltration rates and retard surface runoff. On the other hand, the intense, infrequent and randomly occurring rainfall of central Texas does not promote vegetation or soil development, and the resulting greater overland flow leads to hillslope rills and gullies which eventually become part of the permanent drainage network and dominate slope evolution.

In the moisture-stressed environment of southeast Spain, vegetation cover on slopes is closely correlated with the areas of greatest moisture (Francis et al., 1986). Vegetation occupies the hollows on the slopes where moisture collects, but individual plants compete with each other for the available moisture. As soon as vegetation appears it has a dramatic effect on runoff, whatever runoff there is coming from the bare soils between vegetation clumps. Moisture levels and vegetation quantity are highest in spring, but the litter and soil organic matter peak occurs later. The initial hollows which might have served as the foci of enhanced slope activity become the more densely vegetated and the interfluves the main areas of surface sediment transport. Such feedback relationships vary from climate to climate and will be influenced by slope materials.

14.4.2 Thresholds in Hillslope Evolution

The precise definition of when a hillslope condition will change from one state to another is the goal of much geomorphological enquiry. In modelling, the essential link between components often involved an estimate of the threshold condition, the point at which rill development begins or at which a small hollow will develop into a channel head. In time, the threshold condition is often the movement at which the vegetation cover protecting the soil becomes so degraded or ineffective that surface detachment proceeds at a rate sufficient for rilling to begin.

The hillslope thresholds involved in terms and gully initiation by pipe collapse are more complex than the stream-head threshold model proposed by Kirkby (1980). In the pipeflow stream-head or the tunnel-gully system, the only threshold that is passed at the channel head is that of roof stability (Jones, 1981). The threshold between flow through porous media (throughflow) and concentrated channel flow (pipeflow) has occurred at the head of the pipe. Just as discrete segments of water in a surface channel may combine together during a storm as saturation zones extend (Day, 1978), so may segments of pipe networks become active, extending the flowing subsurface three-dimensional system. In some circumstances these pipeflow thresholds regulate the amount of hillslope erosion (Jones, 1981).

Ideally a threshold can be portrayed in terms of an energy level. When that level is reached, the system moves into a different state. Such a linear response is not likely to occur in a hillslope system (Thornes, 1980). Each individual gully head, rill line, pipe, hollow will have a threshold for extension or collapse related to its dimensions, materials, biotic activity and topographic position.

Other thresholds are related to the exposure of the materials of hillslopes. Duplex soils and mantles related to past climatically-related processes, such as the periglacial slope deposits and colluvium which cover so many hillsides in the temperate zone, offer the same type of resistance thresholds as the successive rock outcrops used in Kirkby's model of slope evolution with resistant hard rock bands (1985). However, where these mantles or strata are almost parallel with the surface, the less permeable or more resistant strata lead to lateral planation at the resistant surface. Such a situation occurs with the erosion of peat above grit stones in the southern Pennines of England and in the gullying of mudstones and sandstones in central Queensland, Australia. In both cases, gullying cuts vertically down through the less resistant material, and then on reaching the resistant layer, sidewall erosion gradually exposes more and more of the resistant rock. A further threshold may be breached if the resistant band is dissected and underlying weaker rocks are exposed.

14.5 LANDFORM HISTORY LEGACIES FROM THE PAST

Incorporating the past into the present, or evaluating the historical hangover effect, has long concerned students of slopes (Chorley, 1964). Hillslopes of vastly different ages and rates of evolution exist, from those of Papua New Guinea whose form changes rapidly in a few hundred years, to those of the Western Australian shield which have changed little for 100 million years. Many geomorphic features relate to past conditions and others have been exhumed by the stripping of overlying sediments.

In terms of modelling slope evolution, geomorphologists are working at a difficult time in earth history, with evidence almost everywhere of considerable morphorgenetic changes in the last 20,000 years and legacies of Pleistocene cold or aridity mantling many parts of the land surface. The rapid growth of the human population has so profoundly modified the vegetation that in few places is it possible to assert that the present ground cover represents what would have existed 200 or 300 years ago. Moreover, the impact of Neolithic people in clearing the forests over hundreds of years—up to 40,000 in the case of the Australian aborigines and probably longer elsewhere in low latitudes—is so great that the vegetation of post-Neolithic activity now taken to be natural vegetation, such as the wet sclerophyll forests adjoining the remaining rainforests of northeastern Australia or the peat-covered moorlands of the North York Moors and Pennines of England.

Generally where slopes are most active, uplift is proceeding more rapidly than denudation. Where slope evolution is slow, rainfall energy is low and slope processes are often weathering limited. While models often envisage effective removal of debris at the base of the slope, the material actually being removed is often that which accumulated in the valley floor under some past climatic conditions. The supply of material by weathering, the denudation of the slope and the work of the river are often out-of-phase with one another.

In the Macclesfield Forest on the western edge of the southern Pennines in England, for example, the present-day geomorphic processes are controlled by a planted-pine forest dating from 1929–1935; the major sediment source is a wind-blown periglacial loess; the chemistry of the stream water is regulated by buffering reactions in a Devensian glacial till which tend to neutralize slightly acidic precipitation; and the general relief of the area reflects either or both exhumation and reactivation of the post-Carboniferous Red Rock and other marginal faults at the eastern edge of the Cheshire-Worcester graben. Some 20 km east of this area the slopes of the limestone knolls of the Earl Sterndale area reflect morphogenesis at the margin of the Lower Carboniferous Woo Dale block when reef-bank nearshore deposits were elevated above sea level and dissected into hills before being buried by further marine sediments, all within the Lower Carboniferous. Broadhurst (1985) writes of the Castleton area, 20 km north of Earl Sterndale: 'the virtual coincidence of the slopes of present-day hillslides with bedding planes in the limestone show that the present-day topography is a resurrection of what was once a Lower Carboniferous sea floor.'

Near Cape Wrath in northwest Scotland is a planate surface in weathered Lewisian Gneiss, largely overlain and protected by late Precambrian Torridonian sediments, the bedding of which parallels the inclination of the underlying unconformity. Where still buried, the surface is gently sloping and remarkably smooth, being disturbed by only one hill of gneiss. Its character suggests a pediment surface eroded in the weathered crystalline rock and subsequently buried beneath a series of coalescing Torridonian alluvial fans (Twidale et al., 1976). Legacies from all episodes in the history of the earth's surface persist and need to be incorporated in general geomorphological history. Perhaps the tendency to start with the wearing down of the landscape emphasizes the most recent to the detriment of the hangovers from the past. Perhaps identification of the hangovers from the past is a necessary precursor to slope modelling.

The references to legacies in Britain should also warn against the growing trend to assume that slope evolutionary models can be successfully applied to such temperate latitudes. Such slopes are so beset with legacies from Quaternary climatic changes that their evolution cannot realistically be modelled as the product of unchanging processes. A start in a more appropriate direction might be made by examining areas of recent uplift, such as Japan or New Guinea, and modelling slope evolution as the result of uplift and downwearing. The ideas of Ohmori (1983) and Yoshikawa (1985) on the lowering of relief should be developed to incorporate lithological and climatic changes as well as tectonics and denudation. Even with such additions the problem of spatial variability remains.

14.5.1 Spatial Variability in Geomorphic Changes

Landforms are so varied that an example can be found to illustrate almost any hypothetical combination of circumstances. The goal of much modelling of slope evolution is to be able to simulate and forecast the process of landform development. Many land surfaces show little change for long periods of time, as duricrusted plateaus, unaltered drumlins and the general expanse of mature soil profiles suggest. Spatially, geomorphic work done per unit time is unequal. River, gully and rill channels are active, abandoned terraces less so. Valley-side slopes evolve rapidly in some circumstances, but in other instances they change little. Summit surfaces and interfluves may change little over long periods of time. The most important foci of denudation in a fluvial landscape may be the channels of the drainage network and the variable source areas and mass movement zones on hillslopes.

14.5.2 Hillslope Evolution as the Modification of Past Forms

The ultimate goal of hillslope modelling must be to predict the future geomorphology of the continents and oceans. Just as geologists have suggested the likely future distribution of the continents in 50 m million years' time, so geomorphologists ought to be able to describe the relief of those continents at that time. The problem for the geomorphologists is far more complicated than for the geologists, for the latter are concerned almost entirely with endogenetic forces, the earth's internal heat engine, while the geomorphologists are concerned with both endogenetic and exogenetic forces. Climatic change, the essential consequence of those exogenetic forces, is currently the subject of great modelling efforts which are beginning to cope with a few hundred or thousand years. Geomorphologists need to combine the climatic and the plate tectonics models to cope with long-term landform evolution.

Plate tectonic models have become more elegant as the theory has been applied in greater detail to both young and ancient continents. Now fragments of continental and oceanic coast, or terranes, are shown to have been brought together to form the present land masses. In a similar manner, landforms comprise remnants of differing age and origin. Hillslope and landform models need to cope with these legacies from the past, particularly as, in many places, present-day processes seem to be doing little to modify past forms.

Rates of erosion may be less than the landforms suggest. Dissected gullied landscapes may become virtually stable, with gully floors aggrading and acquiring vegetation, even though bare gully walls remain. In southeast Spain some sites in gullied areas give the impression of having changed little in the past 4000 years (Thornes and Gilman, 1983). Present-day processes in this area are doing little to modify hillslopes or the drainage network.

The age and persistence of many hillslopes is demonstrated by successive soils and mantles incompletely covering slopes little affected by Quaternary glaciations. Multiple soils and hillslope mantles, with unconformable contacts between variably-truncated remnants of the lower soil mantle and an overriding colluvial mantle, occur in the Canberra and Nowra districts of New South Wales (Butler, 1963). The discontinuous and unconformable contacts between soils and hillslope mantles indicate an alternation of phases of soil development and soil erosion, the change between biostasy and rhexistasy described by Erhart (1956) from African experience. However, in New South Wales, the strong profile differentiation, especially in the older soils, suggests long intervals of biostasy on the hillslopes. Erosive activity has not removed the older soils completely before a new period of soil development has begun. Five such hillside mantles are exposed on the hillsides around Canberra (Van Dijk, 1959) with present-day processes attacking hillslope materials originally weathered and accumulated in five alternations of biostasy and rhexistasy. Chemically varied, with differing porosities, clay minerals, and particle sizes, these materials offer spatial and vertical contrasts of erodibility which vary laterally over short distances. Present-day hillslope evolution thus resembles the uneven dissection and/or stripping of layers of wrapping or protective covering from an insulated cable or complex motor vehicle tyre. Such soil and colluvial mantles, widespread in Mediterranean and subtropical environments, pose particular problems for hillslope modelling as studies of palaeosols have seldom been linked to empirical observations of hillslope processes.

Often hillslope mantles and superimposed profiles are considered solely in terms of pedogenesis in and after the last glaciation (Devensian or Wisconsinian), but many land surfaces are for older, especially those capped by duricrusts. The ferricrete mantled palaeosurface, of which remnants are widespread in the Mount Lofty Ranges, Kangaroo Island and the southern Eyre Peninsula of South Australia, is probably of early Mesozoic age and has been continuously exposed since its uplift by faulting (Twidale *et al.*, 1974). Such duricrusted land surfaces, widespread in low latitudes, persist for long periods. Once any A-horizon has been washed or blown away, the exposed indurated zone, hardened by dissection, suffers little from surface attack. Hillslope evolution at the margins undermines the duricrust and the remnant is reduced in size, but not lowered.

The duricrusted remnant is but one example of the stable landforms which persist through many changes of climate. To such almost unchanging features must be added those which are reinforced by their influence on runoff and the distribution of material. Reinforcement mechanisms, such as the shedding of water from bare rock surfaces onto adjacent weathered material, tend to accentuate the contrast between the resistant features and the more easily weathered and eroded adjacent material.

Many hillslope processes are simply modifying landforms created under past conditions. The material on which those processes work may be quite different from that which would be weathered from the underlying parent rock under the present climate. The contrasted chemical and physical properties of the hillslope mantles make simple general models of weathering and soil profile inadequate for these environments. Past landforms pose enough problems for short-term process, or equilibrium, models, but they add greater difficulties for landform evolution models. For much of the world, Büdel's climatogenetic geomorphology (1982) provides a reasonable general view of landform evolution under changing climates. The European landscape north of the Alps, the type of environment used for many hillslope evolution models, is dominated, according to Büdel, by the effects of past relief generations, especially the Tertiary period of etchplanation, the earliest Pleistocene phase of valley development, the intense glacial and/or periglacial activity and valley incision of the middle and late Pleistocene, and the limited modification of the Holocene. Essentially, in this view, present day slope processes are almost entirely on a relict periglacial landscape. However, as the broad features of the landscape date from the Tertiary, an evolutionary model would have to account for the influence of the first and second relief generations on future landform change.

Most models are uniformitarian, believing that the present is the key to the past, and that the past is the key to the future. Such assumptions run into some of the key difficulties of geomorphology, particularly that of equifinality. If a model using certain assumptions about processes through time can be seen, after many iterations, to produce a sequence of forms similar to those found in the field, it can be argued that those are the processes that produced those forms. Equally, they could have developed by another set of processes.

Distinctive, apparently simple features such as inselbergs evolve through differing paths, as Ahnert's elegant field and modelling study (1982) in Kenya shows. The difficulties of modelling these relatively uncomplicated landforms indicate that modelling long-term landform evolution is still in its infancy. Whether or not such modelling will survive infancy depends on the intellectual effort and research resources to be devoted to it. Geomorphologists, partly through their association with agricultural and civil engineers, can show the economic significance of soil erosion, mass movement, and related hillslope processes. The economic significance of long-term landform evolution, of relief generations, is yet to be demonstrated, beyond the pre-planning, pre-site investigation stage of construction. The huge manpower and computer resources that would be needed to collect sufficient field data to produce adequate parameters and trials of a general landform evolution model are beyond geomorphologists at present. Whether there could be sufficient unity of purpose in such an individualistic discipline to achieve a landform parallel to the Deep Sea Drilling Programme is hypothetical. Perhaps long-term landform evolution modelling is a holy grail which inspires geomorphologists, but is still a little unrealistic and, for many, unsatisfying.

14.5.3 Hillslope Models, Science and Trans-science

One of the major dissatisfactions with hillslope modelling is that the models fail to live up to the complexity of most real hillslope situations. The problem is that of the student preparing a dissertation, who first reads in the textbooks and periodical literature about how things work in the evolution of landforms to create the geomorphic features of the landscape. On leaving the library and entering the field to select a site for detailed study, the student finds the accounts, or verbal models, in the literature are a simplification and generalization of reality. The selection of both field sites and field methods requires some subjective judgement. Scientific rigour of controlled experimentation and measurement begins to be threatened.

In making mathematical models of hillslopes, much the same problem is encountered. Flow diagrams can be drawn to show the linkages between components, but are all the linkages actually incorporated in the diagrams? Can the surface and subsurface variability of vegetation cover, animal activity, soil depth, soil chemistry, soil type, slope angle and substrate, be adequately linked with the temporal variability in the character and energies of precipitation, the state of vegetation, litter layer, animal activity, rill development and mass movement? Is it possible that the full problem enters the realm of what Weinberg (1972) has termed trans-science?

Weinberg (1972) introduced the adjective 'trans-scientific' to describe 'questions which can be asked of science and yet cannot be answered by science.' Although such questions 'are epistemologically speaking, questions of fact and can be stated in the language of science, they are unanswerable by science; they transcend science.' Science as used by Weinberg refers to 'the content and/or methodology of the natural sciences.' Although perhaps a little woolly at the edges, this concept of trans-science is useful in warning scientists and others about the limits to accuracy and precision in understanding what happens in highly complex systems. These situations particularly arise in risk assessment where the impact of physicochemical reactions on organic systems is involved (Crouch, 1986), but it may also apply to studies of earth surface processes.

Discussing problems of hydrological modelling, Philip (1975) asks whether there are elements of trans-science in catchment prediction. He suggests that, for a genuinely stationary catchment, the task of performing an accurate detailed physical characterization and using this in reliable predictive calculations is scientifically feasible; but it would almost always demand for its proper performance the expenditure of resources out of all proportion to the benefits. Such a situation, Philip believes, would be classified by Weinberg as trans-science, *par excellence*. However, the situation becomes more complex if the catchment is subject to precisely known changes; if it is subject to random changes, particularly to irregular human intervention, the incertitudes multiply even more and trans-science dominates.

In the two-dimensional slope profile models great progress has been made in incorporating variations in soil depth, rock resistance, energy inputs and long-term climatic change. Slower progress has been made with three-dimensional slope models, particularly in dealing with the spatially irregular impact of rare, but large events. Weinberg (1972) specifically sees problems involving the probability of highly unlikely, but crucial events as lying in the realms of trans-science. We cannot collect field data for 1000 or 10000 years to be scientifically precise about such events (Philip, 1975). In slope modelling, limitations loom larger the closer this trans-science situation is approached. However, the practical approach geomorphologists should adopt is to work towards a situation where studies become a sceptical science with a rigorous, coherent intellectual content firmly based on the real phenomena. The marriage of field measurement, modelling, field testing and verification must be sustained.

14.6 CONCLUSIONS

Hillslope modelling has some parallels with weather forecasting—many complex physical interactions are involved, but on hillslope there are biological and chemical reactions as well. While for some types of hillslopes the magnitude, rates and nature of some individual processes are well known, others are far less understood. For many processes the questions have yet to be asked. Brave attempts to model pedogenetic processes have been hampered by the inadequacy of field observations. Field observations are usually empirical and opportunistic rather than systematically designed to support work on a deductive model.

Lack of understanding of the linkages between processes, especially how acceleration of one process may advance or retard the operation of another is a major restriction on hillslope modelling. The problem of scale is usually pushed aside rather than tackled head on. The validity of the assumption that processes acting on timescales of an order of magnitude different are effectively independent (Phillips, 1986) requires further checking.

While models of hillslope processes and changes in the short term, 10^4–10^5 s, are becoming satisfactory, those for longer time spans face many difficulties. A valid long-term landform evolution model may demand more intellectual and material effort than is presently available. Much hillslope modelling may still be in the realms of 'trans-science.'

REFERENCES

Ahnert, F. (1976). Brief description of a comprehensive three-dimensional model of landform development. *Z. Geomorph. N. F. Supplbd.*, **25**, 29–49.

Ahnert, F. (1980). A note on measurements and experiments in geomorphology *Z. Geomorph. N. F. Supplbd.*, **35**, 1–10.

Ahnert, F. (1982). Untersuchungen aber das Morphoklima und die Morphologie des Inselberggebiets von Machakos, Kenia. *Catena Suppl.*, **2**, 1–72.

Ahnert, F. (1987). Approaches to dynamic equilibrium in theoretical simulations of slope development. *Earth surf. processes landf.*, **12**, 3–16.

Armstrong, A. C. (1976). A three-dimensional simulation of slope forms. *Z. Geomorph. N. F. Supplbd.*, **25**, 20–28.

Armstrong, A. C. (1980). Simulated slope development sequences in a three-dimensional context. *Earth surf. processes*, **5**, 265–270.

Armstrong, A. C. (1987). Slopes, boundary conditions, and the development of convexo-concave forms—some numerical experiments. *Earth surf. processes landf.*, **12**, 17–30.

Baker, V. R. and Milton, D. J. (1974). Erosion by catastrophic floods on Mars and Earth. *Icarus*, **23**, 27–41.

Beven, K. J. and Kirkby, M. J. (1979). A physically based, variable contributing area model of basin hydrology. *Hydrol. Sci. Bull.*, **24**, 43–69.

Bonell, M., Coventry, R. J., and Holt, J. A. (1986). Erosion of termite mounds under natural rainfall in semi-arid tropical northeastern Australia. *Catena*, **13**, 11–28.

Bonell, M., Gilmour, D. A., and Cassells, D. S. (1983). A preliminary survey of the hydraulic properties of rainforest soils in tropical north-east Queensland and their implications for the runoff process. *Catena Suppl.*, **4**, 57–78.

Broadhurst, F. M. (1985). The geological evolution of North-West England. In Johnson, R. H. (Ed.), *The Geomorphology of North-West England*, (Ed. R. H. Johnson) Manchester University Press, Manchester.

Bruijnzeel, L. A. (1982). *Hydrological and biogeochemical aspects of man-made forests in south-central Java, Indonesia*, Academisch proefschrift, Vrije Universteit te Amsterdam.

Büdel, J. (1957). Die 'Doppelten Einebungsflachen' in den feuchten Tropen. *Z. Geomorph NF*, **1**, 201–228.

Büdel, J. (1982). *Climatic Geomorphology*, Princeton University Press, Princeton.

Butler, B. E. (1963). The place of soils in studies of Quaternary chronology in southern Australia. *Rev. Geomorph. dyn.*, **14**, 160–170.

Carson, M. A. and Kirkby, M. J. (1972). *Hillslope Form and Process*, Cambridge University Press, Cambridge.

Chorley, R. J. (1964). The nodal position and anomalous character of slope studies in geomorphological research. *Geogl. J.*, **130**, 70–73.

Church, M. and Mark, D. M. (1980). On size and scale in geomorphology. *Prog. phys. geogr.*, **4**, 342–390.

Cox, N. J. (1980). On the relationship between bedrock lowering and regolith thickness. *Earth surf. processes*, **5**, 271–274

Crouch, D. (1986). Science and trans-science in radiation risk assessment: child cancer around the nuclear fuel reprocessing plant Sellafield, UK. *Sci. total Env.*, **53**, 201–216.

Culling, W. E. H. (1983). Slow particulate flow in condensed media as an escape mechanism: mean translation distance. *Catena Suppl.*, **4**, 161–190.

Day, D. G. (1978). Drainage density changes during rainfall. *Earth surf. processes*, **3**, 319–326.

Douglas, I. (1980). Climatic geomorphology. Present day processes and landform evolution. Problems of interpretation. *Z. Geomorph. N.F. Supplbd.*, **36**, 27–47.

Dury, G. H. (1980). Neocatastrophism? A further look. *Prog. Phys. Geogr.*, **4**, 391–413.

Erhart, H. (1956). *Le genese des sols, en taut que phenomène géologique. Esquisse d'une theorie géologique et géochimique. Biostasie et rhexistasie.* Masson, Paris.

Francis, C. F., Thomas, J. B., Romero, Diaz, A. Lopez Bermudesz, F., and Fisher, G. C. (1986). Topographic control of soil moisture, vegetation cover and land degradation in a moisture stressed Mediterranean environment. *Catena*, **13**, 211–225.

Freeze, R. A. (1978). Mathematical models of hillslope hydrology. In Kirkby, M. J. (Ed.), *Hillslope Hydrology*, Wiley, Chichester, 177–225.

Hathout, S. A. and Bunting, B. T. (1975). A tentative re-examination of the soil catena in the type area of Kware, north-west Tanzania. *Catena*, **2**, 351–364.

Hazelhoff, L., Van Hoof, P., Imeson, A. C. and Kwaad, F. J. P. M. (1981). The exposure of forest soil to erosion by earthworms. *Earth surf. processes Landf.*, **6**, 235–250.

Herwitz, S. R. (1986). Infiltration-excess caused by stemflow in a cyclone-prone tropical rainforest. *Earth surf. processes Landf.*, **11**, 401–412.

Herwitz, S. R. (1987). Rainfall totals in relation to solute inputs along an exceptionally wet altitudinal transect. *Catena*, **14**, 25–30.

Huggett, R. J. (1985). *Earth Surface Systems*, Springer, Heidelberg.

Imeson, A. C. (1985). Geomorphological processes, soil structure, and ecology. In Pitty, A. F. (Ed.), *Themes in Geomorphology*, Croom Helm, Beckenham, 72–84.

Imeson, A. C. and Kwaad, F. J. P. M. (1976). Some effects of burrowing animals on slope processes in the Luxemburg Ardennes. Part 2: The erosion of animal mounds by splash under forest. *Geogr. Annlr. A.* **58**, 317–328.

Jones, J. A. A. (1981). *The nature of soil piping: a review of research*, (BGRG Research Monograph 3) Geobooks, Norwich.

Kirkby, M. J. (1980). The stream head as a significant geomorphic threshold. In Coates, D. R. and Vitek, J. D. (Eds.), *Thresholds in Geomorphology*, Allen & Unwin, London, 53–73.

Kirkby, M. J. (1985). A model for the evolution of regolith-mantled slopes. In Woldenberg, M. J. (Ed.), *Models in Geomorphology*, Allen & Unwin, London, 213–228.

Komanda, A. (1978). Le role des termites dans la mise en place des sols de plateau dans le Shaba meridional. *Geo-Eco-Trop*, **2**, 81–93.

Millington, A. C. (1981). Relationship between three scales of erosion measurement on two small basins in Sierra Leone. *Int Assoc Hydrol Scient Publ*, **133**, 126–143.

Ohmori, H. (1983). A three-dimensional model for erosional development of mountains on the basis of relief structure. *Trans Japan Geomorph Un*, **4**, 107–120.

Ollier, C. D. (1959). A two cycle theory of tropical pedology. *Soil Sci.*, **10**, 137–148.

Ollier, D. D. (1976). Catenas in different climates. In Derbyshire, E. (Ed.), *Geomorphology and Climate*, Wiley, Chichester, 137–169.

Overrein, L. N., Seip, H. M., and Tolan, A. (1981). *Acid precipitation—effects on forest and fish: Final Report of the SNSF-project 1972–1980*, SNSF project, Oslo.

Patton, P. C. and Baker, V. R. (1976). Morphometry and floods in small drainage basins subject to diverse hydrogeomorphic controls. *Wat. Resour. Res.*, **12**, 941–952.

Philip, J. R. (1975). Some remarks on science and catchment prediction. In Chapman, T. G. and Dunin, F. X. (Eds.), *Prediction in Catchment Hydrology*, Australian Academy of Science, Canberra, 23–30.

Phillips, J. D. (1986). Sediment storage, sediment yields and time scales in landscape denudation studies. *Geogrl. Anal.*, **18**, 161–167.

Pity, A. F. (1971). *Introduction to Geomorphology*, Methuen, London.

Radwanski, S. A. and Ollier, C. D. (1959). A study of an East African catena. *J. Soil Sci.*, **10**, 149–168.

Raynal (1960). Les éboulis ordonnés au Moroc. *Biul. Peryglac.*, **8**, 21–30.

Reich, P. B. and Borchert, R. (1984). Water stress and tree phenology in a tropical dry forest in the lowlands of Costa Rica. *J. Ecol.*, **72**, 61–74.

Remmelzwaal, A. (1978). Soil genesis and Quaternary landscape development in the Tyrrhenian coastal area of south-central Italy. *Publs Fys-Geogr Bodemk Lab Univ Amsterdam*, **28**, 311pp.

Scheidegger, A. E. (1970). *Theoretical Geomorphology*, 2nd Ed., Springer-Verlag, New York.

Simonett, D. S. (1967). Landslide distribution and earthquakes in the Bewani and Torricelli Mountains, New Guinea. In Jennings J. N. and Mabbutt, J. A. (Eds.), *Landform Studies from Australia and New Guinea*, ANU Press, Canberra, 64–84.

Starkel, L. (1976). The role of extreme (catastrophic) meteorological events in contemporary evolution of slopes. In Derbyshire, E. (Ed.), *Geomorphology and Climate*, Wiley, Chichester, 203–246.

Stromquist, L. (1985). Geomorphic impact of snow melt on slope erosion and sediment production. *Z. Geomorph. N.F.*, **29**, 129–138.

Tallis, J. H. (1985). Erosion of blanket peat in the southern Pennines: new light on an old problem. In Johnson, R. H. (Ed.), *The geomorphology of North-west England*, Manchester University Press, Manchester, 372–406.

Tallis, T. H. and Johnson, R. H. (1980). The dating of landslides in Longdendale, north Derbyshire, using pollen-analytical techniques. In Cullingford, R. A., Davidson, D. A., and Lewin, J. (Eds.), *Timescales in Geomorphology*, Wiley, Chichester, 189–205.

Taltasse, F. (1957). Les Cabeças de Jacare et le role des termites. *Rev Géomorph dyn*, **8**, 167–170.

Thornes, J. B. (1980). Conservation practices in erosion models. In Morgan, R. P. C. (Ed.), *Soil Conversation Problems and Prospects*, Wiley, Chichester, 265–271.

Thornes, J. B. and Gilman, A. (1983). Potential and actual erosion around archaeological sites in south-east Spain. *Catena Suppl*, **4**, 91–113.

Tricart, J. (1957). Observations sur le role ameublisseur des termites. *Rev Géomorph dyn*, **8**, 170–179.

Tricart, J. (1974). Phenomènes demesurés et regime permanent dans des bassins montagnards. *Rev Géomorph dyn*, **23**, 99–114.

Tricart, J. (1978). Le sol dans l'environment écologique. *Rev Géomorph dyn*, **27**, 113–128.

Tricart, J. (1979). Paysage et ecologie. *Rev. Géomorph dyn*, **28**, 81–95.

Twidale, C. R., Bourne, J. A., and Smith, D. M. (1974). Reinforcement and stabilisation mechanisms in landform development. *Rev Géomorph dyn*, **23**, 115–125.

Van Dijk, D. C. (1959). Soil features in relation to erosional history in the vicinity of Canberra. *CSIRO Soils Publ*, **13**.

Walker, P. H. (1966). Postglacial environments in relation to landscape and soils. *Iowa Agric Expt Sta Bull*, **549**, 838–875.

Ward, R. C. (1985). Hypothesis testing by modelling catchment response, II. An improved model. *J Hydrol*, **81**, 355–373.

Weinberg, A. M. (1972). Science and trans-science. *Minerva*, **10**, 209–222.

Whitmore, T. C. (1978). Gaps in the forest canopy. In Tomlinson, P. B. and Zimmerman, M. H. (Eds.), *Tropical Trees as Living Systems*, Cambridge University Press, Cambridge, 639–655.

Williams G. P. and Guy, H. P. (1973). Erosional and depositional aspects of hurricane Camille in Virginia 1969. *U.S. Geol Sur Prof Pap.*, **804**.

Yair, A. and Levee, H. (1981). An investigation of the source ares of sediment and sediment transport by overland flow along arid hill-slopes. *Int Assoc Hydrol Scient Publ*, **133**, 433–446.

Yair, A. and Rutin, J. (1981). Some aspects of the regional variation in the amount of available sediment produced by isopods and porcupines, northern Negev, Israel. *Earth surf processes Landf.*, **6**, 221–234.

Yoshikawa, T. (1984). Geomorphology of tectonically active and intensely denuded regions. *Geographical Review of Japan Series A*, **57**, 691–702.

Yoshikawa, T. (1985). Landform development by tectonics and denudation. In Pitty, A. F. (Ed.), *Themes in Geomorphology*, Croom Helm, Beckenham, 194–210.

Modelling Geomorphological Systems
Edited by M. G. Anderson
©1988 John Wiley & Sons Ltd.

Chapter 15

Computer simulation in geomorphology

S. HOWES
Logica Space and Defence Systems Limited

and

M. G. ANDERSON
Department of Geography,
University of Bristol

15.1 INTRODUCTION

Geomorphology comprises an inquiry into the relationship between forms and process. To aid this inquiry, the geomorphological environment is visualized as a system. Geomorphological systems are complex and relationships are rarely straightforward or apparent. These natural environmental systems exhibit a number of characteristics which render description, explanation, and prediction, very intricate tasks. These characteristics include:

1. Infinitely complex structures. It is thus difficult to define meaningful system boundaries which would enable the subject of investigation to be subdivided into manageable and relevant entities.
2. Inseparable and intricate processes, inherent spatial and temporal variability, and significant measurement error associated with even the most carefully designed field measurement program. It is thus impossible to measure accurately and hence to describe system structure and behaviour.
3. The manner of operation of geomorphological systems is scale dependent. Depending upon the scale at which any environmental system is viewed, there are quite different sets of laws which operate. Geomorphological theory has been mostly formulated at the small scale. At this scale, validation with data from the laboratory or field plot scale has been feasible. It has been assumed that behaviour of the system at a larger scale, for example the catchment, can be understood by the simple aggregation of this theory for larger areas. This simple aggregation disregards the very continuity of the real system.

4. Randomness is an inherent feature of all natural environmental systems. The search for explanation and prediction in geomorphology cannot therefore be restricted to purely deterministic approaches.

The choice of a suitable methodology for understanding and predicting the nature of geomorphological processes and their relationship to landform is therefore a very interesting problem. Within the last fifteen years, certain geomorphologists have selected computer simulations as a possible methodology for exploring and developing these relationships. Potentially, computer simulation provides an efficient means to rapidly examine the validity of a range of possible relationships under a variety of assumptions. In addition, the application of a computer simulation methodology by geomorphologists is a useful pursuit as it demands clarity of thought and precise specification and commitment.

Consequently, there has been an increase in the popularity of computer simulation in geomorphology and at this point, it is interesting to consider possible reasons which may have contributed towards this trend. Geomorphology has not always been held in high regard by the wider scientific community and the adoption of a computer simulation methodology can be viewed as an attempt to meet the 'scientific standards' to which these other more highly regarded disciplines aspire. In addition, the more recent pursuits of the applied geomorphologist have to date mostly contributed very basic descriptive skills to contemporary and practical issues. A more interesting and desirable contribution could comprise increased involvement in the provision of prediction and control. Computer simulation can be considered to be a suitably formal and rigorous methodology for the provision of these capabilities. For example, computer simulation models have been developed to provide avalanche warning capability (Judson et al., 1980) and to predict the forces and pressures of avalanche impact upon structures and obstacles (Lang and Dent, 1980). There are also a very large number of computer simulation models used in hydrological and water resource applications. However, despite the potential suitability of computer simulation to applications in geomorphology, it must be stressed that computer simulation is now an extensive and highly advanced discipline in its own right. A simulation exercise involves the geomorphologist not only in the theory of his or her own discipline but also demands familiarity of subjects such as computer science, statistics, mathematics, and operations research. Indeed Bratly et al. (1983) have warned of the problems and difficulties which are involved in simulation. They stressed that a successful simulation program is neither as cheap nor as easy to implement as might initially be perceived.

To embark upon a simulation exercise is clearly an important decision and it is therefore opportune to examine the requirements of computer simulation and to discuss the relevance of this methodology to geomorphology. It is important to establish whether simulation in geomorphology can attain the standards of logicality demanded of scientific methods in other disciplines.

In order to examine the relevance of computer simulation to geomorphology, this paper will be divided into the following three sections:

1. An examination of the nature of computer simulation will be provided with one intention of illustrating the very wide base of knowledge and experience which is required successfully to formulate and implement a computer simulation model.
2. A discussion on model evaluation will emphasize that it is of crucial importance for a suitable methodology and for sufficient data to be provided to effect a comprehensive evaluation of all aspects of the computer simulation model. Simulation models should

be regarded as scientific statements which require falsification. Formal evaluation is necessary to establish the status and relative contribution of simulation models within the context of the proposed application.

3. The requirements of a scientific computer simulation application will be presented and the following discussion will attempt to evaluate the degree to which these requirements can be met in geomorphological studies.

15.2 THE NATURE OF COMPUTER SIMULATION

In order to illustrate the nature of computer simulation and to draw attention to the range of skills involved, this section will concentrate upon the following issues. Firstly, a definition of computer simulation will be provided. Secondly, two fundamental approaches to the development of a computer simulation model, the realist and the functionalist approach, will be considered. Thirdly the need for a clear definition of model application will be discussed. Finally certain aspects of software design and implementation are presented.

15.2.1 Definition of Computer Simulation

Computer simulation is here envisaged as comprising two major stages: firstly the development of a suitable mathematical model, and secondly the design and implementation of the computer simulation model.

A mathematical model which describes the structure, processes and interrelationships of the geomorphological system which are relevant to the aims of the analysis must initially be defined. To effect these mathematical statements as a computer simulation model, it is necessary to provide a computer program which implements the mathematical model in a form which, when driven by suitable inputs and for comparable situations, can imitate the dynamic behaviour of the real system. The changes which the model undergoes are determined by the manner in which the mathematical definition of the system operates on the inputs.

15.2.2 Approaches to the Development of a Computer Simulation Model

Two fundamentally different approaches are considered here, and after Bennett and Chorley (1978), these are termed realist and functionalist. Figure 15.1 serves to summarize and clarify the major differences between these two approaches.

The realist approach is based on the requirement to describe the real system in terms of the fundamental laws of science, for example, continuum mechanics: the conservation of mass, energy, and momentum. This rigorous scientific approach provides the potential to describe the mechanics of all relevant processes, the nature of their interaction, their spatial variability in three dimensions, and their temporal variability. The geomorphologist must also be able to define the manner in which processes interact with landform. The parameters utilized in this mathematical description of the system have physical interpretation, i.e. they represent characteristics such as dimension, velocity, or temperature, which can be measured in the context of the real system.

These physically based models are necessarily complex and in many cases, the description of the dynamics of the system involves the use of differential and non-linear equations.

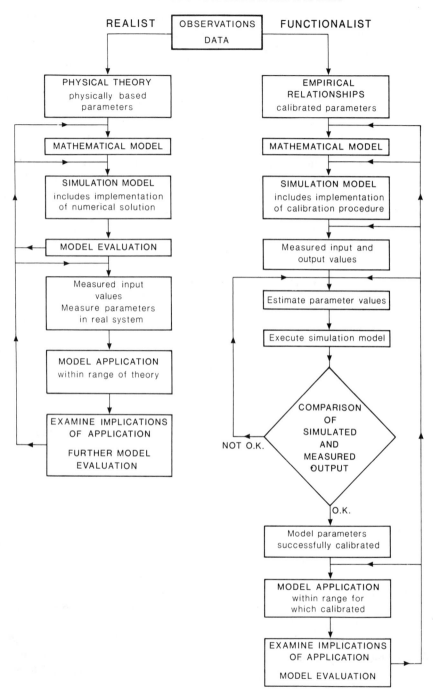

Figure 15.1. Realist and functionalist approaches (after Bennett and Chorley, 1978)

The development of the simulation model therefore requires the selection of one of the appropriate numerical methods for solving these equations over a range of realistic initial and boundary conditions. An appropriate simulation technique is required depending on whether the system is discrete, continuous, or combines elements of each.

Once the simulation model has been implemented and tested to satisfactory standards, application requires that suitable initial parameter values and input values for the system be specified. These values are derived from measurement of the prototype system. It is important that the model should only be applied within the range of the theory from which it has been developed. The implications of the results and further testing can then be considered. The crucial issue of model testing is considered in the following section of this chapter.

Certain examples of the application of this realist approach to simulation modelling which have relevance to geomorphology include the following. Lang *et al.* (1979) provided a two-dimensional simulation model of snow avalanche flow which was based upon transient fluid dynamics theory. The model was designed to provide predictions of avalanche speed, location and depth of avalanche debris, and the runout distance. Dent and Lang (1980) also provided an avalanche simulation model where snow flow was represented as a transient process based upon the numerical solution of the Navier–Stokes equations. Kvasov and Verbitsky (1982) documented a simulation model for the evolution of ice sheets which was based upon the solution of the complete non-linear hydrodynamical equations (see, too, Chapter 1).

These computer simulation models are most useful for investigating and interpreting the detailed causal mechanisms of the various geomorphological process and their controls on form. Richards (1978) documented the application of hydraulics and fluid mechanics theory based on gradually varying floor assumptions to streams with stable riffle pool topography. The simulation model was designed and applied to explore the controls on the hydraulics of flow. Brugnot and Pochat (1981) used the computer simulation of avalanches for isolating and examining the range of parameters and range of snow flow important for understanding avalanches. These parameters cannot be isolated in field studies.

The functionalist approach, which is also illustrated in Figure 15.1 and which contrasts with the realist, describes the system in terms of statistical, or empirical, relationships. Such empirical models may range in complexity from a simple equation involving a single parameter which itself represents an index of the net effect of a range of characteristics and processes, to more complex suites of equations involving a much larger number of parameters.

Models developed from empirical equations may support a conceptual structure, i.e. they may be divided into subsystems which have a physical basis, but the relationships within these subsystems remain empirical. For example Davidson-Arnott (1981) documented a two-dimensional computer simulation model of nearshore bar formation on gently sloping sandy coastlines. This was developed specifically from empirical relationships derived from field data collected in Kouchibougnac Bay, New Brunswick. Ahnert (1976, 1977) developed a three-dimensional slope development computer simulation model which was used to determine the slope forms which result from the application of various geomorphological processes (see, too, Chapter 14). The basis of the processes was conceptual, although their operation was described by empirical relationships.

For the functionalist approach, exact replication of the real system is not considered to be a critical issue. In contrast, therefore, to the realist approach, these models provide little insight into the internal mechanisms of the system and are therefore not designed to aid explanation. Due to their inherent simplicity, they sacrifice scientific rigour and run the risk of not fully representing the most important characteristics of the system.

As indicated in Figure 15.1, the parameters involved in empirical relationships do not always have physical interpretation. In this case, values must be derived by a procedure of calibration. The parameter values are estimated by comparing a series of measured and model outputs which correspond to a given series of inputs. The parameter values are adjusted until the predicted output fits the measured.

A procedure for calibration must therefore be provided when developing the simulation model based upon empirical relationships. In addition, when the simulation model is applied, historical data is required to operate the calibration procedure.

The major aim in calibrating a model is to obtain a unique parameter set which is in some sense physically realistic. The main problem which is experienced is that of an inability to obtain such a solution. In practice, a unique solution cannot be attained; several combinations of parameter values can produce the same result. Calibrated parameter values are also not always physically realistic, but may operate to balance out errors which occur in the data or in the model structure.

Three reasons can be proposed which contribute towards this problem:

1. A poor quantity and/or quality of the data used for calibration can inhibit the realization of a unique and conceptually realistic parameter set. These data will contain measurement error. Additionally, the data may be poor 'activating' data in that if the data do not adequately represent all conditions for which the model is designed, certain processes may not be activated when the model is run. The parameters which describe this non-activation process cannot therefore be calibrated.
2. Error in the model structure also inhibits the attainment of the unique solution. Imperfect representation of the physical processes, and a large number of interacting parameters will adversely affect the success of the calibration procedure.
3. The geomorphological system is almost always indeterminate; the number of model parameters is greater than the number of equations describing the system. There are a large number of degrees of freedom and consequently an infinite number of solutions is possible. The derivation of the optimum parameter set relies far too heavily in the calibration procedure utilized: the selection of the objective function and the search procedure.

The search procedure may be analytical, automatic, or manual. For complex models in which many parameters (which are not necessarily independent) require calibration, the procedure is not always systematic or objective. The ability to adjust parameter values to derive that combination which provides the best estimates is largely a function of the skill and experience of the operator. An objective evaluation is replaced by a subjective assessment, implicit in the operator's knowledge of the system and model.

This discussion of simulation models which are based upon empirical relationships has therefore revealed the following characteristics:

1. An empirically-based model is only applicable to those conditions for which the parameters have been calibrated. We have stressed that geomorphological systems are inherently complex, non-linear, spatially and temporally variable. There exists no physical

justification for the extrapolation of empirical relationships based on the assumption that behaviour will be similar.

2. There are a number of serious problems associated with the procedure of calibration. It is generally considered that insufficient emphasis is placed upon the search for a unique and conceptually realistic parameter set. Many calibration efforts settle on that parameter set which provides the best fit. This has the result that the model has a very poor forecasting ability and thus little confidence can be placed in the operation of the model.

3. Empirical models cannot, by definition, be used to explore or to understand the internal operation of geomorphological systems as their parameters do not have a physical basis. In addition, they cannot be applied beyond the range of conditions for which they are calibrated. These two facts severely limit the potential application of these models. Indeed they are restricted to the position of prediction for that range of conditions for which data is already available. Their application to these conditions is largely academic.

A case has therefore to be made which argues for the philosophical superiority of physically based computer simulation models. These are compatible with our conceptual knowledge of geomorphological systems. These models have physically based parameters which do not require calibration. These models therefore have application to conditions which have not been measured and are consistent with the idea of a unique response. Physically based computer simulation models also have a utility in the determination of the internal mechanisms of geomorphological systems. In this regard, it is important to acknowledge the inevitable extension of the research time frame for the acceptance of such a research philosophy. Parallels in physical hydrology currently suggest, at best, a cautious optimism for such an approach. Given the current emphasis on modelling the interaction of hydrological and geomorphological systems, it is thus important for geomorphologists to avoid isolationism in their modelling approach.

15.2.3 Definition of Model Application

Prior to the initial design of a computer simulation model, it is essential that a precise definition be provided of the objectives and scope of the intended application. A computer simulation model must be designed towards and applied within very specific applications. This is important for three reasons:

1. All natural environmental systems are far too complex and intricate to describe in totality. Therefore, in any modelling exercise, various approximations to reality have to be made, restrictions are placed upon model operation, and certain factors believed to be unimportant have to be neglected. This degree of simplification is necessary in order that the problem remains tractable. It is important that the model only reflects the essential features and behaviour of the real system which are relevant for the application in mind. For example Hey (1979) simulated the development of channels through space and time using a preliminary and highly simplified mathematical model. Insufficient theory regarding all of the relevant process equations restricted the derivation of a fully deterministic mathematical model. The application of the model was therefore restricted to straight gravel bed rivers (as there were no suitable equations to define the meander process or bed forms in sand bed channels), to average bankfull stage (as it is not possible to predict channel response to every flow), and to the initial phase of channel development (where the river is controlled by transport rather than the weathering process) — see, too, Chapter 5.

2. Once implemented, a computer simulation model must be tested and thereby validated. Detailed attention will be paid to this most important aspect of computer simulation modelling in the following section of this chapter. However, it is germane to consider at this point, that as such models are complex, there is not always the time nor the facilities to examine and to validate every mode of behaviour under all possible combinations of conditions. Indeed one of the major problems in the evaluation of such models is deciding what should be evaluated. The assessment of a model must necessarily be made from a limited number of experimental frames. Therefore the selection of exframes will influence the utility of the comments which can be made. Where a clearly defined application has been provided for a model, it is possible and more meaningful to select those frames which enable tests to be undertaken of those particular forms of behaviour which are relevant to the specific application.

3. Once a computer simulation model has been satisfactorily tested, it could be applied to almost an infinite combination of conditions. Very extensive volumes of information may thus be created. Once again, it is essential that an appropriate research design, which meets the aims, requirements, and slope of the desired application be well structured and defined. In the absence of such an experimental design, the interpretation and assessment of results becomes meaningless.

One final point which must be stressed in the context of a well defined application for computer simulation models is that the model should not be applied beyond its design limits. All applications should be consistent with the theoretical and design basis of the model.

15.2.4   Software Design and Implementation

The software which implements the computer simulation model should be understandable, maintainable, and reliable. However, the program will often be very long and complex. In order to ensure these three characteristics, certain aspects of computer science become relevant.

There are a number of considerations which must be made before system design. These involve the selection of a suitable machine for implementation. As far as the hardware is concerned, the Central Processing Unit must have sufficient speed and range of functionality for the required application. Sufficient storage must be available, and suitable peripherals for input and output should be supported. For example the input of graphical information may be desirable. The operating environment provided by the machine should support the language which is to be utilized, and also an appropriate programming environment. This should include a suitable editor, debugging tools, and access to any required subroutine libraries, for example mathematical or graphics libraries.

In design of the computer program, there are certain features which should be considered. Firstly, a programming language should be selected. It has been traditional in disciplines such as Hydrology or Ecology, for General Purpose Languages such as FORTRAN-77 to be used for simulation exercises. However, implementation in a high level simulation language should be considered. Simulation languages have been written which, for example, enable non-linear partial differential equations to be programmed without detailed familiarity with numerical methods and programming. Examples of simulation languages include CSMP (Continuous Simulation Modelling Program) for continuous models (those described by differential equations), GPSS (General Purpose Simulation System) for discrete models (where the state of the system changes at given points), and Simscript, for combined models. Simulation languages hold certain advantages over general purpose languages in that they provide code which is often more convenient, transparent, and concise. Hence the code is easier to assimilate and to evaluate. The sensible use of simulation languages can be highly cost effective as they are less demanding in terms of involvement of the analyst with programming details. It is thus easier and quicker to commit a mathematical model to a computer program. However, as Bratley *et al.* (1983) have stated, where simulation languages are used, convenience may be traded for control. The resultant model may include behaviour which is neither entirely understood, nor predictable.

The design of the computer program should also include a definition of an appropriate structured and modular design for the code and a statement regarding data structures which are to be utilized. If the simulation exercise is to involve a large amount of data, a database structure must be conceived, and an interface between both the simulation program and the user must be designed.

As the operation of the simulated system should not deviate under any circumstances from the functional specification provided by the mathematical definition of the system, it is advantageous, where code becomes necessarily long and complex, to provide error recovery control structures and error tracking facilities which monitor the system, and which will prevent incorrect operations from occurring.

When the code has been implemented, and the performance of the model has been evaluated and accepted, many designers might be tempted to 'optimize' the code. As a general principle, however, it is suggested that this procedure should not be undertaken. The only reason for optimization is to make a simulation program execute faster. However, in so doing, the code is rendered more difficult to maintain and to understand. When the code has been successfully evaluated, the best policy is to leave it alone as any change always has more serious implications than are initially intended. The only reason for optimization is where run time is critical, for example for real-time operation or where charges for Central Processing Unit seconds are critical.

This section has illustrated that certain aspects of computer science experience may be necessary in order to undertake a significant simulation exercise.

15.3 MODEL EVALUATION

In geomorphology, as in other areas, the relative amount of research undertaken in differing techniques related to modelling stages has been somewhat distorted. Figure 15.2 suggests that perhaps the greatest effort has been paid to parameter estimation problems whilst other important issues have been neglected. A central objective of this chapter is thus to propose that a suitable strategy must be adopted to effect a comprehensive evaluation of *all* aspects of any computer simulation model. If the geomorphologist ignores the requirement for such a scheme, then a computer simulation methodology cannot be accepted as a rigorous tool of scientific investigation in geomorphology.

This section will examine the role and importance of evaluation. A three stage computer simulation model evaluation strategy will also be presented which illustrates the scope and depth of evaluation which is required in order to demonstrate the simulation exercise to be a reasonable and worthwhile scientific pursuit.

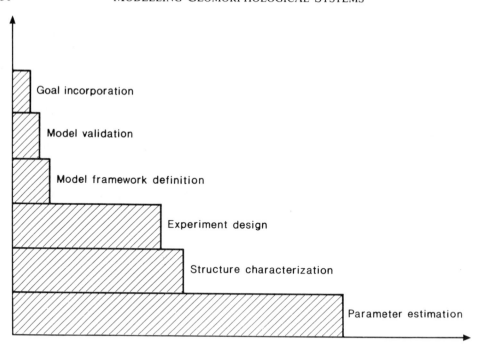

Figure 15.2. Effort currently expended on techniques for the different stages of modelling (after Van Steenkiste and Spriet, 1982)

15.3.1 Importance of Model Evaluation

Leimkuhler (1982) discussed certain methodological problems in modelling in the context of energy systems. The following quotation illustrates some of the major problems which he envisaged. It is suggested here that those problems could also be encountered in computer simulation in geomorphology.

> 'The circumstances are familiar: every time one attends a conference concerned with progress in the modelling business, a number of new or "almost new" models (taking into account the habit of "scientific recycling") with promising features and, hopefully, great explanatory and prognostic or decision-aiding power is presented. The procedures of such a presentation are mutually congruent: in front of experts (or what has to be taken as such) a "model" is announced, and goals and aims of a project for which a model and the use of a model were thought to be of value are shortly mentioned. Then, the model conception is given a brief outline, some important structural features are exposed, most of the underlying assumptions are not mentioned, some results are presented, more of them are promised for the not-too-distant future; and under mild applause one modelling expert gives way to another, who proceeds in a most similar manner.' (Leimkuhler, 1982, p. 61).

Leimkuhler stressed that energy models are rarely fully evaluated. Evaluation is a methodology for establishing the suitability and relevance of a computer simulation model for a particular application and for assessing the level of confidence associated with the information derived from the model. The provision of such information must be considered to be important for the following three reasons:

1. Simulation models can be considered to be scientific statements which require falsification. Model evaluation should establish the relative, effective contribution of simulation models thereby enabling them to be accepted or rejected. Any discipline which utilizes computer simulation models will not advance if vast numbers of untested models are allowed to accumulate.
2. It is the responsibility of the model designer to clarify and to communicate the intended range of application of a computer simulation model. This enables a user correctly to apply the model and to interpret results. Frenkiel and Goodall (1978) have stressed that failure to fully discuss a model's limitations and its consequent application to inappropriate conditions will lead ultimately to a lack of faith in modelling.
3. Physically based computer simulation models are commonly highly sensitive to parameter and input values, or to the numerical methods which are utilized. Even for a model firmly based in theory, significant derivations from reality can be experienced. Evaluation is consequently an important and highly necessary checking procedure.

One possible reason for the notable lack of model evaluation in many disciplines is that there is no commonly accepted methodology as to how this process should be carried out, other than that the procedure must be objective, and applied within the context of the proposed application. It is the proposed application which conditions the appropriate level of detail and precision which can be accepted. Van Horn (1971) considered that model evaluation is not a procedure which can be generalized, but is unique to the specific model and application. Gilmour (1973), however, has documented an attempt to provide a more generalized procedure.

In some modelling studies, it is interesting to note that the term validation has been used specifically to refer only to the process of statistically comparing model output to independent historical series. This process has been applied when model formulation has been completed, and all of the data collected. The sole criterion used to determine a valid model is based upon empirical testing. It is suggested here that model evaluation should encompass a broader series of techniques some of which can be applied during model design, and before significant resources have been wasted.

15.3.2 Mathematical Model Validation

This is the first stage of model evaluation and refers to the process of establishing the model's face validity. It is basically a subjective procedure, based on discussion, aimed at establishing that the assumptions made about the real system by the model are reasonable and that the model adequately reflects the essential features and behaviour of the real system which are relevant to their application in mind.

The need for mathematical model valuation arises since in any modelling exercise, various approximations to reality must be made. No model can be completely comprehensive. As a model's performance is conditional upon the authenticity of its assumptions, it is important that the model be accurately defined and disparities between the model and real world clearly specified. Thus this part of evaluation assesses the model's realism and logical structure to establish that it is internally consistent. It is designed to be applied at a very early stage during model specification and initial development.

15.3.3 Computerized Model Verification

It is important to establish that the computer program which implements the mathematical model is reliable. The majority of software errors are known to occur during translation of a mathematical model into computer code and it is important therefore that the mathematical model be correctly specified and that care be exercised during translation. Hence the second stage of model evaluation involves a series of techniques which are designed to ensure that the computer program actually carries out the logical processes expected of it, that all geomorphological procedures act and interreact rationally, and that the program is consistent with the functionality of the mathematical model.

The literature suggests several aspects of the program operation which should be checked. It is important for example to demonstrate that if the model inputs are held constant, that over several runs of the model there is no variance of the output. This is referred to by Hermann (1967) as establishing the model's internal validity. It should also be established that the continuity condition is satisfied during operation of the model. Bratley *et al.* (1983) also suggested that at a basic level, results derived from a short computer simulation be compared to the results of a hand calculation. They also suggested that the parameter values should be stressed to indicate whether or not the model provides sensible output for infrequent events or conditions. There are many errors or hidden modes of behaviour in a program which may only appear under stressed conditions. The period of time for which the model remains both numerically and physically stable should be established, beyond this point errors may accumulate and predictions become unreasonable.

A very important diagnostic procedure which may be utilized during computerized model verification is a sensitivity analysis. This can be used to examine the effect of variations in model input and parameter values upon model behaviour or output. Such an analysis is achieved by comparing the rate of change of model output with the rate of change of the model input or parameter value.

In the context of model evaluation, a sensitivity analysis can be used to:
1. Demonstrate that in response to representative variation of model input and parameter values, theoretically realistic model behaviour is experienced.
2. Illustrate the model to be sufficiently sensitive to represent actual variation in the prototype system.
3. Identify those model parameters or inputs to which the model is most sensitive.

There are two possible approaches to sensitivity analysis. The first follows a deterministic, and the second, a stochastic methodology. These are outlined only very briefly.

The deterministic sensitivity analysis considers the influence of a small change in parameter value on the model output. This is achieved by differentiation or by factor perturbation (McCuen, 1973, 1976). The general definition of sensitivity is given by:

$$S(F_i) = \frac{\partial F_o}{\partial F_i} \tag{1}$$

where:

$S(F_i)$—sensitivity function of factor F_i
F_o—model output
F_i—model input parameter.

For each factor, the sensitivity function can be derived, which estimates quantitatively the effect of that parameter upon output. This sensitivity function is not independent of the magnitude of the factor. To assess the relative importance of each factor therefore, the relative sensitivity function (R_S) has to be defined.

$$R_S = \frac{\partial F_o}{\partial F_i} \cdot \frac{\partial F_i}{\partial F_o} \qquad (2)$$

The second approach to the sensitivity analysis is a stochastic methodology. This is based upon the assertion that uncertainties in the model structure and data allow a meaningful analysis to deal with probabilities of behaviour. Model input parameters are randomly selected from probability distributions, which are a measure of the relative likelihood of different parameter values, according to a mean and standard deviation. The standard deviation is a measure of the amount of error associated with the specification of that parameter. Variation in model output relating to a much wider spread of data uncertainty can therefore be evaluated. This range typically covers the entire spread of physically realistic parameter values.

Miller *et al.* (1976) have pointed out that model evaluation has traditionally been postponed until such a stage has been reached in research when observed and simulated output can be compared. At this stage, however, large amounts of resources have already been committed to model formulation. Several authors have recently stressed the importance of applying a sensitivity analysis at a very early stage during model formulation (McCuen, 1973, 1976; Miller *et al.*, 1976; Hornberger and Spear, 1981). For a physically based computer simulation model, the sensitivity analysis provides an excellent quantitative method of verifying a simulation model which does not require comparison to a specific or extensive data set. Realistic data may be derived from the literature and used to examine the behaviour of the model throughout the many stages of model development. However, it should be noted that for calibrated, empirically based models, a sensitivity analysis would require that an adequate data base already be established for calibration and independent historical data to be gathered for model validation. This is illustrated in Figure 15.1.

One final point with respect to the sensitivity analysis should be made. The most successful analysis will have been well organized and structured to examine those very specific elements of behaviour and range of parameter and input variability which are relevant in the context of the proposed model application. The importance of research design in the implementation of the sensitivity analysis must be stressed.

In concluding this illustration of the range of techniques which may be devised in order to verify a computer simulation model, it must be stressed that despite the most diligent attempts, it is highly improbable that all software errors will be discovered and remedied. The presence of hidden errors in such complex computer programs should be an expectation of the user. The information provided by a computer simulation model should not be accepted blindly or uncritically. However, the improbability of locating all errors should not provide an excuse for the model builder not to attempt verifying the code. In addition, it should be emphasized that many software errors which do occur are most commonly a function of the manner in which the program is used rather than an inherent property of the program. Many users will take little time fully to appreciate the specification of the model which is provided. They will consequently apply the program outside its design limits.

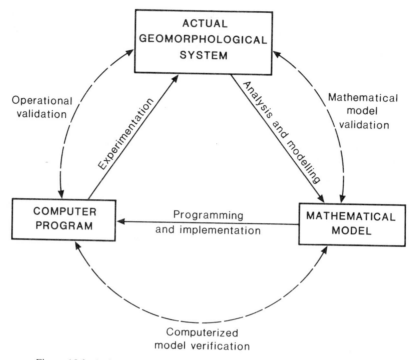

Figure 15.3. A three stage model evaluation strategy (after Sargent, 1982)

The errors which will occur cannot therefore be attributed to programming errors. However, those users who do fully acquaint themselves with the model specification may be unfortunate enough to have been provided with a specification which is inaccurate or insufficiently detailed. Once again, emphasis is placed upon the model developer to provide an appropriate model specification.

15.3.4 Operational Validation

This final stage of model evaluation serves to establish a measure of the extent to which the model and the program implementing it represent an accurate representation of reality. It aims, therefore, to assess the practical or theoretical significance of disparities between the behaviour of the real system and the computer simulation model. This is achieved by a comparison of measured and predicted outputs or behaviour for that range of conditions which are relevant in the context of the proposed model application. In association with these comparisons, those conditions beyond the model's range of application must also be clearly defined. During operational validation, it is impossible to investigate the behaviour of the model for all possible conditions. Assessment of the computer simulation model must therefore be made from a limited number of experimental frames. The design of these frames is therefore crucial.

This stage in model evaluation primarily involves a comparison of calculated and measured values. It is necessary to offer an assessment of the goodness of fit of the model's behaviour to the measured system. This may be achieved by the application of various graphical and numerical comparisons. It is suggested that a better impression of any model's performance is gained by the use of a number of indices. However, in addition, an important part of operational diagnostics includes specific examination of the nature of model error (measured minus predicted behaviour). This is essential as further model improvement can only be effected when the source of error, should this prove to be of significance, has been correctly identified (Weber *et al.*, 1973).

There are two distinct sources of model error: random and systematic. If the errors are random, and provided that they are small, then a model can be considered to be satisfactory in the context of the application for which it is evaluated. The error in this case may be attributed to random measurement error with respect to model input, parameter values, or output. However, where error is systematic, model modification or restructuring is required to remove the source of this error. The view we have presented of model verification and validation as consisting of mathematical model validation, computerized model validation and operational validity is shown in Figure 15.3.

Having illustrated the depth required of model evaluation, we now examine the detailed requirements of a computer simulation model.

15.4 REQUIREMENT OF COMPUTER SIMULATION

This chapter has argued for the desirability of a computer simulation model which is based upon sound theoretical principles. It is appropriate to outline more specifically both the conceptual and technical requirements of these physically based models, and to discuss the degree to which these requirements can be met in geomorphology.

Theory is a major conceptual requirement of these models. Both the temporal and spatial dimensions of the geomorphological system must be resolved in terms of sound theoretical principles rather than empirical generalizations. There are two specific technical requirements. Firstly, a sufficient quality and quantity of data is required to evaluate and to apply the model. Secondly, suitable techniques for implementing the computer simulation and numerical solution must be determined. These must be relevant to the dynamic and spatial behaviour of the prototype system.

15.4.1 Conceptual Requirements

It is to be proposed that in geomorphology, as indeed in any other natural environmental discipline, the conceptual requirements of these computer simulation models cannot be met. This proposal is based upon the following notion. The procedures for mathematical modelling and simulation were designed for use in physical, deterministic systems, or in the 'hard' sciences. These are clearly definable systems, in which the general theoretical principles are well known, for example, electrical network systems, weapon systems, or industrial systems. Simulation is based upon an assumption that mathematics can clearly and distinctly describe the system and is therefore the only basis for sound knowledge. The geomorphological system has been described in the introduction to this chapter as a complex, inaccessible, scale dependent and inherently random natural system. The following discussion aims to illustrate that due to these characteristics, the requirement for precise theoretical description cannot fully be met.

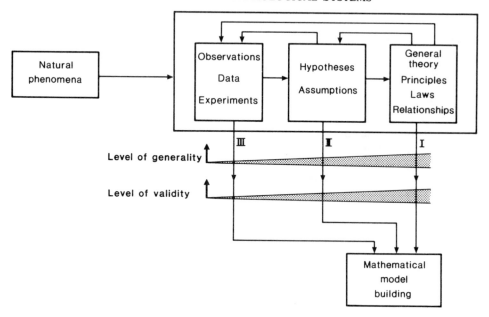

Figure 15.4. The progression of scientific investigation and three possible routes for mathematical model building (after Van Steenkiste and Spriet, 1982)

Geomorphology does contain a few established theories, but it mostly consists of a large number which are vague and questionable. As Haines-Young and Petch (1986) have emphasized, landforms are extremely difficult objects of study. Direct experimentation is usually impossible and this inhibits the development of theories which are sufficiently well formulated to be tested critically. They claim that geomorphology is 'intrinsically unscientific'.

Van Steenkiste and Spriet (1982) considered that a very similar condition of unsatisfactory theoretical description exists for biological systems. They determined this to be one cause of certain mathematical modelling exercises in biology being considered unsatisfactory. It is interesting to consider their argument in that a claim can also be made for geomorphological modelling. Figure 15.4 is based upon a figure which Van Steenkiste and Spriet (1982) produced, and illustrates all scientific investigation as beginning with observations and progressing iteratively, rather than consecutively, to a stage where general theory can be developed. Mathematical modelling may be attempted at any of these evolutionary stages, but as suggested by the figure, the models derived from each route are associated with varying degrees of generality and validity. Biological systems are not considered to be sufficiently well defined to allow modelling to be approached via Route I. This could be considered to be the case for theoretical geomorphology. In these environmental systems, mathematical models are typically approached via routes II and III.

Implicit in any physically based explanation of the real world, is the assumption that natural processes are deterministic. However, this search for exact deterministic explanation may not be entirely appropriate to the geomorphological system. Klemes (1978) warned that, in the context of hydrological modelling, such is our deterministic view of the world, we attribute our inability to apply physically based models successfully to natural hydrological systems to failure in our theoretical basis and to an inability to determine the correct parameter values. Few would consider that this failing may be due to randomness, an inherent feature of the system. Yevjevich (1974) has also argued with the assumption in deterministic hydrology that the ratio of signal (which can be explained by physical laws) to unexplained noise, will increase in time, presumably as hydrology moves to the right in Figure 15.4. He argues that deterministic models are not applicable to the natural environment and therefore cannot hope to succeed.

The consideration of scale is also crucial for mathematical modelling in geomorphology. It must be appreciated that depending on the scale at which an environmental system is viewed, there are different sets of laws which operate. Scale is a function of the real system and cannot be imposed upon the system by the scientist. Geomorphological theory has mostly been validated for a small spatial and temporal scale. This theory will not necessarily be appropriate for the larger range of scales with which the geomorphologist might be interested.

There have been criticisms of geomorphology for not developing a sound theoretical basis. It has been illustrated that there are certain features of geomorphological systems which inhibit such description. However, there are those who have argued that in part, this failure may be attributed to weakness in our methods of study.

15.4.2 Technical Requirements

A number of technical problems in the application of physically based models to environmental systems have been experienced. In general these complex models have not enjoyed the success in prediction and explanation which 'theoretically' should have been theirs.

Physically based models characteristically require a large quantity and good quality of information for their application. Very often, their application is not sufficiently supported by data. For many environmental systems, it is not possible to assign representative parameter values due to inherent variability, measurement error, and inseparable processes (Anderson and Rogers, 1987). Problems are imposed by the application of theory to scales for which it has not been derived. For example, by increasing the scale of application, parameters which originally referred to points, now refer to areas. In many cases to derive the parameter values relevant to the scale of interest, the models have to be calibrated. This defeats the point of designing physically based models in the first place.

Where the required degree of data resolution cannot be obtained, it is met with synthetically derived data and thus the model cannot be fully validated. Such models do have a tendency to be highly sensitive to data error and so there is a danger that they may possess behaviour which is uncharacteristic of the system. McPhearson (1975, p. 247) has commented:

'Strident advocates of complex models appear to reason that the application of a larger number of guesses to a large number of unknowns is somehow better than a smaller number, on the assumption that the more complicated the procedure the better the results.'

The application of physically based models also suffers additional technical problems; those involved in the implementation or simulation of the mathematical model. Natural processes are both continuous and simultaneous. There are both impossible features to

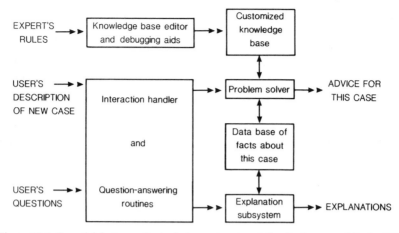

Figure 15.5. Essential features of a typical expert system (after Buchanan and Duda, 1983)

incorporate in simulation models. The dynamic behaviour of such systems can only be programmed at successive, but discrete intervals. Only a degree of 'quasi-simularity' can be achieved. In a simulation model, only one process can be allowed to operate at any one time (Hillel, 1977). Simultaneously occurring events must be assumed to be independent. Each event is controlled only by the conditions at the start of each time step. Processes may in reality affect variables describing the system, but their values are not updated until the beginning of the next time step. It must therefore be assumed that the order in which each process is considered is not critical. No computer simulation model, even where based upon the most complete theory, would exactly reproduce the behaviour of the prototype system in this respect.

Technical problems are also involved, in the development of the simulation model, and with the numerical solution of the mathematical model. Model error is introduced by the selection of numerical method, and the assumptions made of the system which are necessary to keep the solution tractable. Frequently, these methods exhibit stability and convergence problems which can only be removed by reducing the space and time discretion to such a small scale that the time for computation becomes impractical. Typically, these models require significant computer capacity.

At this point, it might be reasonable to consider that the nature of geomorphological systems sets it apart from the more traditional simulation problem in the 'hard' sciences. There is an upper limit to the degree of modelling which is possible for such complex and intricate systems. There is certainly little procedural information for mathematical modelling and simulation in ill-defined systems. It may be necessary to consider alternative modelling strategies and techniques for use in these conditions.

Modelling ill-defined systems has been the subject of a limited amount of recent research. Van Steenkiste and Spriet, for example, argue that too much emphasis is currently given to the mechanics of building a sophisticated computer simulation model, with too little emphasis on ensuring the model has been adequately identified, estimated, and validated. They therefore perceive a simulation layout for ill-defined system modelling which throws attention on the key position data bases will play in its architecture. Of course, in the geomorphological field such a structure is not yet apparent—data, testing, and program availability are all generally each restricted to just a handful of investigators. The more general availability of both geographical information systems and remote sensing imagery may, in the long term, serve to better integrate data sources amongst users.

We are arguing that computer simulation as a discipline within geomorphology needs greater professional recognition. It becomes a vehicle by which, often unwittingly, theoreticians within the subject are judged. As we have outlined, computer simulation in geomorphology is presented with potentially greater problems of articulation than in many other branches of science—a fact that surely makes rigour in development and application all the more necessary. Such a recognition as we have sought to achieve in this chapter, should help to clarify the objectives of mathematical modelling. In particular, more detailed attention to computer simulation concepts will aid the evaluation of models as predictors of complex dynamic and partially understood geomorphological systems.

In addition, such an effort would more likely result in a significant move towards the exploration of expert systems in geomorphological modelling. Figure 15.5 shows the effective division into expert and user interaction as perceived by Buchanan and Duda (1983). The severe limitation of geomorphological model availability, resulting in the near absence of the link between 'customized knowledge' and 'problem solver' in Figure 15.5, not only limits model application (see Figure 15.3) but effectively prevents any evaluation of 'cautious artificial intelligence' (Searle, 1987) with geomorphology. This point was made as a major conclusion to Chapter 1, albeit in a slightly different context. The intervening chapters, combined with the more general views expressed in this chapter, have hopefully helped those with an interest in geomorphological modelling to appreciate the challenges and stimuli that remain in aspiring to improve the level of professional computer simulation in geomorphology.

REFERENCES

Ahnert, F. (1976). Brief description of a comprehensive three dimensional process-response model of landform development. *Zeit. Geomorph. Suppl.*, **25**, 29–49.

Ahnert, F. (1977). Some comments on the quantitative formulation of geomorphological processes in a theoretical model. *Earth Surface Processes*, **2**, 191–201.

Anderson, M. G. and Rogers, C. C. M. (1987). Catchment scale distributed models: a discussion of future research directions. *Prog. in Phys. Geog.*, **11**, 28–51.

Bennett, R. J. and Chorley, R. J. (1978). *Environmental Systems; Philosophy, Analysis and Control*, Methuen, London.

Bratley, P., Fox, B. L., and Schrage, L. E. (1983). *A Guide to Simulation*, Springer-Verlag, 383 pp.

Brugnot, G. and Pochat, R. (1981). Numerical simulation study of avalanches. *Journal of Glaciology*, **27**, 77–88.

Buchanan, B. G. and Duda, R. O. (1983). Principles of rule-based expert systems, *Advances in Computers*, **22**, 164–216.

Davidson-Arnott, R. G. D. (1981). Computer simulation of nearshore bar formation. *Earth Surface Processes*, **6**, 23–34.

Dent, J. D. and Lang, T. E. (1980). Modelling of snow flow. *Journal of Glaciology*, **26**, 131–140.

Frenkiel, F. N. and Goodall, D. W. (1978). *Simulation Modelling of Environmental Problems*, John Wiley and Sons, Chichester, 112 pp.

Gilmour, P. (1973). A general validation for computer simulation models. *Australian Computer Journal*, **5**, 127–131.

Haines-Young, R. H. and Petch, J. R. (1986). *Physical Geography. its Nature and its Methods*, Harper and Row, London, 230 pp.

Hermann, C. F. (1967). Validation problems in games and simulations with special reference to models of international politics. *Behavioural Science*, **12**, 216–231.

Hey, R. D. (1979). Dynamic process–response model of river channel development. *Earth Surface Processes*, 59–72.

Hillel, D. (1977). *Computer Simulation of Soil Water Dynamics*, Ottawa, International Development Research Centre, Canada, 214 pp.

Hornberger, G. M. and Spear, R. C. (1981). An approach to the preliminary analysis of environmental systems. *Journal of Environmental Management*, **12**, 7–18.

Judson, A., Leaf, C. F., Brink, G. E. (1980). A process-oriented model for simulating avalanche danger. *Journal of Glaciology*, **26**, 53–63.

Klemes, V. (1978). Physically based stochastic hydrologic analysis. *Advances in Hydroscience*, **11**, 285–356.

Kvasov, D. D. and Verbitsky, M. Ya. (1982). Numerical simulation of the evolution of ice covers using the Scandinavian and Laurentide ice sheets as examples. *Journal of Glaciology*, **28**, 267–272.

Lang, T. E. and Dent, J. D. (1980). Scale modeling of snow-avalanche impact on structures. *Journal of Glaciology*, **26**, 189–196.

Lang, T. E., Dawson, K. L., and Martinelli, M. (1979). Application of numerical transient fluid dynamics to snow avalanche flow. Part I. Development of computer program AVALNCH. *Journal of Glaciology*, **22**, 107–115.

Leimkuhler, K. (1982). Some methodological problems in energy modelling. In Cellier, F. E. (Ed.), *Progress in Modelling and Simulation*, Academic Press, London, 61–75.

McCuen, R. H. (1973). Role of sensitivity analysis in hydrologic modeling. *Journal of Hydrology*, **18**, 37–53.

McCuen, R. H. (1976). The anatomy of the modelling process. In Brebbin, C. A. (Ed.), *Mathematical Models for Environmental Problems*, Proceedings of the International Conference held at University of Southampton, September 8–12, 1975, London, Pentach Press, 401–412.

McPhearson, M. B. (1975). Special characteristics or urban hydrology. In Chapman, T. G. and Dunin, F. X. (Eds.) *Prediction in Catchment Hydrology*, Australian Academy of Science, Griffin Press, 240–255.

Miller, D. R., Butler, G. and Bramall, L. (1976). Validation of ecological system models. *Journal Environmental Management*, **4**, 383–401.

Richards, K. S. (1978). Simulation of flow geometry in a riffle–pool stream. *Earth Surface Processes*, **3**, 345–354.

Sargent, R. G. (1982). Verification and validation of simulation models. In Cellier, F. E. (Ed.), *Progress in Modelling and Simulation*, Academic Press, London, 159–169.

Searle, J. R. (1987). Minds, brains and programs. In Born, R. (Ed.), *Artificial Intelligence — the Case Against*, Croom Helm, London, 18–40.

Van Horn, R. L. (1971). Validation of simulation results. *Management Science*, **17**, 247–258.

Van Steenkiste, G. C. and Spriet, J. A. (1982). Modelling ill-defined systems. In Cellier, F. E. (Ed.), *Progress in Modelling and Simulation*, Academic Press, London, 11–38.

Weber, J. E., Kisiel, C. C. and Duckstein, L. (1973). On the mismatch between data and models of hydrologic and water resource systems. *Water Resources Bulletin (Journal of the American Water Resource Associations)*, **9**, (6), 1075–1088.

Yevjevich, V. (1974). Determinism and stochasticity in hydrology. *Journal of Hydrology*, **22**, 225–238.

Author Index

441

Subject Index